Fungi and Insect Interactions: Pathogenicity, Immune Defenses and Biocontrol

Fungi and Insect Interactions: Pathogenicity, Immune Defenses and Biocontrol

Guest Editors

Jiaqin Xie
Nicolás Pedrini

MDPI

Basel • Beijing • Wuhan • Barcelona • Belgrade • Novi Sad • Cluj • Manchester

Guest Editors
Jiaqin Xie
School of Life Sciences
Chongqing University
Chongqing
China

Nicolás Pedrini
Instituto de Investigaciones
Bioquímicas de La Plata
(INIBIOLP)
National University of La
Plata (UNLP)
La Plata
Argentina

Editorial Office
MDPI AG
Grosspeteranlage 5
4052 Basel, Switzerland

This is a reprint of the Special Issue, published open access by the journal *Journal of Fungi* (ISSN 2309-608X), freely accessible at: https://www.mdpi.com/journal/jof/special_issues/265WR9CQ1W.

For citation purposes, cite each article independently as indicated on the article page online and as indicated below:

Lastname, A.A.; Lastname, B.B. Article Title. *Journal Name* **Year**, *Volume Number*, Page Range.

ISBN 978-3-7258-6742-4 (Hbk)
ISBN 978-3-7258-6743-1 (PDF)
https://doi.org/10.3390/books978-3-7258-6743-1

Contents

Journal of *Fungi*

MDPI

Editorial

Fungi and Insect Interactions: Pathogenicity, Immune Defenses and Biocontrol

Jiaqin Xie [1,*] and Nicolás Pedrini [2,*]

[1] School of Life Sciences, Genetic Engineering Research Center, Chongqing University, Chongqing 400044, China
[2] Instituto de Investigaciones Bioquímicas de La Plata (INIBIOLP), CCT La Plata Consejo Nacional de Investigaciones Científicas y Técnicas (CONICET), Universidad Nacional de La Plata (UNLP), Calles 60 y 120, La Plata 1900, Argentina
* Correspondence: xiejqin@cqu.edu.cn (J.X.); npedrini@med.unlp.edu.ar (N.P.)

1. Introduction

Insects are the largest animal group in nature, with significant effects on ecology, human health, and indigenous flora and fauna. Naturally abundant insect-pathogenic (i.e., entomopathogenic) fungi play a crucial role in regulating insect populations and have complex interactions with them. Entomopathogenic fungi infect insects by attaching to, germinating on, and penetrating the host's exoskeleton, ultimately proliferating within the insect's body and tissues. In response, host insects have evolved physical barriers, immune responses, microbiota, and behavioral defenses against fungal infections. Furthermore, insect-pathogenic fungi are valuable in biological pest management, as they often lead to the death of their hosts. They can also act as facultative saprophytes in the soil and enhance plant resistance to herbivores as endophytes. While the general mechanisms of the interaction between fungi and insects are well understood, many aspects at the molecular, physiological, and behavioral levels remain unclear. Therefore, studying the interaction between fungi and insects in this specific context could provide valuable and extensive insights.

These topics were addressed in the Special Issue "Fungi and Insect Interactions: Pathogenicity, Immune Defenses and Biocontrol", comprising 13 original studies focused on the biological and molecular aspects of the interaction between entomopathogenic fungi and their wide array of hosts, including arthropods and plants, which we briefly summarize in the following paragraphs and encourage readers to explore in full.

2. An Overview of Published Articles

The study by Chacón-Fuentes et al. (Contribution 1) reported a countermeasure strategy employed by the Chilean pest *Chilesia rudis* against peramine, often produced by fungi. This pest larva has evolved the ability to adapt to endophyte-infected ryegrass and develop counter-adaptation mechanisms to mitigate the side effects of such alkaloids. In their research, Wang et al. (Contribution 2) aimed to enhance the efficiency of the entomopathogenic fungus *Metarhizium rileyi* in targeting insect pests, addressing its slow growth rate, low resistance to abiotic stress, and slow killing speed by incorporating various lipids. Their findings revealed that exogenous oleic acid and linoleic acid could boost stress tolerance and virulence of *M. rileyi* by safeguarding conidial germination and facilitating cuticle infection. Su et al. (Contribution 3) identified the catalase gene *MrCat1* as a key player influencing various functions of the entomopathogenic fungus *M. rileyi*, including oxidative stress tolerance, microsclerotia formation, and virulence. Furthermore, Song et al. (Contribution 4) and Hong et al. (Contribution 5) reported on the negative regulation of UV- and thermo-tolerances by the forkhead box gene

MaSep1 and the role of a new transcription factor *MaAzaR* in influencing virulence to *Locusta migratoria manilensis* through cuticle penetration in *Metarhizium acridum*, respectively. Chen et al. (Contribution 6) highlighted the significant potential of two isolated *Beauveria bassiana* strains in controlling rice pests *Nilaparvata lugens* and *Sogatella furcifera*. A subsequent study suggested that combining these strains with chlorfenapyr or dinotefuran could enhance control efficacy against these pests. The study by Mao et al. (Contribution 7) examined the effects of three insecticides and three fungicides on the mycelial growth and spore germination of *Cordyceps javanica*. They also investigated the impact of combining these chemical pesticides with the fungus against *Myzus persicae* to enhance the efficiency of mycoinsecticides in biocontrol strategies. Preisegger et al. (Contribution 8) utilized a conidial hydrogel formulation of GFP-tagged *B. bassiana* to study per os infection and track the conidial movement through the alimentary canal of the red flour beetle, *Tribolium castaneum*. They found that ingested conidia are unable to germinate within the midgut but have a measurable impact on the insects' fitness, increasing their susceptibility to fungal infection and eventual mortality. Wang et al. (Contribution 9) utilized homologous recombination to create deletion mutants and complementation strains of the nitrate transporter MaNrtB in *M. acridum*. They found that the disruption strain was more vulnerable to various stressors and exhibited impaired fungal virulence, characterized by reduced appressorium formation on the cuticle and attenuated growth in the hemolymph of *Locusta migratoria manilensis*. The study by Boaventura et al. (Contribution 10) provides a comprehensive understanding of the effects of both constant and fluctuating temperatures on the growth and virulence of the entomopathogenic fungus *C. javanica* against the whitefly *Bemisia tabaci*. Their results, based on fungal efficacy and spatial predictions under fluctuating temperatures, indicated that *C. javanica* is suitable for use in different latitudes, although its performance varied at constant temperatures. Wakil et al. (Contribution 11) evaluated the acaricidal efficacy of *B. bassiana* and *M. robertsii* combined with spinosad in both laboratory and greenhouse assays against a tomato pest, the two-spotted spider mite *Tetranychus urticae*. They found a synergistic effect of the combined treatments on tomato plants under greenhouse conditions, which can provide enhanced control of *T. urticae* life stages compared to each treatment applied alone. Liu et al. (Contribution 12) explored the priming effects of *B. bassiana* on tomato plants. They found that the fungus activates phenylpropanoid metabolic pathways, with this modulation being influenced by jasmonate, thereby improving the plant's resistance to herbivore stress caused by *B. tabaci*. Finally, Almeida et al. (Contribution 13) investigated whether *M. robertsii* can reduce the feeding of the neotropical brown stink bug, *Euschistus heros* using electropenetrography on soybean plants. They found that fungus-treated adults spent significantly less time on probing plants compared to untreated insects and the number of waveform events decreased in the former.

3. Conclusions

Entomopathogenic fungi have significant potential as biocontrol agents for managing insect pests through direct infection or the production of indirect metabolites, exhibiting complex interactions with the hosts. This Special Issue covers a diverse range of research, highlighting the richness of the field. Four articles concentrate on enhancing the efficiency of fungi through interactions with lipids and chemical insecticides. Another four articles delve into the functions of genes in *Metarhizium*, aiming to uncover the regulatory mechanisms governing resistance, virulence, or conidiation. Additionally, four articles assess the potential of fungi in insect control, exploring the underlying mechanisms. Finally, one paper specifically examines the infection dynamics of *Beauveria bassiana* in the gut of *Tribolium castaneum*. Overall, each paper introduces novel insights that underscore the significance of this Special Issue, offering readers research focused on multi-level contexts beyond the usual scope, providing a more diverse view of the field.

Author Contributions: Conceptualization, J.X. and N.P.; writing—review and editing, J.X. and N.P.; funding acquisition, J.X. and N.P. All authors have read and agreed to the published version of the manuscript.

Funding: J.X. obtained funding from the Natural Science Foundation of Chongqing (project no. CSTB2024TIAD-GPX0024) and the Fundamental Research Funds for the Central Universities with projects no. 2023CDJKYJH043 and 2024CDJXY016. N.P. obtained funding from the National Agency for Science and Technology Promotion of Argentina, with grant PICT 2020 series A 02932, and from the Williams Foundation in Argentina.

Conflicts of Interest: The authors declare no conflicts of interest.

List of Contributions:

1. Chacón-Fuentes, M.; Martínez-Cisterna, D.; Lizama, M.; Asencio-Cancino, V.; Matamala, I.; Bardehle, L. A Countermeasure Strategy against Peramine Developed by *Chilesia rudis* in the Endophyte–Ryegrass–Herbivore Model. *J. Fungi* **2024**, *10*, 512. https://doi.org/10.3390/jof10080512.
2. Wang, G.; Zhang, X.; Du, G.; Wang, W.; Yao, Y.; Jin, S.; Cai, H.; Peng, Y.; Chen, B. Oleic Acid and Linoleic Acid Enhances the Biocontrol Potential of *Metarhizium rileyi*. *J. Fungi* **2024**, *10*, 521. https://doi.org/10.3390/jof10080521.
3. Su, Y.; Wang, X.; Luo, Y.; Jiang, H.; Tang, G.; Liu, H. The Catalase Gene MrCat1 Contributes to Oxidative Stress Tolerance, Microsclerotia Formation, and Virulence in the Entomopathogenic Fungus *Metarhizium rileyi*. *J. Fungi* **2024**, *10*, 543. https://doi.org/10.3390/jof10080543.
4. Song, T.; Li, C.; Jin, K.; Xia, Y. The Forkhead Box Gene, *MaSep1*, Negatively Regulates UV- and Thermo-Tolerances and Is Required for Microcycle Conidiation in *Metarhizium acridum*. *J. Fungi* **2024**, *10*, 544. https://doi.org/10.3390/jof10080544.
5. Hong, G.; Wang, S.; Xia, Y.; Peng, G. MaAzaR Influences Virulence of *Metarhizium acridum* against *Locusta migratoria manilensis* by Affecting Cuticle Penetration. *J. Fungi* **2024**, *10*, 564. https://doi.org/10.3390/jof10080564.
6. Chen, Z.; Mu, H.; Peng, Y.; Huo, R.; Xie, J. The Susceptibility of Two *Beauveria bassiana* Strains on Rice Pests *Nilaparvata lugens* and *Sogatella furcifera*. *J. Fungi* **2025**, *11*, 128. https://doi.org/10.3390/jof11020128.
7. Mao, R.; Cai, X.; Wang, T.; Liu, Z.; Xing, P.; Zhang, G.; Zhou, W.; Diao, H.; Ma, R. In Vitro Response of Two Strains of *Cordyceps javanica* to Six Chemical Pesticides. *J. Fungi* **2024**, *10*, 852. https://doi.org/10.3390/jof10120852.
8. Preisegger, L.; Flecha, J.C.; Ghilini, F.; Espin-Sánchez, D.; Prieto, E.; Oberti, H.; Abreo, E.; Huarte-Bonnet, C.; Pedrini, N.; Mannino, M.C. Visualizing Oral Infection Dynamics of *Beauveria bassiana* in the Gut of *Tribolium castaneum*. *J. Fungi* **2025**, *11*, 101. https://doi.org/10.3390/jof11020101.
9. Wang, J.; Zou, Y.; Xia, Y.; Jin, K. MaNrtB, a Putative Nitrate Transporter, Contributes to Stress Tolerance and Virulence in the Entomopathogenic Fungus *Metarhizium acridum*. *J. Fungi* **2025**, *11*, 111. https://doi.org/10.3390/jof11020111.
10. Boaventura, H.A.; Queirós, L.A.; Silva, J.F.A.; Goble, T.A.; Pazolini, K.; Marciano, A.F.; Quintela, E.D. Temperature-Dependent Modeling and Spatial Predictions for Identifying Geographical Areas in Brazil Suitable for the Use of *Cordyceps javanica* in Whitefly Control. *J. Fungi* **2025**, *11*, 125. https://doi.org/10.3390/jof11020125.
11. Wakil, W.; Boukouvala, M.C.; Kavallieratos, N.G.; Naeem, A.; Ntinokas, D.; Ghazanfar, M.U.; Avery, P.B. The Inevitable Fate of *Tetranychus urticae* on Tomato Plants Treated with Entomopathogenic Fungi and Spinosad. *J. Fungi* **2025**, *11*, 138. https://doi.org/10.3390/jof11020138.
12. Liu, M.; Xiang, D.; Hokkanen, H.M.T.; Niu, T.; Zhang, J.; Yang, J.; Wei, Q.; Chen, H.; Liu, H.; Li, Y. *Beauveria bassiana* Induces Strong Defense and Increases Resistance in Tomato to *Bemisia tabaci*. *J. Fungi* **2025**, *11*, 141. https://doi.org/10.3390/jof11020141.
13. Almeida, A.C.d.S.; Rodrigues, M.A.; Boaventura, H.A.; Vieira, A.S.; e Silva, J.F.A.; de Jesus, F.G.; Quintela, E.D. Can *Metarhizium anisopliae* Reduce the Feeding of the Neotropical Brown Stink Bug, *Euschistus heros* (Fabricius, 1798), and Its Damage to Soybean Seeds? *J. Fungi* **2025**, *11*, 247. https://doi.org/10.3390/jof11040247.

Journal of *Fungi*

MDPI

Article

Can *Metarhizium anisopliae* Reduce the Feeding of the Neotropical Brown Stink Bug, *Euschistus heros* (Fabricius, 1798), and Its Damage to Soybean Seeds?

André Cirilo de Sousa Almeida [1,*], Mayara Alves Rodrigues [1], Heloiza Alves Boaventura [2], Angélica Siqueira Vieira [2], José Francisco Arruda e Silva [2], Flávio Gonçalves de Jesus [1] and Eliane Dias Quintela [2,3]

[1] Instituto Federal Goiano—Campus Urutaí, Rodovia Professor Geraldo Silva Nascimento, Km 2,5, Urutaí 75790-000, GO, Brazil; mayara-@live.com (M.A.R.); flavio.jesus@ifgoiano.edu.br (F.G.d.J.)

[2] Embrapa Arroz e Feijão, Rodovia GO-462, Km 12, Fazenda Capivara, Zona Rural, CP 179, Santo Antônio de Goiás 75375-000, GO, Brazil; boaventuraheloiza@gmail.com (H.A.B.); vieira.angelica29@gmail.com (A.S.V.); jose.arruda-silva@embrapa.br (J.F.A.e.S.); eliane.quintela@embrapa.br (E.D.Q.)

[3] Escola de Agronomia, Universidade Federal de Goiás, Goiânia 74690-900, GO, Brazil

* Correspondence: andre.almeida@ifgoiano.edu.br

Abstract: The fungus *Metarhizium anisopliae* is under development as a bioinsecticide for *Euschistus heros*. To further elucidate the effect of this fungus on *E. heros* behavior, we monitored the feeding activities of adults treated with the fungus at 1×10^8 conidia mL^{-1} on soybean in the pod-filling stage (R5) through electropenetrography (EPG) AC-DC. We also determined the virulence of *M. anisopliae* to adults and its damage to soybean seeds. *M. anisopliae* displayed high levels of virulence to adults even at low concentrations of 5×10^6 conidia mL^{-1} (98% mortality). *E. heros* females were more susceptible to *M. anisopliae* than males, exhibiting a lower LT_{50} for mycosed adults (7.1 and 9.7 days, respectively). The EPG experiment showed that fungus-treated adults spent significantly less time on probing activities (reduced by 86% at day four and ceased at day five) than untreated insects, and the number of waveform events per insect significantly decreased. This information is valuable for managing stink bugs at the field level, as it shows that even though the insect is alive, its feeding is compromised, consequently minimizing the damage inflicted to the crop. This study paves the way for further research employing entomopathogenic fungi in pest control.

Keywords: biological control; electropenetrography; entomopathogenic fungi; EPG

1. Introduction

The Neotropical brown stink bug *Euschistus heros* (Fabricius, 1798) (Hemiptera: Pentatomidae) can seriously damage commodity crops, including soybean [*Glycine max* (L.) Merr.], corn (*Zea mays* L.), and cotton (*Gossypium* spp.) in Brazil [1,2]. This sucking species is also widely distributed in Argentina, Paraguay, and Uruguay [3,4].

On average, it develops from egg to adult in 25 to 32 days (5 to 7 days in the egg stage and 20 to 25 days in the nymphal stage), with the adult lifespan ranging from 75 to 116 days [5,6]. The Neotropical brown stink bug begins feeding from the second instar, and significant damage to soybean seeds occurs only after the third instar. The nymphal stages exhibit feeding behaviors similar to those of adult *E. heros*, but with some differences, particularly in the shorter duration of feeding activities compared to adults [7].

The stink bug initiates the colonization of soybean during the vegetative phase and is harmful in the reproductive phase as a pod feeder. Therefore, their feeding can directly affect yield and/or other grain-quality parameters during pod development and seed filling [1,5]. The damage results from their probing activities in soybean plants or pods when the stylet penetrates the plant tissues and injects saliva [8]. The digestive enzymes in their salivary secretions facilitate digestion, degrade tissues, and cause darkening. Feeding on pods and immature seeds can lead to malformation and abscission of pods and seeds, wrinkled or empty seeds, and darkening, with delayed maturation and reduced seed vigor [4]. In addition, the holes left by the stylets can facilitate infection by pathogenic microorganisms [9,10].

The most common method of controlling *E. heros* is application of synthetic insecticides [11]. The insecticides used to reduce stink bugs populations on soybean belong mostly to three chemical groups (neonicotinoids, organophosphates, and pyrethroids), which can be used alone or in formulated mixtures [12]. The limited number of available insecticides, combined with the frequent use of the same mode of action (often requiring 33–34 spray applications per season), has led to the selection of resistant strains [1,13–16]. Moreover, the overuse of synthetic insecticides can be detrimental to the environment, reducing natural biocontrol agents and pollinators [16,17].

Reducing synthetic chemical use in agriculture has become a global goal. Therefore, efficient and environmentally friendly multiple control strategies are needed [18,19]. Biological control [20] is a practical and ecologically friendly strategy for pest control. Entomopathogenic fungi, a proven biocontrol agent for managing stink bugs, can directly infect their host by contact through the integument, making them an important bioinsecticide for sap-sucking insects [21]. The research on stink bug management has mainly involved the fungus *Metarhizium anisopliae* (Metsch.) Sorok. (Hypocreales: Clavicipitaceae) [21–23].

The isolate BRM 2335 of *M. anisopliae* selected for this study has already been shown to be highly virulent against several key stink bug pests [21,23] and is under development as a mycoinsecticide for *E. heros*. Although the mode of action, virulence, and efficiency of *M. anisopliae* in managing *E. heros* have been studied, the probing behavior of this stink bug after infection with entomopathogenic fungi has yet to be elucidated.

The electropenetrography (EPG) technique is used to observe and quantify feeding behaviors otherwise hidden in the opaque food substrate [24]. EPG is an accurate and comprehensive method to evaluate the probing behavior of hemipterans [25,26] and has been employed to evaluate the probing behavior of *E. heros* [8,27]. This technique operates using an electrical circuit connecting the sucking insect to its host plant. The low electrical current flowing through the system generates waveforms that represent different probing activities (stylet penetration, salivation, and ingestion), non-probing activities, and the insect standing still or walking on the plant surface [25–28].

As fungal infection can influence host behavior [22], this technique could improve understanding of the probing behavior of piercing–sucking insects infected by an entomopathogenic fungus. The fungi take a considerable amount of time to kill their hosts—typically ranging from 2 to 10 days—depending on factors such as inoculum concentration, host species, and age. This slow process is adaptive for the pathogen, as it allows ample time to maximize nutrient extraction from the host [27]. Therefore, the probing behavior, as well as when feeding is ceased after infection, are important information for managing stink bugs, as disruptions in stylet activities can reduce the damage to productive structures of the plants. Recently, Maluta et al. [29] reported a significant disruption of the probing activities of *Dalbulus maidis* (DeLong & Wolcott) (Hemiptera: Cicadellidae) on corn after spraying *Cordyceps javanica* (Frieder. & Bally) (Hypocreales: Cordycipitaceae). The stylet activities were altered within 30 h, and these changes lasted until about 96 h.

The probing behavior of *E. heros* after entomopathogenic fungus application is not known. This pioneering study aimed to (1) determine the virulence of different *M. anisopliae* conidia concentrations to *E. heros* adults; (2) monitor the probing behavior of infected *E. heros* adults during the soybean pod-filling stage (R5) through EPG AC-DC; and (3) determine the damage to soybean pods caused by *E. heros*.

2. Materials and Methods

2.1. Stink Bug Rearing

Adult *E. heros* were collected from a soybean field at Embrapa Rice and Beans (Santo Antônio de Goiás, GO, Brazil), (16°28′00″ S, 49°17′00″ W; 823 m.a.s.l), taken to the Laboratory of Entomology, and placed in a plastic box (25 × 17 × 17 cm) lined with moistened filter paper (T 25 ± 2 °C, RH 70 ± 10%, 14-h photoperiod). Fresh green bean pods (*Phaseolus vulgaris* L.), okra fruits (*Abelmoschus esculentus* L.), mature soybean seeds (*Glycine max* L.), and raw shelled peanuts (*Arachis hypogaea* L.) were provided as food source according to Silva et al. [30].

2.2. Fungal Strain and Preparations

The *M. anisopliae* BRM 2335 was obtained from infected adult *Tibraca limbativentris* Stål (Hemiptera: Pentatomidae) collected from rice in a screenhouse at Embrapa Rice and Beans in 1985. The BRM 2335 isolate was identified through the sequence analysis of the elongation factor 1-alpha gene, following the protocol described in Bischoff et al. [31]. The isolate was preserved at −80 °C in the Invertebrate Fungal Collection at Embrapa Genetic Resources and Biotechnology, Brasília, DF, Brazil.

Conidia were grown on parboiled rice for 7–10 days, according to Mascarin and Quintela [32]. The conidial suspension was prepared by washing the colonized rice grains in 10 mL of sterile aqueous solution of 0.01% (v/v) Tween 80, using 50 mL plastic centrifuge tubes. The suspension was vigorously agitated on a vortex mixer for 1 min and filtered through two layers of 30 µm pore-sized nylon cheesecloth. The filtered suspension (10 mL) was vortexed again for 1 min before application, and conidial concentrations were enumerated by hemocytometer (Brightline Improved Neubauer, New Optik®, São Paulo, SP, Brazil) at 400× magnification. Conidial germination exceeded 98% on potato dextrose agar (PDA) after 18 h at 26 °C. Only conidia with germ tubes greater than conidial diameter were considered germinated.

2.3. Virulence of *M. anisopliae* to Adult *E. heros*

The fungus was tested using a control (sterile aqueous solution of 0.01% (v/v) Tween 80), as well as concentrations of 5.0×10^6, 1.0×10^7, 5.0×10^7, and 1.0×10^8 conidia mL^{-1}. Carbon dioxide gas (CO_2) was used to anesthetize adults for 15 s before fungal spraying. One milliliter of the fungal suspension was applied to 10 insects placed together in a single 60 mm Petri dish in a Potter tower calibrated at 20 psi of working pressure. After spraying, adults were kept in Gerbox-type boxes (110 × 110 × 35 mm) lined with moistened filter paper and two green bean pods. The pods were previously surface-sterilized with a sterile aqueous solution of 5% (v/v) sodium hypochlorite (2.5% NaClO) for 15 min and rinsed twice with distilled water. Assessments of live and dead insects were performed daily for 12 days. To confirm mortality resulting from fungal infection (cadavers with fungal sporulation), dead adults were transferred to Petri dishes (60 mm) with a wet cotton and maintained at room temperature. Insects were considered infected by the fungus when mycelial and conidial growth was observed on the insect cadaver.

The experiments were conducted in a completely randomized design with five replicates, each consisting of 10 insects, totaling 50 per treatment. The experiment was repeated three times. The experiments were maintained at room temperature and with a 14 h photoperiod. The temperature and relative humidity in the laboratory were monitored at 1 h intervals by two dataloggers (Hobo® U12-012, Onset Computer Corp., Ltd., Bourne, MA, USA). Small variations were observed for the datalogger measurements, with an average of 25 ± 2 °C and 57 ± 15% RH.

2.4. Damage of *E. heros* Adults Treated with *M. anisopliae* on Soybean Seeds

Groups of five females or males of *E. heros* aged 4 to 6 days were sprayed with *M. anisopliae* at 1×10^8 conidia mL^{-1} in a Potter Tower, similarly to the methodology described above. The control groups were sprayed with a sterile aqueous solution containing 0.01% (v/v) Tween 80. After spraying, each insect was transferred to a Petri dish (60 × 15 mm) with moistened filter paper and a soybean pod containing two seeds. Mortality was assessed daily for 11 days, and dead insects were transferred to Petri dishes with a wet cotton to confirm fungal infection. At 2, 4, 7, 9, and 11 days after spraying, the pod of each repetition was replaced and transferred to paper bags to dry for 7 days at room temperature. After drying, the number of feeding punctures and the damage in soybean seeds were determined by tetrazolium test, according to the methodology described in França-Neto et al. [33]. The experimental design was randomized blocks, with 15 treated males or females and untreated (control). The experiment was repeated three times.

2.5. Electropenetrography Studies

Soybean BRS 7470 IPRO seeds were sown weekly in soil (5 L plastic pots) and kept in a greenhouse (T 32 ± 10 °C, RH 60 ± 20%, 12 h photoperiod). After reaching the R5 stage (pod filling), stems containing pods were collected (the entire plants were too large for the Faraday cages) with a razor and placed into a small pot (0.2 L) filled with substrate [1:1–1 soil (red dystrophic oxisol), 1 sand]. The plant electrode was introduced into the substrate.

EPG data were collected using a four-channel AC/DC monitor [23] (EPG Technologies, Inc., Gainesville, FL, USA), which was connected to a computer. The recordings were made using an input impedance of 10^7 Ohms for each of the four channels, a voltage of 50 mV alternating current (AC) via the plant electrode, and a gain set to 400× (actual gain 400 with 1× multiplier) [34]. An offset control was used to avoid rectifier fold-over and retain native waveform polarity after rectification [24]. Voltage outputs were amplified and captured at a rate of 100 Hz per channel using WinDaq DI-710 equipment (Dataq Instruments, Akron, OH, USA) and recorded on a computer with WinDaq Lite software version 3.11 (Dataq Instruments, Akron, OH, USA). The amplifiers, plants, and insects were kept inside a Faraday cage during the recordings to reduce external electrical noise.

Adults were treated with a concentration of 1×10^8 conidia mL^{-1}, and the control group with 0.01% v/v Tween 80, as described above. Afterward, the insects were kept in the Petri dish with two green bean pods for 48 h. Then, the insects were fasted for 5 h (without water or food) before being submitted to EPG recordings. Adhesive tape was employed to immobilize adults in a Petri dish, and dental sandpaper was used to remove the lipid layer of the insect pronotum [35]. The insects were wired by gluing one end of a gold wire (0.127 mm in diameter and 3.5 cm in length) (Sigma Aldrich, Barueri, SP, Brazil) to a copper wire (3 cm in length) soldered to a brass nail. A small loop was made at the other end of the gold wire to increase the contact area with the insect and improve electrical conductivity [36]. The gold wire was attached to the surface of the insect using a silver glue made with silver flake (Sigma Aldrich, St. Louis, MO, USA), water, and white glue (Cascorez, Jundiai, SP, Brazil) (1:1:1 wt/vol/vol) [36]. The glue dried for approximately 40 min before recording. Each stink bug was placed on a soybean pod inside the Faraday cage (as described above) and EPG-recorded for 72 h under laboratory conditions (25 ± 2 °C) and constant light [37]. Twenty adults were successfully recorded per treatment in a completely randomized design.

Waveforms were identified and named according to the nomenclature proposed by Lucini and Panizzi [6], who characterized an EPG waveform library produced by *E. heros* adult females feeding on soybean pods. The recorded waves and their biological significance are presented in Table 1. Six behaviors represented by waveforms were assessed using the variables non-probing, pathway, xylem ingestion, cell rupture, short ingestion of macerated tissues, and probable phloem sap ingestion. The five variables evaluated were waveform duration per insect (WDI), number of waveform events per insect (NWEI), waveform duration per event per insect (WDEI), percentage of recording time spent in probing (PRTP) [37–40], and final time at the last probe (FTLP).

Table 1. Waveforms recorded using electropenetrography (EPG) of *Euschistus heros* treated and untreated (control) with *Metarhizium anisopliae* fed on soybean pod and their biological significance.

Phase	Family	Type/Subtype	*Euschistus heros* Biological Meaning
Non-probing	-	Z	Insect immobile on plant surface/walking on plant surface
Pathway	P	Eh1	Penetration of stylets and salivary sheath secretion
Ingestion	I	Eh2	Xylem sap ingestion
Salivation	I	Eh3a	Cellular laceration and enzymatic maceration of seed endosperm
Ingestion	I	Eh3b	Short ingestion of macerated tissues of seed endosperm
Ingestion	I	Eh4	Short ingestion from an unknown site (probable phloem sap ingestion)

Source: Adapted from Lucini and Panizzi [7].

2.6. Statistical Analysis

The virulence of *Metarhizium* was expressed and compared for percent of mortality, cadavers with fungal sporulation (% mycosis), and mean lethal time (LT_{50}). Overall mortality and confirmed mortality curves were adjusted according to non-linear models and compared using the Chi-Square test ($p < 0.05$). To estimate the LT_{50}, Gompertz and Weibull non-linear models were fitted, and values were compared by the overlap of their 95% confidence intervals (95% CI) using the Package 'drc' version 3.0-1 [41] in statistical software R version 4.2.2 (R Core Team 2023, available at https://www.r-project.org/ (accessed on 11 November 2024.)). Data on feeding punctures were analyzed using the Student's *t*-test to identify statistically significant differences between groups ($p < 0.05$).

The EPG data were entered in INFEST–Insect Feeding Behavior Statistics software (available at https://arsilva.shinyapps.io/infest/ (accessed on 20 July 2024.)), which provided spreadsheets. Next, a generalized linear model (GLM) for Poisson fitted to the count data (NWEI) and GLM for Gamma were fitted for the duration variables (WDI, WDEI, PRTP, and FTLP). Differences were considered significant at $\alpha = 0.05$. All analyses were performed in INFEST.

3. Results

3.1. Virulence of *M. anisopliae* to Adult *E. heros*

The four concentrations of *M. anisopliae* BRM 2335 caused overall and confirmed mortalities (cadavers with fungal sporulation) of adults that were statistically different from the control (Figure 1A,B; Table 2). At the lowest concentration (5×10^6 conidia mL^{-1}), adult mortalities were similar to those of the other concentrations (Figure 1A,B; Table 2). The fungus at 1×10^8 conidia mL^{-1} killed more adults only when compared with 1×10^7 conidia mL^{-1}. Cadavers with fungal sporulation at 1×10^8 conidia mL^{-1} were significantly higher in number than those at 1×10^7 and 5×10^7 conidia mL^{-1}. The median lethal times (LT_{50})

ranged from 5.8 to 7.6 days for all concentrations, and no differences were observed among them (Table 3).

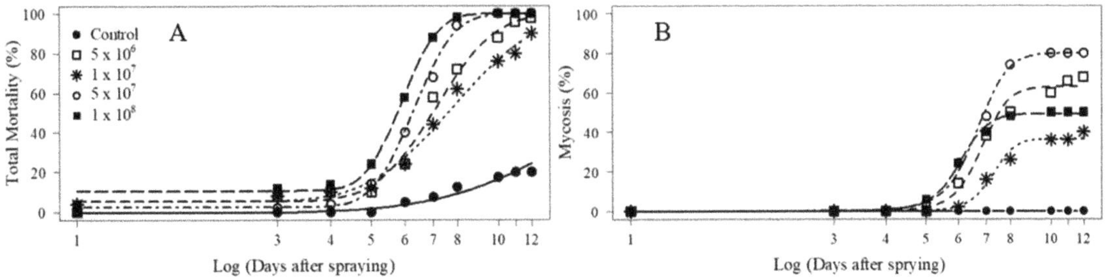

Figure 1. Overall (total) and confirmed mortalities (cadavers with fungal sporulation) means on different days for *Euschistus heros* adults treated with *Metarhizium anisopliae* at different concentrations. Curves were adjusted according to Logistic (**A**) and Brain–Cousens (**B**) non-linear models.

Table 2. *p*-values of the comparisons of overall (total) and confirmed mortalities (cadavers with fungal sporulation) curves for *Euschistus heros* adults after treatment with *Metarhizium anisopliae* at different concentrations. The Chi-Square test was used for *p* value calculation. Curves were considered significantly different at $p \leq 0.05$.

		Mortality		
Treatments	**5.0×10^6**	**1.0×10^7**	**5.0×10^7**	**1.0×10^8**
Control	<0.001	<0.001	<0.001	<0.001
5.0×10^6	-	0.3565	0.2401	0.0663
1.0×10^7	-	-	0.0460	0.0114
5.0×10^7	-	-	-	0.3482
		Confirmed Mortalities		
Treatments	5.0×10^6	1.0×10^7	5.0×10^7	1.0×10^8
Control	<0.001	<0.001	<0.001	<0.001
5.0×10^6	-	0.0196	0.1398	0.2713
1.0×10^7	-	-	0.0001	0.0754
5.0×10^7	-	-	-	0.0224

Table 3. Estimates of parameters of non-linear models and median lethal time (LT_{50}) of *Euschistus heros* adults treated with *Metarhizium anisopliae* at different concentrations (conidia mL^{-1}).

Treatments (Conidia mL^{-1})	**Model Parameters**			**LT_{50} (d) (CI95%)**
	B	**C**	**E**	
5.0×10^6	3.90	0.05	7.11	7.0 (5.8–8.2)
1.0×10^7	2.77	0.05	7.80	7.6 (5.9–9.2)
5.0×10^7	5.67	0.02	6.29	6.3 (5.4–7.1)
1.0×10^8	6.42	0.10	5.89	5.8 (4.9–6.7)

Gompertz model parameters: B is the slope factor around the "E" parameter; C is the lower limit of the curve; E is the inflection point of the curve.

3.2. Damage of E. heros Adults Treated with M. anisopliae on Soybean Seeds

Female adult cadavers with fungal sporulation (mycosis) were significantly higher in number than those of males at 11 days after spraying ($p < 0.001$) (Figure 2). No infected adults were observed in the controls, and the counts differed between male and female groups ($p = 0.005$, $p < 0.0001$, respectively) (Figure 2). The LT_{50} was lower for females (7.1 days) than for males (9.7 days) (Table 4).

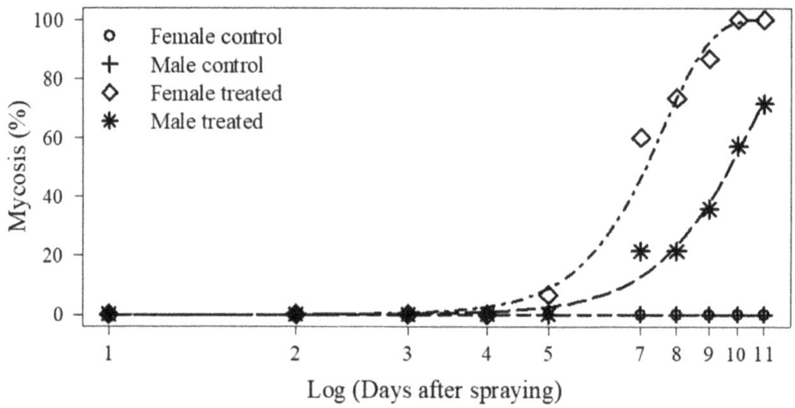

Figure 2. Confirmed mortality (cadavers with fungal sporulation) of female and male adult *Euschistus heros* on different days after treatment with *Metarhizium anisopliae*. Curves were adjusted according to non-linear Weibull model.

Table 4. Estimates of parameters of non-linear Weibull model and median lethal time (LT_{50}) of female and male adult *Euschistus heros* treated with *Metarhizium anisopliae* at 1×10^8 conidia mL^{-1}.

Treatments	Model Parameters				LT_{50} (d) (CI95%)
	b	c	d	e	
Female treated	5.7	0.0	1.0	7.6	7.1 (6.5–7.7)
Male treated	5.0	0.0	1.0	10.4	9.7 (8.6–10.7)

Weibull model parameters: b is the is the slope of the curve; c is the lower limit of the curve; d is the upper limit of the curve; e is the inflection point of the curve.

The number of feeding punctures by *E. heros* females treated with *M. anisopliae* was significantly reduced compared to untreated females (t = 2.28, $p = 0.03$) at 11 days after spraying (Figure 3). However, the number of feeding punctures was similar for treated and untreated males (t = 1.25, $p = 0.22$) (Figure 3).

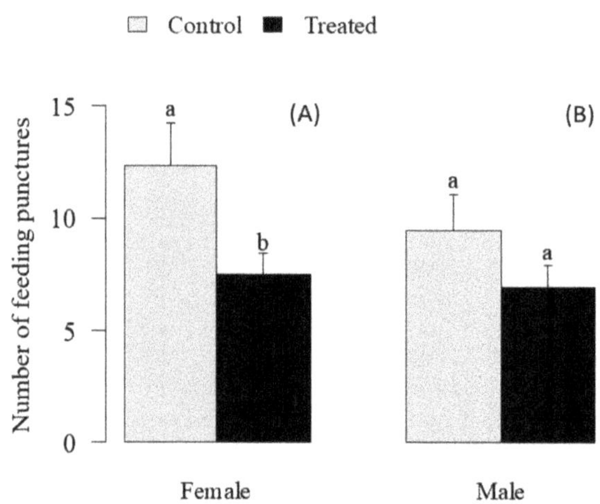

Figure 3. Number of feeding punctures by female (**A**) and male (**B**) adults *of Euschistus heros* in soybean seeds at 11 days after spraying of *Metarhizium anisopliae* at 1×10^8 conidia mL^{-1}. Different letters indicate significant differences ($p < 0.05$) by Student's *t*-test.

3.3. Electropenetrography

Metarhizium anisopliae influenced the probing behavior of *E. heros* adults on soybean pods. The NWEI was significantly reduced on treated stink bugs compared to control (Table 5). We found that *E. heros* treated with *M. anisopliae* spent significantly less time on probing activity. The WDI was shorter in xylem ingestion (waveform Eh2), laceration and maceration of endosperm (CLE–waveform Eh3a), and the short periods of ingestion of macerated tissues of endosperm (SIE–waveform Eh3b). Overall, non-probing duration (Z-Np waveform) was significantly longer in treated than in control insects.

Table 5. Number of waveform events for (NWEI) (mean ± SE) and waveform duration per insect (WDI–min) of *Euschistus heros* treated and untreated (control) with *Metarhizium anisopliae* at 1×10^8 conidia mL^{-1} feeding on soybean pod-filling stage (R5) plants.

Waveform	NWEI				WDI			
	Control	Treated	F	*p* Value	Control	Treated	F	*p* Value
Z + Np-non-probing	15.9 ± 0.9	12.5 ± 0.8	6.66	0.009	3488 ± 115	3924 ± 122	6.65	0.015
Eh1-pathway	19.1 ± 0.8	12.9 ± 1.0	19.6	<0.001	77.6 ± 10.7	49.6 ± 10.3	3.51	0.065
Eh2-xylem sap ingestion	5.9 ± 0.6	1.0 ± 0.2	62.3	<0.001	189.6 ± 46.1	24.8 ± 11.7	11.93	0.016
Eh4-unknown site ingestion	2.7 ± 0.4	0.9 ± 0.4	14.8	<0.001	9.07 ± 4.9	4.23 ± 2.7	0.781	0.399
Eh3a-CLE	26.5 ± 1.2	14.9 ± 0.9	52.3	<0.001	550.2 ± 106.4	239.4 ± 50.4	8.20	0.007
Eh3b-SIE	13.6 ± 0.9	4.6 ± 0.5	74.2	<0.001	5.42 ± 1.60	1.26 ± 0.4	8.08	0.007

CLE–cellular laceration of seed endosperm, SIE–short ingestion of macerated tissues of seed endosperm. $p < 0.05$ is considered different by F-test.

The durations of waveform events per insect (WDEI) (Table 6) were reduced for adults infected with *M. anisopliae.* This was pronounced in non-probing (Z-Np), laceration and maceration of endosperm (CLE–waveform Eh3a), and the short periods of ingestion of macerated tissues of endosperm (SIE–waveform Eh3b). Thus, *E. heros* performed longer Z-Np, CLE, and SIE in control compared to treated insects. The PRTP was also influenced by *M. anisopliae.* The control spent more time on non-probing activities. On the other hand, the recording times in probing activities (Eh2, Eh3a, Eh3b waveforms) were lower in control compared with treated stink bugs.

Table 6. Waveform duration per event per insect (WDEI-min) (mean ± SE) and percentage of recording time spent in probing (PRTP-%) of *Euschistus heros* untreated (control) and treated with *Metarhizium anisopliae* at 1×10^8 conidia mL^{-1} feeding on soybean pod-filling stage (R5) plants.

Waveform	WDEI				PRTP			
	Control	Treated	F	*p* Value	Control	Treated	F	*p* Value
Z + Np-non-probing	356.1 ± 81.2	1115.2 ± 401.8	6.80	0.014	80.8 ± 2.8	90.3 ± 2.6	6.55	0.015
Eh1-pathway	4.44 ± 0.4	3.23 ± 0.4	3.73	0.628	1.8 ± 0.2	1.2 ± 0.2	3.22	0.082
Eh2-xylem sap ingestion	34.5 ± 10.2	15.0 ± 7.8	1.29	0.263	4.4 ± 1.0	0.6 ± 0.2	14.7	0.007
Eh4-Phloem ingestion	2.0 ± 0.4	1.0 ± 0.6	1.29	0.264	0.3 ± 0.1	0.1 ± 0.0	0.825	0.370
Eh3a-CLE	21.4 ± 2.6	13.7 ± 2.2	4.55	0.041	12.8 ± 2.4	5.6 ± 1.1	8.23	0.004
Eh3b-SIE	0.5 ± 0.1	0.2 ± 0.0	4.68	0.038	0.2 ± 0.0	0.1 ± 0.0	8.77	0.005

CLE—cellular laceration of seed endosperm, SIE—short ingestion of macerated tissues of seed endosperm. $p < 0.05$ is considered different by F-test.

Insects treated with *M. anisopliae* showed a reduction in WDI in probing activities (Eh1, Eh2, Eh3a, Eh3b, Eh4 waveforms) (86%) (F = 24.43, $p < 0.001$) at 48 h and (98%) (F = 10.68, $p < 0.001$) at 72 h after recording started (Figure 4).

There was a 45% reduction in the FTLP of treated stink bug adults compared to the control (Figure 5). In *E. heros* treated with *M. anisopliae*, the FTLP was 31.1 h, while in the control insects, the final time was 55.1 h (F = 18.35, $p < 0.001$).

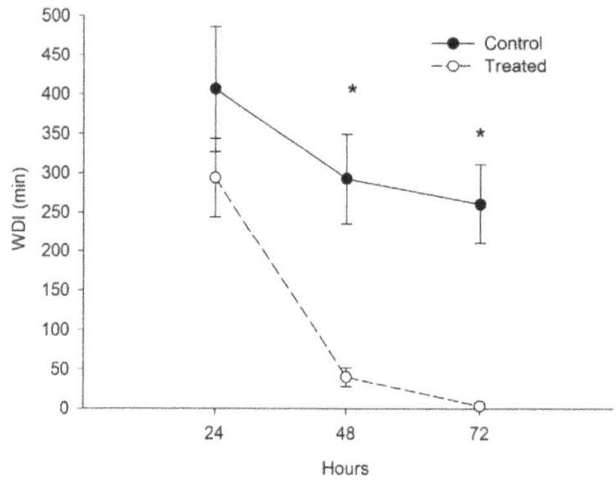

Figure 4. Waveform duration per insect (WDI–min) (mean ± SE) of probing activates (Eh1, Eh2, Eh3a, Eh3b, Eh4 waveforms) of *Euschistus heros* untreated (control) and treated with *Metarhizium anisopliae* feeding on soybean plants 24, 48, and 72 h after the start of recording (corresponding to 72, 96, and 120 h after fungus application, respectively). Treatments within each post-recording start time with an asterisk (*) were statistically different by F-test ($p < 0.05$) within each post-recording time.

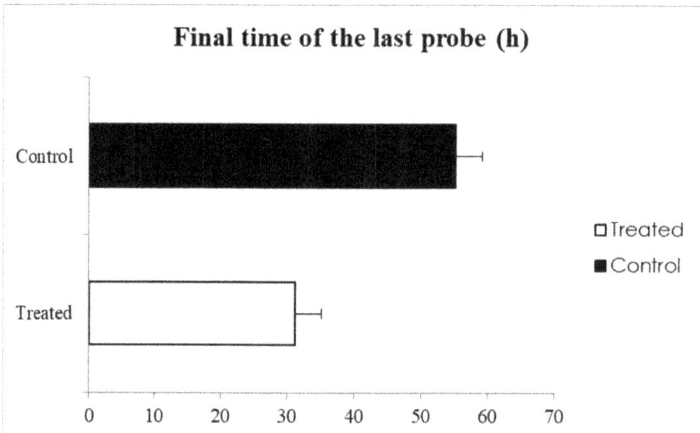

Figure 5. The final time of the last probe (%) of *Euschistus heros* untreated (control) and treated with *M. anisopliae* feeding on soybean plants. $p < 0.05$ is considered different by F-test.

4. Discussion

Among all stink bug species that damage soybean, the Neotropical brown stink bug *E. heros* is considered the most relevant pest species in all Brazil production regions [1,5], and this species is the main target of insecticide applications [42]. A soybean area of approximately 135 million ha was treated with chemical insecticides for stink bug control in the 2023/24 season [43]. Despite the large treated area, control failures or low control efficiencies have become more common [42–44].

The search for sustainable strategies to manage stink bug is underway with biologicals because they are host and prey of several natural enemies [1]. The most abundant egg parasitoid in soybean, *Telenomus podisi* Ashmead, 1893 (Hymenoptera, Platygastridae) is the species most commonly associated with *E. heros* in Brazil [45]. Ten commercial products containing *T. podisi* are available in Brazil for *E. heros* management [12]. Management strategies that favor the beneficial potential of parasitism must be adopted to allow more efficient natural and applied biological control. Entomopathogenic fungi, such as *M. anisopliae*, can be used to complement the egg parasitoid *T. podisi* [46].

Stink bugs, including *E. heros*, are notoriously difficult to control with entomopathogenic fungi due to the deployment of biochemical barriers, such as aldehyde production, that are quite efficient [47–50]. Despite the biochemical defenses against fungal infections, the BRM 2335 isolate of *M. anisopliae* showed high virulence against several stink bug species (Hemiptera: Pentatomidae), *E. heros*, *Oebalus poecilus* (Dallas, 1851), *Oebalus ypsilongriseus* (De Geer, 1773) (Hemiptera: Pentatomidae), *Thyanta perditor* (Fabricius, 1794) (Hemiptera: Pentatomidae), and *Tibraca limbativentris*, when compared to other *M. anisopliae* isolates, as well as *Beauveria bassiana* (Bals.) Vuill. and *Cordyceps javanica* (Hypocreales: Cordycipitaceae) [23,51]. This is supported by several studies showing that variation in host susceptibility to fungal infections depends on both species and genetic variability among isolates [52–55]

Confirming the findings of previous research, the BRM 2335 isolate of *M. anisopliae* tested in our study also displayed high levels of virulence to *E. heros* adults even at low concentrations of 5×10^6 conidia mL^{-1} (98% adult mortality after 10 days). This isolate was also tested for *E. heros* under laboratory, screenhouse, and field conditions [23,56]. A limitation of *Metarhizium* for *E. heros* control is that it requires a long time to kill this pod feeder. The LT_{50} for the isolate BRM 2335 tested at 1×10^8 conidia mL^{-1} (the concentration used in the EPG experiment) was 5.8 days. Despite the longer time to kill the host, the EPG experiment showed that adults treated with *M. anisopliae* spent significantly less time on probing activities than untreated insects. Furthermore, the NWEI was significantly reduced for feeding activities. Probing activities by infected adults were reduced by 86% on day 4 and ceased on day 5 (98%). EPG is an accurate method to evaluate the probing behavior of hemipterans that feed on the host's opaque tissue [25,26]. This tool may also be useful in the applied field of stink bug management, such as to evaluate the action of insecticides affecting their feeding and survivorship [27].

Our results showed that female *E. heros* were more susceptible to *M. anisopliae* than males, exhibiting a lower LT_{50} (7.1 versus 9.7 days, respectively). This may be attributed to the fecundity of females, which drains their energy and nutrients [57]. Studies also showed that female stink bug adults accumulate more lipids than male adults, indicating higher metabolic demands and possibly longer feeding times [58,59]. This behavior explains why EPG experiments are mostly being conducted with female stink bugs. Nevertheless, additional studies are required to elucidate the physiological mechanisms underlying the susceptibility of female *E. heros* to fungal infection.

In our study, *M. anisopliae* reduced female feeding on soybean seeds, resulting in fewer punctures compared to the untreated control. These results confirmed that fungal infection can influence the damage caused by stink bugs in soybean seeds. In addition to the reduction in probing, upon penetration into the host hemocoel, *Metarhizium* fungal cells absorb nutrients and destroy host cells [60]. The fungus also secretes toxic compounds known as secondary metabolites [61]. These substances can facilitate fungal invasion [62,63] or act as immunosuppressants, compounds that resist the host's defense [64].

Several studies have proven that *Metarhizium* can efficiently kill these insects [28]; however, little information is available about what happens to the probing behavior from the time of insect infection until death. Chen et al. [65] showed that infection by the fungus *Pandora neoaphidis* (Entomophthoromycotina: Entomophthorales) decreased stylet pathway activity and increased the time spent in non-probing activities compared with uninfected aphids. The corn leafhopper, *Dalbulus maidis*, treated with the fungus *Cordyceps javanica*, also exhibited disruption of the stylet activities in phloem and non-phloem phases 48 h after spraying [29].

Stink bugs treated with *M. anisopliae* showed a reduction (45%) in the FTLP compared with non-treated insects, i.e., treated stink bugs stopped their feeding activities earlier. This information is valuable for managing stink bugs at the field level, as it shows that even though the insect is still alive, its feeding is compromised, and consequently, the damage caused to the crop is reduced. Therefore, the slower lethal effect of the entomopathogenic fungus compared to synthetic insecticides can be minimized by changing the probing behavior of the insect treated with *M. anisopliae*. Despite the reduced time spent on probing activities by treated *Metarhizium* adults, the insect can still damage soybean seeds. One strategy to reduce the time to kill the host is mixing the entomopathogenic fungi with chemical insecticides. Sublethal or full concentrations of an insecticide can be mixed with *Metarhizium* to avoid damage to soybean by stink bugs. This strategy has proven to enhance the efficiency of entomopathogenic fungi for the control of several insect pests, in addition to reducing the selection of insects resistant to chemical insecticides [66–68]. Currently, this strategy (chemical + EPF) has been widely used by Brazilian producers to manage pests that are difficult to control, such as the corn leafhopper (*D. maidis*) and the whitefly *Bemisia tabaci* (Gennadius) (Hemiptera: Aleyrodidae) [69,70]. The quick action of the chemical insecticide combined with the long-lasting action of the fungi enhances the control efficiency, resulting in fewer field spraying applications [69]. Important aspects such as the compatibility between each species of fungus and the concentration of the synthetic insecticides must be taken into account.

5. Conclusions

In conclusion, our studies showed that Neotropical brown stink bug treated with *M. anisopliae* reduced its feeding activities by 86% at 4 days and ceased at 5 days after spraying; consequently, its damage to soybean seeds was reduced. EPG is a promising technique to evaluate the feeding behavior of stink bugs treated with entomopathogenic fungi. To further support the findings of our research, studies are underway to address the following questions: Can other fungal species and *M. anisopliae* isolates perform similarly to the highly adapted BRM 2335 isolate in reducing feeding and damage to soybean seeds by stink bugs? How do different concentrations of BRM 2335 affect the feeding activities of *E. heros*? Furthermore, this study paves the way for the further use of entomopathogenic fungi for pest control.

Author Contributions: Conceived and designed the experiments: A.C.d.S.A., E.D.Q., F.G.d.J. and H.A.B. Performed the experiments: M.A.R., A.S.V. and H.A.B. Analyzed the data: A.C.d.S.A. and J.F.A.e.S. Wrote the paper: A.C.d.S.A., E.D.Q., M.A.R., H.A.B., A.S.V. and F.G.d.J. All authors have read and agreed to the published version of the manuscript.

Funding: This work was made possible through financial support from the: FAPED/EMBRAPA/TED grant n° SEG 20.23.10.002.00.00, under the project "Generation of biological inputs and study of the bio-inputs production chain as strategies for the sustainable development of agriculture in Goiás", and Centro de Excelência em Bioinsumos (CEBIO) and Fundação de Amparo a Pesquisa de Goiás, grant n° FAPEG 202117647003995.

Institutional Review Board Statement: Not applicable.

Informed Consent Statement: Not applicable.

Data Availability Statement: The original contributions presented in the study are included in the article, further inquiries can be directed to the corresponding author.

Conflicts of Interest: Author Heloiza Alves Boaventura, Angélica Siqueira Vieira, José Francisco Arruda e Silva and Eliane Dias Quintela were employed by the company Embrapa Arroz e Feijão.

The remaining authors declare that the research was conducted in the absence of any commercial or financial relationships that could be construed as a potential conflict of interest.

References

1. Sosa-Gómez, D.R.; Corrêa-Ferreira, B.S.; Kraemer, B.; Pasini, A.; Husch, P.E.; Delfino Vieira, C.E.; Martinez, C.B.R.; Negrao Lopes, I.O. Prevalence, damage, management and insecticide resistance of stink bug populations (Hemiptera: Pentatomidae) in commodity crops. *Agric. For. Entomol.* **2020**, *22*, 99–118. [CrossRef]
2. Hickmann, F.; Cordeiro, E.M.G.; Aurélio, M.S.L.; Saldanha, A.V.; Corrêa, A.S. Host preference and survivorship of *Euschistus heros* strains on cotton and soybean. *Entomol. Exp. Appl.* **2023**, *171*, 564–571. [CrossRef]
3. Saluso, A.; Xavier, L.; Silva, F.A.C.; Panizzi, A.R. An invasive pentatomid pest in Argentina: Neotropical brown stink bug, *Euschistus heros* (F.) (Hemiptera: Pentatomidae). *Neotrop. Entomol.* **2011**, *40*, 704–705. [PubMed]
4. Panizzi, A.R. Growing problems with stink bugs (Hemiptera: Heteroptera: Pentatomidae): Species invasive to the US and potential Neotropical invaders. *Am. Entomol.* **2015**, *61*, 223–233. [CrossRef]
5. Panizzi, A.R.; de Freitas Bueno, A.; Silva, F.A.C. Insetos que atacam vagens e grãos. In *Soja: Manejo Integrado de Insetos e Outros Artrópodes-Praga*; Hoffmann-Campo, C.B., Corrêa-Ferreira, B.S., Moscardi, F., Eds.; Embrapa: Londrina, Brazil, 2012; pp. 335–420.
6. Costa, M.L.; Borges, M.; Vilela, E.F. Biologia reprodutivade *Euschistus heros* (F.) (Heteroptera: Pentatomidae). *An. Soc. Entomol. Bras.* **1998**, *27*, 559–568. [CrossRef]
7. Lucini, T.; Panizzi, A.R. Electropenetrography monitoring of the neotropical brown-stink bug (Hemiptera: Pentatomidae) on soybean pods: An Electrical Penetration Graph-histology analysis. *J. Insect Sci.* **2018**, *18*, 5. [CrossRef]
8. Lucini, T.; Panizzi, A.R.; Semeão, A.A. Electropenetrography study of *Euschistus heros* (F.) (Heteroptera: Pentatomidae) nymphs feeding at different sites of the soybean pod. *Environ. Entomol.* **2023**, *52*, 546–554. [CrossRef]
9. Silva, F.A.C.; da Silva, J.J.; Depieri, R.A.; Panizzi, A.R. Feeding activity, salivary amylase activity, and superficial damage to soybean seed by adult *Edessa meditabunda* (F.) and *Euschistus heros* (F.) (Hemiptera: Pentatomidae). *Neotrop. Entomol.* **2012**, *41*, 386–390. [CrossRef]
10. Tessmer, M.A.; Kuhn, T.M.A.; Appezzato-da-Gloria, B.; Lopes, J.R.S.; Erler, G.; Bonani, J.P. Histology of damage caused by *Euschistus heros* (F.) nymphs in soybean pods and seeds. *Neotrop. Entomol.* **2022**, *52*, 112–121. [CrossRef]
11. Ecco, M.; Borella Junior, C.; Miranda, D.S.; Araújo, M.S.; Almeida, A.C.d.S.; Jesus, F.G. Stink bug control at different stages of soybean development. *Arq. Inst. Biol.* **2020**, *87*, e1132018. [CrossRef]
12. AGROFIT. Sistema de Agrotóxicos Fitossanitários. Available online: http://agrofit.agricultura.gov.br/agrofit_cons/principal_agrofit_cons/ (accessed on 2 June 2024).
13. Guedes, R.N.C. Insecticide resistance, control failure likelihood and the First Law of Geography. *Pest Manag. Sci.* **2017**, *73*, 479–484. [CrossRef] [PubMed]
14. Tuelher, E.; Silva, E.H.; Rodrigues, H.S.; Hirose, E.; Guedes, R.N.C.; Oliveira, E.E. Area-wide spatial survey of the likelihood of insecticide control failure in the neotropical brown stink bug *Euschistus heros*. *J. Pest Sci.* **2017**, *91*, 849–859. [CrossRef]
15. Tibola, C.M.; Silva, L.; Sgubin, F.; Omoto, C. Monitoring resistance of *Euschistus heros* (Fabricius) (Hemiptera: Pentatomidae) to insecticides by using encapsulated artificial diet bioassay. *Insects* **2021**, *12*, 599. [CrossRef]
16. Woodcock, B.A.; Bullock, J.M.; Shore, R.F.; Heard, M.S.; Pereira, M.G.; Redhead, J.; Ridding, L.; Dean, H.; Sleep, D.; Henrys, P.; et al. Country-specific effects of neonicotinoid pesticides on honey bees and wild bees. *Science* **2017**, *356*, 1393–1395. [CrossRef]
17. Torres, J.B.; Bueno, A.F. Conservation biological control using selective insecticides: A valuable tool for IPM. *Biol. Control* **2018**, *126*, 53–64. [CrossRef]
18. Lee, R.; den Uyl, R.; Runhaar, H. Assessment of policy instruments for pesticide use reduction in Europe; learning from a systematic literature review. *Crop Prot.* **2019**, *126*, 104929. [CrossRef]
19. Maciel, R.M.A.; Bueno, A.F. The Role of Integrated Pest Management for Sustainable Food Production: The Soybean Example. In *Biodiversity, Functional Ecosystems and Sustainable Food Production*; Galanakis, C.M., Ed.; Springer Nature: Cham, Switzerland, 2022; pp. 117–139.
20. van Lenteren, J.C.; Bolckmans, K.; Köhl, J.; Ravensberg, W.J.; Urbaneja, A. Biological control using invertebrates and microorganisms: Plenty of new opportunities. *BioControl* **2018**, *63*, 39–59. [CrossRef]
21. Quintela, E.D.; Mascarin, G.M.; Silva, R.A.; Barrigossi, J.A.F.; Martins, J.F.S. Enhanced susceptibility of *Tibraca limbativentris* (Heteroptera: Pentatomidae) to *Metarhizium anisopliae* with sublethal doses of chemical insecticides. *Biol. Control* **2013**, *66*, 56–64. [CrossRef]
22. Boaventura, H.B.; Quintela, E.D. The multifunctionality of the fungus *Metarhizium* spp. and its use in Brazilian agriculture. *Bragantia* **2025**, *in press*.
23. Sousa, L.M.; Quintela, E.D.; Boaventura, H.A.; Silva, J.F.A.; Tripode, B.M.D.; Miranda, J.E. Selection of entomopathogenic fungi to control stink bugs and cotton boll weevil. *Pesqui. Agropecu. Trop.* **2023**, *53*, e76316. [CrossRef]

24. Backus, E.A.; Bennett, W.H. The AC–DC Correlation Monitor: New EPG design with flexible input resistors to detect both R and emf components for any piercing-sucking hemipteran. *J. Insect Physiol.* **2009**, *55*, 869–884. [CrossRef] [PubMed]
25. Walker, G.P. A beginner's guide to electronic monitoring of homopteran probing behavior. In *Principles and Applications of Electronic Monitoring and Other Techniques in the Study of Homopteran Feeding Behavior*; Walker, G.P., Backus, E.A., Eds.; Entomological Society of America: Lanham, MD, USA, 2000; pp. 14–40.
26. Lucini, T.; Panizzi, A.R. Behavioral comparisons of ingestion and excretion by selected species of pentatomids: Evidence of feeding on different food sources supports pest status. *Neotrop. Entomol.* **2017**, *46*, 361–367. [CrossRef] [PubMed]
27. Lucini, T.; Panizzi, A.R. Electropenetrography (EPG): A breakthrough tool unveiling stink bug (Pentatomidae) feeding on plants. *Neotrop. Entomol.* **2018**, *47*, 6–18. [CrossRef] [PubMed]
28. Tjallingii, W.F. Electronic recording of penetration behaviour by aphids. *Entomol. Exp. Appl.* **1978**, *24*, 721–730. [CrossRef]
29. Maluta, N.; Castro, T.; Spotti Lopes, J.R. DC-electrical penetration graph waveforms for *Dalbulus maidis* (Hemiptera: Cicadellidae) and the effects of entomopathogenic fungi on its probing behavior. *Sci. Rep.* **2023**, *13*, 22033. [CrossRef]
30. Silva, C.C.; Laumann, R.A.; Blassioli, M.C.; Pareja, M.; Borges, M. *Euschistus heros* mass rearing technique for the multiplication of *Telenomus podisi*. *Pesqui. Agropecuária Bras.* **2008**, *43*, 575–580. [CrossRef]
31. Bischoff, J.F.; Rehner, S.A.; Humber, R.A. A multilocus phylogeny of the *Metarhizium anisopliae* lineage. *Mycologia* **2009**, *101*, 508–528. [CrossRef]
32. Mascarin, G.M.; Quintela, E.D. Técnica de produção do fungo entomopatogênico *Metarhizium anisopliae* para uso em controle biológico. In *Embrapa Arroz e Feijão-Documentos (INFOTECA-E)*; Embrapa: Brasília, Brazil, 2013. Available online: https://www.embrapa.br/busca-de-publicacoes/-/publicacao/968726/tecnica-de-producao-do-fungo-entomopatogenico-metarhizium-anisopliae-para-uso-em-controle-biologico (accessed on 17 November 2024).
33. França-Neto, J.B.; Krzyzanowski, F.C.; Costa, N.P. *O Teste de Tetrazólio em Sementes de Soja*; Embrapa Soja: Brasília, Brazil, 1998. Available online: https://www.embrapa.br/busca-de-publicacoes/-/publicacao/461306/o-teste-de-tetrazolio-em-sementes-de-soja (accessed on 17 November 2024).
34. Almeida, A.C.d.S.; de Jesus, F.G.; Barrigossi, J.A.F. Unveiling the feeding behavior of *Tibraca limbativentris* (Hemiptera: Pentatomidae) on rice using an electropenetrography waveform library. *J. Insect Sci.* **2020**, *20*, 14. [CrossRef]
35. Lucini, T.; Panizzi, A.R. Waveform characterization of the soybean stem feeder *Edessa meditabunda* (F.) (Hemiptera: Heteroptera: Pentatomidae): Overcoming the challenge of wiring pentatomids for EPG. *Entomol. Exp. Appl.* **2016**, *158*, 118–132. [CrossRef]
36. Cervantes, F.A.; Backus, E.A.; Perring, T.M. EPG waveform library for *Graphocephala atropunctata* (Hemiptera: Cicadellidae): Effect of adhesive, input resistor, and voltage levels on waveform appearance and probing behaviors. *J. Insect Physiol.* **2018**, *109*, 21–40. [CrossRef]
37. Backus, E.A.; Cline, A.R.; Ellerseick, M.R.; Serrano, M.S. *Lygus hesperus* (Hemiptera: Miridae) feeding on cotton: New methods and parameters for analysis of nonsequential electrical penetration graph data. *Ann. Entomol. Soc. Am.* **2007**, *100*, 296–310. [CrossRef]
38. Sarria, E.; Cid, M.; Garzo, E.; Fereres, A. Excel workbook for automatic parameter calculation of EPG data. *Comput. Electron. Agric.* **2009**, *67*, 35–42. [CrossRef]
39. Ebert, T.A.; Backus, E.A.; Cid, M.; Fereres, A.; Rogers, M.E. A new SAS program for behavioral analysis of electrical penetration graph data. *Comput. Electron. Agric.* **2015**, *116*, 80–87. [CrossRef]
40. Cervantes, F.A.; Backus, E.A.; Godfray, L.; Akbar, W.; Clark, T.L. Characterization of an EPG waveform library for adult *Lygus lineolaris* and *Lygus hesperus* (Hemiptera: Miridae) feeding on cotton squares. *Ann. Entomol. Soc. Am.* **2016**, *109*, 684–697. [CrossRef]
41. Gottschalk, P.G.; Dunn, J.R. Measuring parallelism, linearity, and relative potency in bioassay and immunoassay data. *J. Biopharm. Stat.* **2005**, *15*, 437–463. [CrossRef]
42. Sosa-Gómez, D.R.; Silva, J.J.d. Neotropical brown stink bug (*Euschistus heros*) resistance to methamidophos in Paraná, Brazil. *Pesqui. Agropecu. Bras.* **2010**, *45*, 767–769. [CrossRef]
43. Kynetec. Available online: https://www.kynetec.com/agriculture (accessed on 20 December 2024).
44. Roggia, S.; Utiamada, C.M.; Hirose, E.; Stoetzer, A.; Avila, C.J.; Kischel, E.; Marzarotto, F.O.; Tomquelski, G.V.; Guedes, J.V.C.; Arnemann, J.A.; et al. Eficiência de inseticidas no controle do percevejo-marrom (*Euschistus heros*) em soja, na safra 2013/14: Resultados sumarizados de ensaios cooperativos. *Rev. Bras. Entomol.* **2018**, *62*, 45–58.
45. Pacheco, D.J.P.; Corrêa-Ferreira, B.S. Parasitismo de *Telenomus podisi* Ashmead (Hymenoptera: Scelionidae) em populações de percevejos pragas da soja. *An. Soc. Entomol. Bras.* **2000**, *29*, 295–302. [CrossRef]
46. Battisti, L.; Warmling, J.V.; de Freitas Vieira, C.; de Oliveira, D.H.R.; de Lima, Y.R.A.; de Freitas Bueno, A.; Lozano, E.R. Seletividade de *Metarhizium anisopliae* e *Beauveria bassiana* para adultos de *Telenomus podisi* (Hymenoptera: Scelionidae). *Semina Ciênc. Agrar.* **2022**, *43*, 727–738. [CrossRef]
47. Sosa-Gómez, D.R.; Boucias, D.G.; Nation, J.L. Attachment of *Metarhizium anisopliae* to the southern green stink bug *Nezara viridula* cuticle and fungistatic effect of cuticular lipids and aldehydes. *J. Invertebr. Pathol.* **1997**, *69*, 31–39. [CrossRef]

48. Sosa-Gómez, D.R.; Moscardi, F. Laboratory and f ield studies on the infection of stink bugs, *Nezara viridula, Piezodorus guildinii* and *Euschistus heros* (Hemiptera: Pentatomidae) with *Metarhizium anisopliae* and *Beauveria bassiana* in Brazil. *J. Invertebr. Pathol.* **1998**, *71*, 115–120. [CrossRef] [PubMed]
49. Pike, T.J. Interactions Between the Invasive Brown Marmorated Stink Bug, *Halyomorpha halys* (*Hemiptera: Pentatomidae*), and *Entomopathogenic fungi*. Master's Dissertation, University of Maryland: College Park, MD, USA, 2014.
50. Silva, R.A.; Quintela, E.D.; Mascarin, G.M.; Pedrini, N.; Lião, L.M.; Ferri, P.H. Unveiling chemical defense in the rice stalk stink bug against the entomopathogenic fungus *Metarhizium anisopliae. J. Invertebr. Pathol.* **2015**, *127*, 93–100. [CrossRef] [PubMed]
51. Martins, J.F.S.; Lima, M.G.A.; Botton, M.; Carbonari, J.J.; Quintela, E.D. Efeito de isolados de *Metarhizium anisopliae* (Metsch.) Sorok. e *Beauveria bassiana* (Bals.) Vuill. sobre o percervejo-do-colmo do arroz, *Tibraca limbativentris* Stal. *Ana. Soc. Entomol. Bras.* **1997**, *26*, 277–283. [CrossRef]
52. Hajek, A.E.; St. Leger, R.J. Interactions between fungal pathogens and insect hosts. *Annu. Rev. Entomol.* **1994**, *39*, 293–322. [CrossRef]
53. St. Leger, R.J.; Wang, J.B. *Metarhizium*: Jack of all trades, master of many. *Open Biol.* **2020**, *10*, 200307. [CrossRef]
54. Castrillo, L.A.; Roberts, D.W.; Vandenberg, J.D. The fungal past, present, and future: Germination, ramification, and reproduction. *J. Invertebr. Pathol.* **2005**, *89*, 46–56. [CrossRef]
55. Islam, W.; Adnan, M.; Shabbir, A.; Naveed, H.; Abubakar, Y.S.; Qasim, M.; Tayyab, M.; Noman, A.; Nisar, M.S.; Khan, K.A.; et al. Insect-fungal-interactions: A detailed review on entomopathogenic fungi pathogenicity to combat insect pests. *Microb. Pathog.* **2021**, *159*, e105122. [CrossRef]
56. Vieira, A.S. *Fatores relacionados* a *Euschistus heros* (Hemiptera: Pentatomidae) que Afetam a Virulência de *Metarhizium anisopliae* (Hypocreales: Clavicipitaceae). Master's Dissertation, Universidade Federal de Goiás, Goiânia, Brazil, 2023.
57. Panizzi, A.R.; Oliveira, E.D. Performance and seasonal abundance of the neotropical brown stink bug, *Euschistus heros* nymphs and adults on a novel food plant (pigeonpea) and soybean. *Entomol. Exp. App.* **1998**, *88*, 169–175. [CrossRef]
58. Panizzi, A.R. Ecologia nutricional de insetos sugadores de sementes. In *Ecologia Nutricional de Insetos e Suas Implicações no Manejo de Pragas*; Panizzi, A.R., Parra, J.R.P., Eds.; Editora Manole Ltda: São Paulo, Brazil, 1991; pp. 253–287.
59. Piubelli, G.C.; Hoffmann-Campo, C.B.; Arruda, I.C.D.; Lara, F.M. Nymphal development, lipid content, growth and weight gain of *Nezara viridula* (L.) (Heteroptera: Pentatomidae) fed on soybean genotypes. *Neotrop. Entomol.* **2003**, *32*, 127–132. [CrossRef]
60. Qu, S.; Wang, S. Interaction of entomopathogenic fungi with the host immune system. *Dev. Comp. Immunol.* **2018**, *83*, 96–103. [CrossRef]
61. Boucias, D.G.; Pendland, J.C. Entomopathogenic fungi: Fungi imperfecti. In *Principles of Insect Pathology*; Springer: Boston, MA, USA, 1998; pp. 321–364.
62. Ferron, P. Fungal control. In *Comparative Insect Physiology, Biochemistry and Pharmacology*; Kerkut, G.A., Gilbert, L.I., Eds.; Pergamon Press: Oxford, UK, 1985; Volume 12, pp. 313–346.
63. Roberts, D.W.; Hajek, A.E. Entomopathogenic fungi as bioinsecticides. In *Frontiers in Industrial Mycology*; Springer US: Boston, MA, USA, 1992; pp. 144–159.
64. Trienens, M.; Rohlfs, M. Insect–fungus interference competition–the potential role of global secondary metabolite regulation, pathway-specific mycotoxin expression and formation of oxylipins. *Fungal Ecol.* **2012**, *5*, 191–199. [CrossRef]
65. Chen, C.; Ye, S.; Hu, H.; Xue, C.; Yu, X. Use of electrical penetration graphs (EPG) and quantitative PCR to evaluate the relationship between feeding behaviour and *Pandora neoaphidis* infection levels in green peach aphid, *Myzus persicae. J. Insect Physiol.* **2018**, *104*, 9–14. [CrossRef] [PubMed]
66. Quintela, E.D.; McCoy, C.W. Pathogenicity enhancement of *Metarhizium anisopliae* and *Beauveria bassiana* to first instars of *Diaprepes abbreviatus* (Coleoptera: Curculionidae) with sublethal doses of imidacloprid. *Environ. Entomol.* **1997**, *26*, 1173–1182. [CrossRef]
67. Quintela, E.D.; McCoy, C.W. Synergistic effect of imidacloprid and two entomopathogenic fungi on the behavior and survival of larvae of *Diaprepes abbreviatus* (Coleoptera: Curculionidae) in soil. *J. Econ. Entomol.* **1998**, *91*, 110–122. [CrossRef]
68. Quintela, E.D.; McCoy, C.W. Conidial attachment of *Metarhizium anisopliae* and *Beauveria bassiana* to the larval cuticle of *Diaprepes abbreviatus* (Coleoptera: Curculionidae) treated with imidacloprid. *J. Invertebr. Pathol.* **1998**, *72*, 220–230. [CrossRef]
69. Boaventura, H.A. Eficácia de *Cordyceps javanica* Sozinho ou em Combinação com Inseticidas Químicos no Controle de *Bemisia tabaci* MEAM1 e Persistência em Folhas de Soja. Master's Dissertation, Universidade Federal de Goiás, Goiânia, Brazil, 2019.
70. Poletto, T.B. Enhancing *Dalbulus maidis* (Hemiptera: Cicadellidae) Control: Na Integrated Approach Combining *Cordyceps javanica* (Ascomycota: Hypocreales) and Insecticides. Ph.D. Thesis, Universidade de São Paulo, São Paulo, Brazil, 2024.

Journal of
Fungi

MDPI

Article

Beauveria bassiana Induces Strong Defense and Increases Resistance in Tomato to *Bemisia tabaci*

Mengying Liu [1,2,†], Dong Xiang [3,†], Heikki M. T. Hokkanen [4], Tiandi Niu [1,2], Junjie Zhang [1,2], Jinlin Yang [1,2], Qiuyang Wei [1], Hanqiu Chen [3], Huai Liu [1,2,*] and Yaying Li [1,2,*]

[1] Yibin Academy of Southwest University, Southwest University, Yibin 644000, China; liumengying@swu.edu.cn (M.L.); ysdtz1912@email.swu.edu.cn (T.N.); zhangjunjie627@163.com (J.Z.); xzlpsyjl@163.com (J.Y.); weiqy_88@126.com (Q.W.)

[2] Key Laboratory of Agricultural Biosafety and Green Production of Upper Yangtze River, College of Plant Protection, Southwest University, Chongqing 400715, China

[3] Institute of Vegetable, Tibet Academy of Agriculture and Animal Husbandry Sciences, Lhasa 850032, China; xiangd666@126.com (D.X.); chenhanqiu11@126.com (H.C.)

[4] Stockbridge School of Agriculture, University of Massachusetts, Amherst, MA 01003, USA; heikki.hokkanen@ecostack.eu

* Correspondence: liuhuai@swu.edu.cn (H.L.); liyaying@swu.edu.cn (Y.L.)

† These authors contributed equally to this work.

Abstract: Pre-stimulation of plants can change their resistance mechanisms, thereby enhancing their defense responses. *Beauveria bassiana*, a broad-spectrum entomogenous fungi, can also induce plant defenses, but it received little attention. Here, we show that *B. bassiana* can act as a stimulus to prime tomato defense responses, improving resistance in the plant to herbivore stress. The results illustrated that four defense genes (*PIN2*, *PR2*, *PAL*, and *MPK3*) were upregulated in all *B. bassiana* treatments, especially the phenylalanine deaminase (*PAL*) gene, which was highly expressed in tomato plants after *B. bassiana* inoculation. Feeding through *Bemisia tabaci* resulted in a weak upregulation of defense genes. However, in combined fungal inoculation and *B. tabaci* feeding, a total of nine defense genes were upregulated, among which five genes—*PAL*, *PPO*, *PIN2*, *PR2*, and *PR1*—were closely related to the phenol synthesis. The results of tomato plant metabolism showed that *B. bassiana* mainly activates tomato phenylpropane metabolic pathways, with this modulation being influenced by jasmonate. Further explorations revealed a significant enhancement in the antioxidant capacity of the plants, as evidenced by the determination of their antioxidant compounds and the coloration of leaf phenolic substances. Thus, entomopathogenic fungi can act as an exogenous substance to activate the defense responses of tomatoes without damaging the plant, indicating a good potential for developing applications using *B. bassiana* to promote resistance in tomatoes for pest management.

Keywords: phenylalanine deaminase; induced resistance; phenolics; antioxidant

1. Introduction

Plants are exposed to a wide range of pests and pathogens to which they respond by activating a variety of defense signaling pathways [1]. The defense responses constitute localized responses that reduce the outside attack at the sites of stress, as well as systemic responses that impart resistance to other parts of the plant on a subsequent attack [2]. This inducible defense could be activated by some beneficial microbes, such as rhizosphere microorganisms and entomogenous fungi [3].

Beauveria bassiana is an entomogenous fungus widely used in biological control. The role of pathogenic mechanisms in insect infection has been described [4], but previous studies have predominantly focused on pest pathogenicity, ignoring the interactions among plants, insect-pathogenic microorganisms, and pests. Microbes are not only associated with insects but also can profoundly influence plant–insect interactions [5–7]. Indeed, it has been confirmed that entomogenous fungi *Verticillium lecanii* and *Metarhizium anisopliae* could stimulate plants by altering plant metabolic components, induce attraction or repulsion to insects, and may be toxic to pests [8–10].

Plants were subject to feeding stress by many pests, evolving a series of complex defense systems against insect herbivores, such as *Bemisia tabaci* [11]. Researches showed that *B. tabaci*, a generalist herbivore, damages crops by feeding on phloem sap and may suppress induced plant responses, including reducing the induction of the jasmonic acid (JA)-mediated defense signaling pathway via its salivary protein, the ferritin BtFer1 [12,13]. Therefore, the importance of enhancing plant defenses and eliminating pest interference effects is indisputable.

Since the publication of the review article 'Priming: getting ready for battle', the importance of defense priming as an adaptive trait for the adjustment of plant defense in unpredictable environments has been well established [14]. Defense priming has been proposed as an adaptive and cost-effective defensive strategy, characterized by a subtle and transient response initiated by specific stimuli. In primed plants, defense responses are activated more rapidly, robustly, or sustainably, enabling them to better address challenges compared to unprimed plants [15,16]. Moreover, much of this type of resistance can be passed down through generations [14]. Defense priming can be triggered by various factors, including pest herbivores, pathogens, and chemical compounds, as well as beneficial edaphons, rhizosphere microorganisms, and even entomogenous fungi [17,18].

Plant stimulation can enhance their resistance by eliciting a stronger defense response, serving as an adaptive mechanism to cope with external stress. Defensive traits may encompass a range of alterations, including changes in defense-related processes, signaling compounds, and actual defense responses [19–21]. Secondary metabolites in plants are compounds synthesized by plants themselves without influencing normal plant growth and development, serving as direct indicators of plant defense functions [22]. Among the secondary metabolites, phenolic substances are among the most prevalent and widely distributed components, playing a crucial role in enhancing plant defense mechanisms [23,24]. When plants respond to insect feeding attacks, the phenylpropane metabolic pathways are activated, mediating the production of phenolic substances, and subsequently, the content of reactive oxygen species increases to resist pests [25,26].

Jasmonic acid (JA) is a central regulator of plant defense responses and is a lipid-derived compound commonly found in plants, belonging to the oxylipin family of lipoxygenase products [27,28]. It can be obtained from α-linolenic acid as a substrate through the catalysis of lipoxygenase (LOX) and β-oxidation reaction [29,30]. JA can respond to pathogen infection or herbivore feeding, thereby activating the expression of defense genes such as pathogenesis-related genes (PRs), LOXs, and vegetative storage proteins (VSPs), and simultaneously regulating the increase in the activities of defense enzymes such as protease inhibitor (PI), polyphenol oxidase (PPO), and catalase (CAT), playing a crucial role in plant growth, development, and defense [31,32].

Some studies indicate systemic protection of *Papaver somniferum* against *Iraella luteipes* by an endophytic strain of *B. bassiana* [33]. Entomogenous fungi can colonize corn, weakening the drill collar activity of lepidopteran larvae on stalks [34]. These findings suggest that entomopathogenic fungi can act as inducers to enhance plant defense, but research on the mechanisms of enhancement is incomplete.

The aim of this study was to check the hypothesis that *B. bassiana* can stimulate the plant to release a faster and stronger response to *B. tabaci* stress. To assess this, quantitative real-time PCR and gas chromatograph mass spectrometry were used to detect plant primary defense pathways induced by *B. bassiana*, and plant antioxidant tests were performed to evaluate the plant's defensive ability. This study contributes to our understanding of how *B. bassiana* continually interacts with plant defenses and enhances plant resistance to the generalist herbivore *B. tabaci*.

2. Materials and Methods

2.1. Insect, Fungal, and Plant Materials

The insect used in the experiments was the whitefly *Bemisia tabaci* (biotype: MED) adults, which were collected from vegetables in the Chongqing Tongnan district, China, and maintained at 26 ± 1 °C, relative humidity (RH) 75 ± 5%, and light (L):dark (D) = 14:10, using tomato (*Solanum lycopersicum*) as a host plant for more than 30 generations. The fungal strain was *Beauveria bassiana* Bb252, stored in the Biotechnology Center of Southwest University at −80 °C. It was originally isolated from *Chilo suppressalis* on maize in Yongchuan District, Chongqing, China, and was separated into a monosporic culture. The fungus was grown on potato dextrose agar (PDA) (pH = 6.5–6.7) at 28 ± 1 °C in a temperature-controlled chamber under darkness. Tomato seeds were purchased from Shanghai Changzhong Tomato Co., Ltd., Shanghai, China, without germination tests. Before sowing, they were sterilized by soaking in 1% NaClO for 5 min, then rinsed three times with sterile distilled water [35,36]. The seedlings were then planted in sterilized nutrient soils (dry sterilization: 75 °C for 24 h) and transferred to an artificial climate chamber at 26 ± 1 °C, RH = 75 ± 5%, and L:D = 14:10. The test started when the tomato plants reached the height of 20 cm.

2.2. Experimental Design

There were a total of five treatments in this study. Five plant treatments were included: (1) plants were foliar sprayed with 10 mL of 1×10^8 conidia/mL concentration of spore suspension and tested after 3 days; (2) plants were foliar sprayed with 10 mL of 1×10^8 conidia/mL concentration of spore suspension and tested after 7 days; (3) plants were foliar sprayed with 10 mL of 1×10^8 conidia/mL concentration of spore suspension and tested after 14 days; (4) whiteflies were released on plants at a density of 30 adults per plant to feed for 7 days; (5) fungal inoculation consistent with treatment (2), combined with inoculation with *B. tabaci* 7 days later, when 30 whitefly adults were released for per plant and were allowed to feed for 7 days before test [20,37]; and (6) plants spraying with 0.1 mL of methyl jasmonate ($C_{13}H_{20}O_3$ assay ≥ 95% Coolaber Science & Technology, Beijing, China). Each group contained four replicates. The subsequent experimental treatments were all numbered accordingly.

2.3. Tomato Plant Defense Gene Amplification and Characterization

Total RNA from *Solanum lycopersicum* was isolated from fresh lamina tissue using pBIOZOL Plant Total RNA Extraction Reagent (BIOER Technology Co., Ltd., Hangzhou, China). The RNA concentration and purity were determined using a NanoDrop™ Spectrophotometer ND-2000 (Thermo Scientific, Waltham, MA, USA). Rapid amplification of cDNA with the PrimeScript™ RT reagent Kit (TaKaRa, Otsu-shi, Japan) was performed according to the manufacturer's instructions. Based on the RNA-seq data, the transcript abundance of multiple defense pathway key genes was quantified. The primers used in this experiment are listed in Table S1. Regarding gene expression, a two-step method was used for quantitative real-time PCR detection with the following thermal cycling conditions: 95 °C 1 min, followed by 40 cycles of 20 s at 95 °C, and 1 min at 60 °C. All reactions were run in duplicate technical replicates, and average values were used in the subsequent analysis. The gene expression was quantified by using the $2^{-\Delta\Delta Ct}$ method with *β-actin* as an endogenous control gene and by drawing a heat map based on genes' relative expression multiples. In this experiment, the treatments (1), (2), (3), (4), and (5) and a control group (plants without treatment) were used.

2.4. Plant Metabolite Extraction

In this experiment, the treatments (2), (4), and (5) and a control group were used. Fresh tomato leaf samples (1 g) were initially freeze-dried at −50 °C for 48 h to remove moisture, then ground and placed into headspace vials with 1 mL acetone (HPLC) for 24 h [38,39]. The obtained extracts were filtered through a 0.2 mm polytetrafluoroethylene (PTFE) filter to eliminate impurities. The supernatant fractions were collected and stored at −20 °C (not more than 2 days) until used. The samples were evaporated in a vacuum freeze dryer and derivatized with 200 μL N, O-Bis (trimethylsilyl) trifluoroacetamide (BSTFA) prior to GC–MS analysis [40].

2.5. GC–MS Analysis

In this experiment, the treatments (2), (4), and (5) and control group were used. The samples were analyzed using an Agilent 7890B GC-MS system (Agilent Technologies, Santa Clara, CA, USA) equipped with a DB-5MS fused-silica capillary column (30 m × 0.25 mm, 0.25 μm). Helium (grade 5.0, 1.5 mL min^{-1}) was used as carrier gas. The injector was operated in the splitless mode for all chromatographic runs. The extracted sample using the M/C technique was analyzed under the following oven temperature program: injector temperature was 250 °C, 35 °C for 1 min, followed by 2 °C/min oven temperature ramp to 70 °C, then by 5 °C/min to 250 °C, and concluding with a 24 min isothermal hold at 280 °C. Mass spectra were acquired using the full scan monitoring mode with a mass scan range of 40–650 m/z. Qualitative analysis was performed by comparing the obtained MS spectra with the NIST library and ChEBI library. Plant metabolites were annotated for pathways through the Kyoto encyclopedia of genes and genomes (KEGG) pathway. The MetaboAnalyst 6.0 tool (https://www.metaboanalyst.ca/, 20 December 2024) was employed to conduct pathway enrichment analysis on the mass spectrometry results.

2.6. Plant Total Antioxidant Activity

In this experiment, the treatments (2), (4), (5), and (6) and a control group were used. The tomato plant samples were prepared using 0.5 g of fresh plant tissue homogenized in 2 mL of sterile water. The total antioxidant capacity was determined using a T-AOC Assay Kit (Nanjing Jiancheng Bioengineering Institute, Nanjing, China, A015-2-1). According to the kit instructions, serially dilute the standard Trolox (an antioxidant), then add the ABTS working solution and let it react for 6 min. Read the OD_{405} and fit the standard curve. Subsequently, in a similar manner, replace the standard with the sample to be tested, add it to the ABTS working solution, react for 6 min, and read the OD_{405}. Calculate the Trolox-Equivalent Antioxidant Capacity according to the standard curve.

2.7. Phenolic Substance Staining

In this experiment, the treatments (4) and (5) and a control group were used. A 0.5% toluidine blue staining was used for the phenolic substance of plant leaf staining via the characteristic that phenolic substances can combine with dyes and produce distinct blue-purple colors. Wash the leaves with PBS buffer 3 times, add the dye dilution 200 times, and dye for 30 min. Then, wash the leaves twice with PBS buffer and take pictures using an Olympus optical microscope (Olympus Corporation, Tokyo, Japan).

2.8. Statistical Analysis

All data are presented as mean ± SE. Statistical analysis was performed using ANOVA and Tukey's test, with significance reported when $p < 0.05$. IBM SPSS Statistics V21.0 software was used for all data calculations, and the figures were produced using GraphPad Prism 8.

3. Results

3.1. Tomato Plant Defense Gene Expression

Tomatoes rely on a complex defense system, and the expression of defense genes can be conveniently detected by quantitative PCR. The expression of defense genes in tomatoes was significantly upregulated after 7 days of *B. bassiana* treatment. Compared with 3 or 14 days of treatment, the number of upregulated defense genes at this time was the largest, and the upregulation fold was higher, showing a higher defense level. Four genes (*PIN2*, *PR2*, *PAL*, and *MPK3*) were concurrently upregulated at all three time points, with the phenylalanine deaminase gene (*PAL*) showing particularly high expression in tomato plants following *B. bassiana* inoculation. Its expression reached an 18.5-fold level with a peak at 7 days after the fungal treatment ($F_{10,43} = 8.181$, $p < 0.001$, Figure 1A).

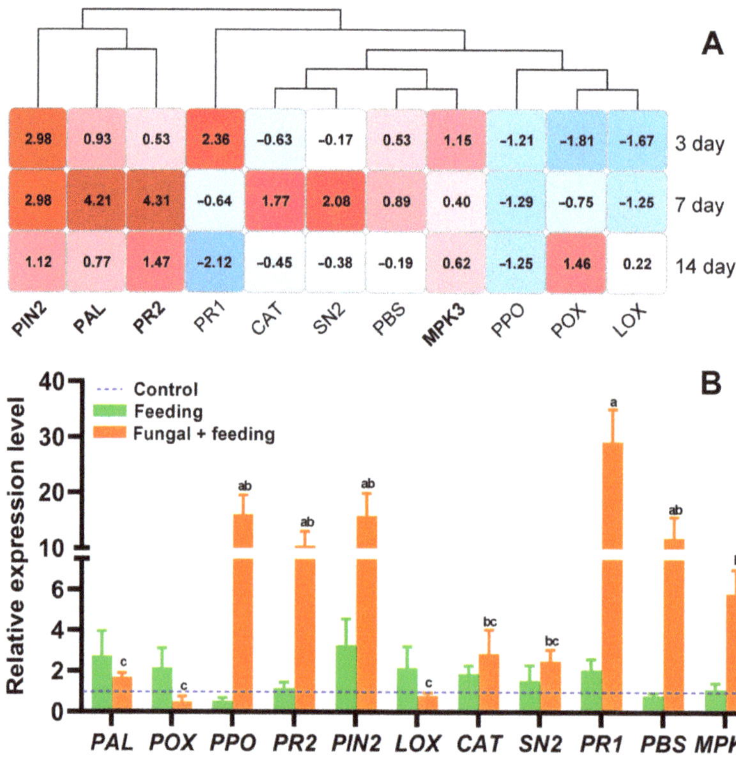

Figure 1. The expression patterns of defense genes in tomato. (**A**) Red represents upregulated expression; blue represents downregulated expression. The numbers in the figure represent the differential expression levels (the fold change) of various genes compared with the control. The bolded genes in the figure refer to genes that were upregulated for expression at all three times. (**B**) Feeding means only feeding by *B. tabaci*; fungal + feeding means *B. bassiana* treatment combined with *B. tabaci* feeding. The bars in the figure show the expression levels of each gene after different treatments. Different letters indicate significant differences (Tukey's multiple comparison test).

Compared with fungal inoculation, *B. tabaci* feeding induced gene expression at a lower level, and only a few defense genes were upregulated, including *PAL, POX, PIN2, LOX, CAT*, and *PR1*. The average gene expression level was two-fold. However, the gene expression profiles in plants with combined fungal inoculation and insect feeding produced remarkable results: 9 genes were upregulated, and the increases were over 10-fold ($F_{10,43} = 3.566$, $p = 0.003$, Figure 1B). In this case, defense response genes were mainly in the primary and secondary metabolic pathways, of which four upregulated genes (*PPO, PIN2, PR2*, and *PR1*) were identified as being involved in phenolic compound synthesis.

3.2. Identification of Plant Metabolites

The GC–MS results for the four treatment groups were different. The mass spectrometry results were compared based on the KEGG Pathway database. The results showed that the plants always take nutrient anabolism as the primary approach (Figure 2). The nutritional pathways obtained by comparing metabolites focus on fatty acid and amino acid synthesis and gluconeogenesis.

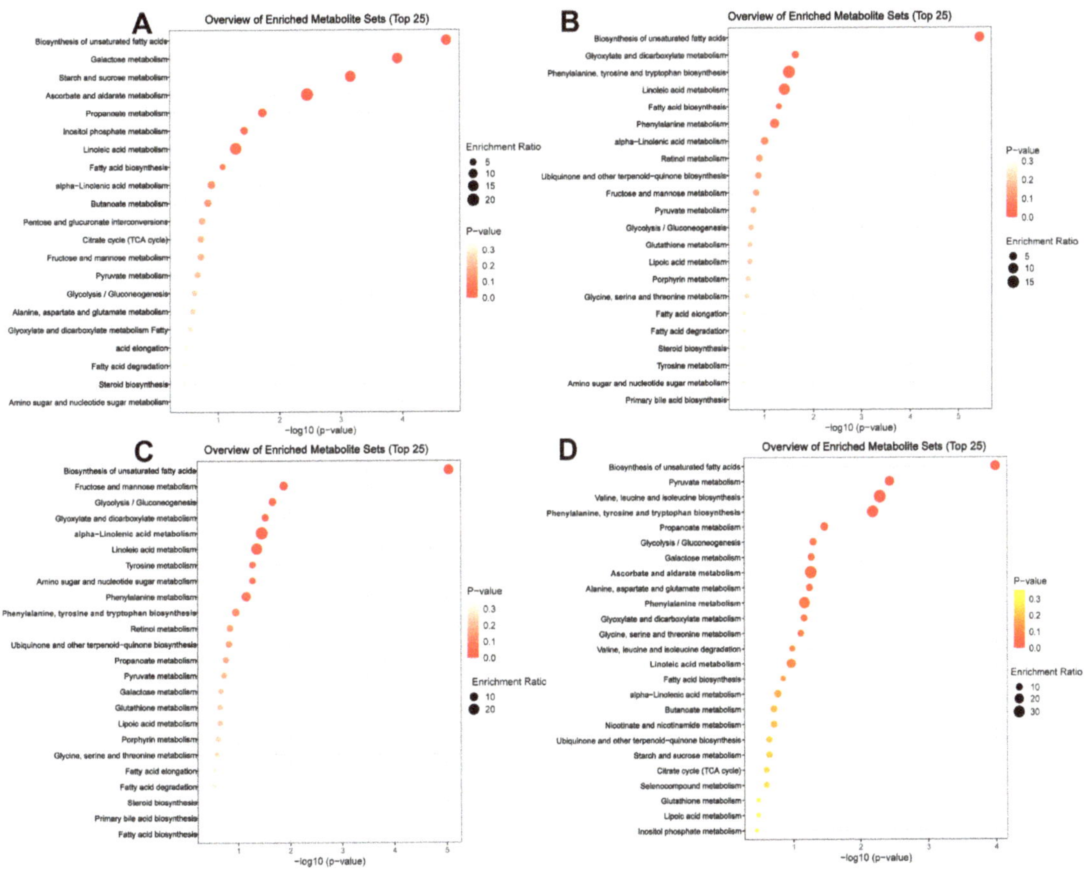

Figure 2. Kyoto encyclopedia of genes and genomes (KEGG) pathway enrichment analysis. (**A**) gives results for the control; (**B**) gives results by only spraying *B. bassiana*; (**C**) gives results for *B. tabaci* feeding; and (**D**) gives results for *B. bassiana* treatment combined with *B. tabaci* feeding.

Specific defense metabolites were produced only after being subjected to external stress signals. The plant secondary metabolite compounds increased sharply in *B. bassiana*-induced plants. Through compound database comparison, these belong to defense pathways and regulation of the hormone network, such as phenylalanine metabolism, linoleic acid metabolism, and alpha-linolenic acid metabolism (Figure 2B). Treatment with feeding by *B. tabaci* also induces phenylalanine metabolism, alpha-linolenic acid metabolism, glycolysis/gluconeogenesis, and glyoxylate and dicarboxylate metabolism (Figure 2C). Particularly striking was the result that the treatment combining fungal inoculation with insect feeding resulted in the activation of more defense-related pathways, with phenylpropaine metabolism mediated by *PAL* being abundant and prominent. In addition, phenylalanine, a crucial precursor involved in the synthesis of defensive substances such as phenols, had its biosynthetic pathway significantly enriched (Figure 2D). Furthermore, the phytohormone involved in plant defense regulation, jasmonate, along with its synthetic precursors, α-linolenic acid, and I-isoleucine, were detected. For the normalized peak-area calculations, their contents slightly increased as a result of *B. bassiana* inoculation (Figure 3). Based on the measurement of peak area, the jasmonate area was 2.8×10^4 prior to the fungal treatment and increased rapidly to 6.6×10^4 after the fungi and feeding treatment (2.4-fold, $p < 0.001$).

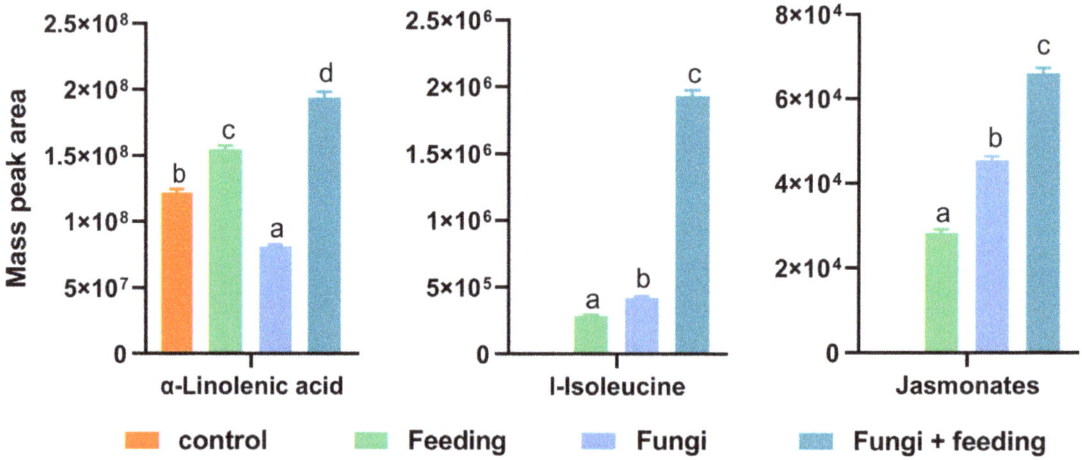

Figure 3. The relative content of the jasmonates and their precursors. The peak area of the mass spectrum was used to indicate the relative content of the substance. Different letters indicate significant differences ($p < 0.05$; Tukey's multiple comparison test).

3.3. Total Antioxidant Capacity of Plants

The strength of the antioxidant capacity of plants is directly related to the levels of phenolic metabolites. Inoculation by *B. bassiana* can slightly increase the antioxidant capacity of tomatoes. When the whitefly was feeding on the plant, the antioxidant content increased only to 0.47 ± 0.0115 mM. However, the methyl jasmonate (MeJA) (0.81 ± 0.0036 mM) and *B. bassiana* inoculation combined with pest feeding (0.79 ± 0.0075 mM) significantly enhanced the antioxidant concentrations ($F_{2,17} = 6.394$, $p = 0.002$, Figure 4), and the staining test of a combination of fungi and insects treating leaves also showed the deepest color (Figure S1A); the antioxidant concentrations of the control groups were only 0.47 ± 0.0115 Mm, and the leaves showed little color (Figure S1C).

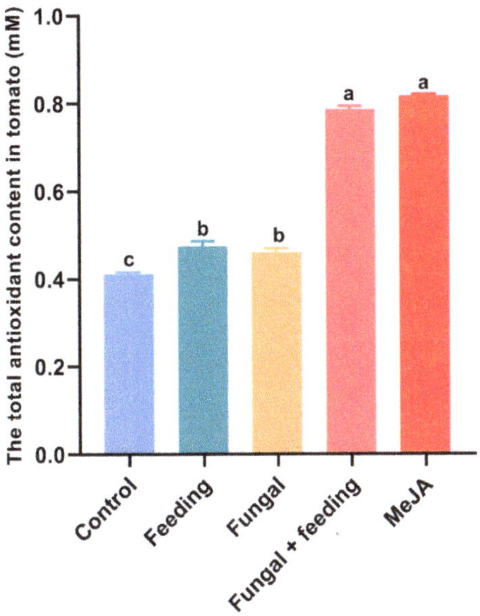

Figure 4. Comparison of tomato plant total antioxidant capacity. Methyl jasmonate (MeJa) means spraying with methyl jasmonate. Fungal + feeding means *B. bassiana* treatment combined with *B. tabaci* feeding. Fungal means *B. bassiana* treatment. Feeding means feeding by *B. tabaci* only. Different letters indicate significant differences ($p < 0.05$; Tukey's multiple comparison test).

4. Discussion

External stimulation of plants can induce a high defense gene expression and produce a variety of defensive metabolites, mainly involving phenylpropanoids that mediate the synthesis of phenolic substances, thereby evoking an enhanced response to subsequent pest stress. An activated plant defense system using entomogenous fungi as the stimulus has interesting potential for plant protection [34]. *B. bassiana* and *Metarhizium anisopliae* have also been found to be able to colonize as endophytes in many plant species, such as tomato, corn, and cotton [41]. Over the past few decades, some studies have shown that *B. bassiana* can be artificially introduced into tomato plants by various inoculation methods, such as leaf spraying, root irrigation, stem injection, and seed treatments [42–44]. Each of these methods has a different influence on the colonization efficiency of *B. bassiana* in tomato tissues. Our previous research found that when tomatoes were inoculated with *B. bassiana* by leaf spraying, the colonization efficiency reached 100% within 14 days [6]. In recent years, more in-depth insights have been obtained. After the foliar application of *B. bassiana*–GFP, fluorescence was detected in the root and stem tissues of tomatoes [45]. *B. bassiana* was mainly located in the intercellular spaces and vascular tissue [46]. In addition, *B. bassiana* can move within tomato tissues, migrating from the inoculated sites to non-inoculated areas [45,47]. These results suggest that *B. bassiana* can successfully colonize multiple tissues and achieve a systemic distribution in tomato plants.

In this study, we illustrated that *B. bassiana* can act as an exogenous stimulant for tomato plants, activate its own defense system, and even in the mode of pre-induction by the fungus, it can promote plants to respond to subsequent feeding attacks by *B. tabaci*, producing a faster and stronger defensive response. The interaction between *B. bassiana* and tomato plants in our study resulted in the upregulation of the expression patterns of several defense genes, explained to be primarily due to plant recognition of the change in the electrical potential of the plasma membrane [48]. This causes reversible phosphorylation, allowing fast and specific signal transduction of extracellular stimuli to the cytosol and nucleus, namely the mitogen-activated protein kinase (MAPK) cascade [49], an important signaling pathway involved in stress signal transduction in plants.

During the plant defense, the primary signal transduces to the secondary signal process, rapidly increasing the content of the plant-regulating hormone jasmonic acid. This trend was confirmed in our study during both the fungal stimulation and whitefly feeding. Jasmonic acid has been proven to be the most closely related hormone to plant insect resistance [50]. However, plants often use the same defense system when dealing with different stresses. Part of the plant defense intermediates could participate in multiple pathways for insect and microbial resistance [51]. In cotton infected with black spot disease or when fed upon by cotton bollworm, it promotes the increase in chitinase, β-1,3-glucanase, phenylalanine ammonia-lyase, and lipoxygenase enzyme activities [52]. Varying the content of these intermediate products could feed back the signal path in JA [50]. Eventually, this leads to the phosphorylation of transcription factors, which in turn activate an array of defense enzymes (POD, PPO, and PAL) and antioxidant enzymes (CAT, SOD, APx, and GPx) [34,53].

Phenylalanine ammonia lyase (PAL) was highly expressed in our study both as a response to whitefly feeding or following *B. bassiana* treatment. PAL is a key gene, representing an important subset of secondary metabolites in plant defense responses [54]. PAL plays a key role in the resistance to plant diseases. Its synthesis can be enhanced by pathogenic microorganisms as well as some beneficial microorganisms [25,55,56]. Typically, external stimulation leads to plant accumulation of pathogenesis-related (PR) protein, one of the major sources of plant-derived allergens [57]. PAL is more important in the early stages of induction since it is the first enzyme in the phenylpropanoid pathway that leads to the production of phenolic substances [58]. Similarly, the expression of PAL increased significantly during feeding by *B. tabaci*, which has also been reported for feeding by *Phenacoccus peruvianus* on *Bougainvillea* [56]. Phenolic enhancement is regarded as the result of an increase in the activity of PAL in wounded tissues [59].

Phenolic compounds are the largest category of plant-derived secondary metabolites [60], mainly formed via the malonic and shikimic acid pathways. In the shikimic acid pathway, phenylalanine is the most common intermediate for glycolysis and pentose phosphate conversion, being further catalyzed by PAL to form phenolic metabolites [61]. Most phenolic compounds have a crucial role in plant growth and defense against microbes and herbivore attacks. In general, large amounts of phenolic compounds can be detected, and phenylalanine ammonia-lyase activity will significantly increase in the parts of the plant that are subjected to insect feeding [62]. Particularly in plant resistance processes, phenolics have an antioxidant activity as they are hydrogen donors and scavenge reactive oxygen species (ROS) [48]. Here, we showed that the PAL expression level of plants treated by the fungus was the highest, and the phenolic compounds stained the deepest, which was consistent with this statement.

After suffering from insect stress, ROS content in the plant increased rapidly in our study. ROS is an important activator of plant defense. Under normal conditions, the level of cell toxic ROS is kept in homeostasis by effective enzymes and the non-enzymatic antioxidant system [63]. However, under stress conditions, the production of ROS exceeds the plant's antioxidant capacity, thereby resulting in oxidative stress, which also induces the secretion of antioxidant active substances [64]. Jasmonates, methyl salicylate, and other hormones play a pivotal role as transmission signals in this response to oxidative stress [50,65]. In addition, due to the accumulated level of hormone in plant hosts caused by ROS, higher expression may be induced for a defensive response against external challenges [66].

Based on the above results, our findings have great positive implications for agricultural production. Firstly, understanding the role of *B. bassiana* in activating plant defense mechanisms can offer a new approach to pest management. Instead of relying solely on chemical pesticides, farmers could consider using *B. bassiana* as an alternative or complementary method. For example, in tomato cultivation, introducing *B. bassiana* to the plants could enhance their natural defenses against pests like *B. tabaci* [45]. This could potentially reduce the amount of chemical pesticides used, which is beneficial for both the environment and the cost-effectiveness of farming operations. However, our study currently lacks field experiments, which is a significant limitation. Laboratory and greenhouse studies, while valuable, cannot fully replicate the complex and variable field conditions. In the field, factors such as weather fluctuations, soil heterogeneity, and the presence of other pests and diseases can significantly influence the performance of *B. bassiana* [41].

Field experiments are essential to validate the effectiveness of *B. bassiana* in real-world farming scenarios [67]. They could help determine the optimal application rates, frequencies, and timings for different regions and crop varieties. For example, in different soil types, the colonization ability of *B. bassiana* may vary, which could affect its overall effectiveness in pest control. Field experiments would also provide data on the long-term impacts of *B. bassiana* on soil health, water quality, and biodiversity, which are important considerations for sustainable agriculture [68]. Therefore, future research should also focus on conducting field experiments to further confirm the potential of *B. bassiana*.

5. Conclusions

In conclusion, *B. bassiana* can stimulate the synthesis of defense substances in tomato plants, eliciting a response similar to that induced by feeding from *B. tabaci*. However, further studies are required to fully understand the mechanisms of the interaction between entomogenous fungi and plants and their potential for managing the interaction between the pest insect and the plant. Subsequent research can be based on the interaction analysis between pathogen-associated molecular patterns such as chitin and β-glucan and plant pattern-recognition receptors to further clarify the defensive recognition mechanism of plants against entomogenous fungi. Although our study has limitations in terms of this defensive recognition mode, the application potential shown by this approach is still considerable.

Supplementary Materials: The following supporting information can be downloaded at https://www.mdpi.com/article/10.3390/jof11020141/s1, Figure S1: Phenolic substance staining; Table S1: Sequence of primers used in this study.

Author Contributions: Conceptualization, D.X., H.C., H.L. and Y.L.; data curation, M.L., D.X., T.N., J.Z., J.Y., Q.W. and Y.L.; formal analysis, M.L., D.X., Q.W. and Y.L.; funding acquisition, H.C., H.L. and Y.L.; methodology, M.L., D.X., H.M.T.H., H.C., H.L. and Y.L.; project administration, H.C. and Y.L.; software, M.L.; visualization, M.L.; writing—original draft, M.L., D.X. and Y.L.; writing—review and editing, M.L., D.X., H.M.T.H., H.C., H.L. and Y.L. All authors have read and agreed to the published version of the manuscript.

Funding: This research was supported by the Shuangcheng cooperative agreement research grant of Yibin (XNDX2022020014) and the Key R&D Program of Tibet, China (XZ202301ZY0014N).

Institutional Review Board Statement: Not applicable.

Informed Consent Statement: Not applicable.

Data Availability Statement: The original contributions presented in the study are included in the article, further inquiries can be directed to the corresponding author.

Acknowledgments: We thank Ingeborg Menzler-Hokkanen (Stockbridge School of Agriculture, University of Massachusetts, Boston, MA, USA) for assistance with improving the manuscript.

Conflicts of Interest: The authors declare no conflicts of interest.

References

1. Nair, A.; Kolet, S.P.; Thulasiram, H.V.; Bhargava, S. Systemic jasmonic acid modulation in mycorrhizal tomato plants and its role in induced resistance against *Alternaria alternata*. *Plant Biol.* **2015**, *17*, 625–631. [CrossRef] [PubMed]
2. Halim, V.A.; Vess, A.; Scheel, D.; Rosahl, S. The role of salicylic acid and jasmonic acid in pathogen defence. *Plant Biol.* **2010**, *8*, 307–313. [CrossRef]
3. Jogawat, A.; Meena, M.K.; Kundu, A.; Varma, M.; Vadassery, J. Calcium channel CNGC19 mediates basal defense signaling to regulate colonization by *Piriformospora indica* in Arabidopsis roots. *J. Exp. Bot.* **2020**, *71*, 2752–2768. [CrossRef]
4. Sánchez-Rodríguez, A.R.; Del Campillo, M.C.; Quesada-Moraga, E. *Beauveria bassiana*: An entomopathogenic fungus alleviates *Fe chlorosis* symptoms in plants grown on calcareous substrates. *Sci. Hortic.* **2015**, *197*, 193–202. [CrossRef]
5. Po-Wen, C.; Singh, P.; Zimmerli, L. Priming of the *Arabidopsis* pattern-triggered immunity response upon infection by necrotrophic *Pectobacterium carotovorum* bacteria. *Mol. Plant Pathol.* **2012**, *14*, 58–70. [CrossRef] [PubMed]
6. Wei, Q.-Y.; Li, Y.-Y.; Xu, C.; Wu, Y.-X.; Zhang, Y.-R.; Liu, H. Endophytic colonization by *Beauveria bassiana* increases the resistance of tomatoes against *Bemisia tabaci*. *Arthropod-Plant Interact.* **2020**, *4*, 1–12. [CrossRef]
7. Lin, P.-A.; Peiffer, M.; Felton, G.W. Induction of defensive proteins in Solanaceae by salivary glucose oxidase of *Helicoverpa zea* caterpillars and consequences for larval performance. *Arthropod-Plant Interact.* **2020**, *14*, 317–325. [CrossRef]
8. Benhamou, N.; Brodeur, J. Evidence for antibiosis and induced host defense reactions in the interaction between *Verticillium lecanii* and *Penicillium digitatum*, the causal agent of green mold. *Phytopathology* **2000**, *90*, 932–943. [CrossRef]
9. Chairin, T.; Petcharat, V. Induction of defense responses in longkong fruit (*Aglaia dookkoo* Griff.) against fruit rot fungi by *Metarhizium* guizhouense. *Biol. Control* **2017**, *111*, 40–44. [CrossRef]
10. Ownley, B.; Gwinn, K.; Vega, F. Endophytic fungal entomopathogens with activity against plant pathogens: Ecology and evolution. *Biocontrol* **2010**, *55*, 113–128. [CrossRef]
11. Bruce, T.J.A. Interplay between insects and plants: Dynamic and complex interactions that have coevolved over millions of years but act in milliseconds. *J. Exp. Bot.* **2015**, *66*, 455–465. [CrossRef] [PubMed]
12. Ali, J.G.; Agrawal, A.A. Specialist versus generalist insect herbivores and plant defense. *Trends Plant Sci.* **2012**, *17*, 293–302. [CrossRef] [PubMed]
13. Su, Q.; Peng, Z.; Tong, H.; Xie, W.; Wang, S.; Wu, Q.; Zhang, J.; Li, C.; Zhang, Y. A salivary ferritin in the whitefly suppresses plant defenses and facilitates host exploitation. *J. Exp. Bot.* **2019**, *70*, 3343–3355. [CrossRef] [PubMed]
14. Conrath, U.; Beckers, G.J.; Flors, V.; Garcia-Agustin, P.; Jakab, G.; Mauch, F.; Newman, M.A.; Pieterse, C.M.; Poinssot, B.; Pozo, M.J.; et al. Priming: Getting ready for battle. *Mol. Plant Microbe. Interact.* **2006**, *19*, 1062–1071. [CrossRef]
15. Conrath, U.; Beckers, G.J.; Langenbach, C.J.; Jaskiewicz, M.R. Priming for enhanced defense. *Annu. Rev. Phytopathol.* **2015**, *53*, 97–119. [CrossRef]
16. Hilker, M.; Schwachtje, J.; Baier, M.; Balazadeh, S.; Baurle, I.; Geiselhardt, S.; Hincha, D.K.; Kunze, R.; Mueller-Roeber, B.; Rillig, M.C.; et al. Priming and memory of stress responses in organisms lacking a nervous system. *Biol. Rev. Camb. Philos. Soc.* **2016**, *91*, 1118–1133. [CrossRef]
17. Mauch-Mani, B.; Baccelli, I.; Luna, E.; Flors, V. Defense priming: An adaptive part of induced resistance. *Annu. Rev. Plant Biol.* **2017**, *68*, 485–512. [CrossRef]
18. Pescador, L.; Fernandez, I.; Pozo, M.J.; Romero-Puertas, M.C.; Pieterse, C.M.J.; Martinez-Medina, A. Nitric oxide signalling in roots is required or MYB72-dependent systemic resistance induced by *Trichoderma* volatile compounds in Arabidopsis. *J. Exp. Bot.* **2022**, *73*, 584–595. [CrossRef] [PubMed]
19. Elliot, S.L.; Sabelis, M.W.; Janssen, A.; Geest, L.P.S.V.D.; Beerling, E.A.M.; Fransen, J. Can plants use entomopathogens as bodyguards? *Ecology* **2010**, *3*, 228–235. [CrossRef]
20. Hodge, S.; Bennett, M.; Mansfield, J.W.; Powell, G. Aphid-induction of defence-related metabolites in *Arabidopsis thaliana* is dependent upon density, aphid species and duration of infestation. *Arthropod-Plant Interact.* **2019**, *13*, 387–399. [CrossRef]
21. Sorokan, A.V.; Burkhanova, G.F.; Benkovskaya, G.V.; Maksimov, I.V. Colorado potato beetle microsymbiont Enterobacter *BC-8* inhibits defense mechanisms of potato plants using crosstalk between jasmonate-and salicylate-mediated signaling pathways. *Arthropod-Plant Interact.* **2020**, *14*, 161–168. [CrossRef]
22. Kadam, S.B.; Pable, A.A.; Barvkar, V.T. Mycorrhiza induced resistance (MIR): A defence developed through synergistic engagement of phytohormones, metabolites and rhizosphere. *Funct. Plant Biol.* **2020**, *47*, 880–890. [CrossRef] [PubMed]

23. Dinis, L.T.; Bernardo, S.; Conde, A.; Pimentel, D.; Ferreira, H.; Felix, L.; Geros, H.; Correia, C.M.; Moutinho-Pereira, J. Kaolin exogenous application boosts antioxidant capacity and phenolic content in berries and leaves of grapevine under summer stress. *Plant Physiol.* **2016**, *191*, 45–53. [CrossRef]
24. Mechri, B.; Tekaya, M.; Cheheb, H.; Attia, F.; Hammami, M. Accumulation of flavonoids and phenolic compounds in olive tree roots in response to mycorrhizal colonization: A possible mechanism for regulation of defense molecules. *J. Plant Physiol.* **2015**, *185*, 40–43. [CrossRef]
25. Geetha, N.P.; Amruthesh, K.N.; Sharathchandra, R.G.; Shetty, H.S. Resistance to downy mildew in pearl millet is associated with increased phenylalanine ammonia lyase activity. *Funct. Plant Biol.* **2005**, *32*, 267–275. [CrossRef]
26. Adams, Z.P.; Ehlting, J.; Edwards, R. The regulatory role of shikimate in plant phenylalanine metabolism. *J. Theor. Biol.* **2019**, *462*, 158–170. [CrossRef]
27. Anfang, M.; Shani, E. Transport mechanisms of plant hormones. *Curr. Opin. Plant Biol.* **2021**, *63*, 102055. [CrossRef]
28. Ullah, A.; Manghwar, H.; Shaban, M.; Khan, A.H.; Akbar, A.; Ali, U.; Ali, E.; Fahad, S. Phytohormones enhanced drought tolerance in plants: A coping strategy. *Environ. Sci. Pollut. Res.* **2018**, *25*, 33103–33118. [CrossRef] [PubMed]
29. Ruan, J.; Zhou, Y.; Zhou, M.; Yan, J.; Khurshid, M.; Weng, W.; Cheng, J.; Zhang, K. Jasmonic Acid Signaling Pathway in Plants. *Int. J. Mol. Sci.* **2019**, *20*, 2479. [CrossRef]
30. Singh, P.; Arif, Y.; Miszczuk, E.; Bajguz, A.; Hayat, S. Specific Roles of Lipoxygenases in Development and Responses to Stress in Plants. *Plants* **2022**, *11*, 979. [CrossRef]
31. Wang, Y.; Mostafa, S.; Zeng, W.; Jin, B. Function and Mechanism of Jasmonic Acid in Plant Responses to Abiotic and Biotic Stresses. *Int. J. Mol. Sci.* **2021**, *22*, 8568. [CrossRef]
32. Yang, J.; Duan, G.; Li, C.; Liu, L.; Han, G.; Zhang, Y.; Wang, C. The Crosstalks Between Jasmonic Acid and Other Plant Hormone Signaling Highlight the Involvement of Jasmonic Acid as a Core Component in Plant Response to Biotic and Abiotic Stresses. *Front. Plant Sci.* **2019**, *10*, 1349. [CrossRef] [PubMed]
33. Quesadamoraga, E.; Muñozledesma, F.J.; Santiagoálvarez, C. Systemic protection of *Papaver somniferum* L. against *Lraella luteipes* (Hymenoptera: Cynipidae) by an endophytic strain of *Beauveria bassiana* (Ascomycota: Hypocreales). *Environ. Entomol.* **2009**, *38*, 723–730. [CrossRef]
34. Martinez-Medina, A.; Flors, V.; Heil, M.; Mauch-Mani, B.; Pieterse, C.M.J.; Pozo, M.J.; Ton, J.; van Dam, N.M.; Conrath, U. Recognizing Plant Defense Priming. *Trends Plant Sci.* **2016**, *21*, 818–822. [CrossRef]
35. Rondot, Y.; Reineke, A. Endophytic *Beauveria bassiana* in grapevine *Vitis vinifera* (L.) reduces infestation with piercing-sucking insects. *Biol. Control* **2018**, *116*, 82–89. [CrossRef]
36. McKinnon, A.C.; Saari, S.; Moran-Diez, M.E.; Meyling, N.V.; Raad, M.; Glare, T.R. *Beauveria bassiana* as an endophyte: A critical review on associated methodology and biocontrol potential. *BioControl* **2016**, *62*, 1–17. [CrossRef]
37. Lin, Y.; He, S.-Q.; Lu, Z.-H.; Gao, Y.-L.; Duan, Y.-R.; Li, Z.-Y.; Chen, B.; Gui, F.-R. Influence of *Aphis gossypii* feeding on defense strategy of native and introduced populations of *Ageratina adenophora. J. Arthropod-Plant Interact.* **2020**, *14*, 345–356. [CrossRef]
38. Bagavathi Perumal, E.; Ramasamy, N. GC-MS analysis of phytocomponents in the ethanol extract of *Polygonum chinense* L. *Pharmacogn. Res.* **2011**, *4*, 11–14. [CrossRef]
39. Selvamangai, G.; Bhaskar, A. GC-MS analysis of phytocomponents in the methanolic extract of *Eupatorium triplinerve. Asian Pac. J. Trop. Biomed.* **2012**, *2*, S1329–S1332. [CrossRef]
40. Cardenas-Manríquez, G.; Vega-Muñoz, I.; Villagómez-Aranda, A.L.; León-Galvan, M.F.; Cruz-Hernandez, A.; Torres-Pacheco, I.; Rangel-Cano, R.M.; Rivera-Bustamante, R.F.; Guevara-Gonzalez, R.G. Proteomic and metabolomic profiles in transgenic tobacco (*N. tabacum* xanthi nc) to CchGLP from *Capsicum chinense* BG-3821 resistant to biotic and abiotic stresses. *Env. Exp. Bot.* **2016**, *130*, 33–41. [CrossRef]
41. Greenfield, M.; Gomez-Jimenez, M.I.; Ortiz, V.; Vega, F.E.; Kramer, M.; Parsa, S. *Beauveria bassiana* and *Metarhizium anisopliae* endophytically colonize cassava roots following soil drench inoculation. *Biol. Control* **2016**, *95*, 40–48. [CrossRef] [PubMed]
42. Sinno, M.; Ranesi, M.; Di Lelio, I.; Iacomino, G.; Becchimanzi, A.; Barra, E.; Molisso, D.; Pennacchio, F.; Digilio, M.C.; Vitale, S.; et al. Selection of endophytic *Beauveria bassiana* as a dual biocontrol agent of tomato pathogens and pests. *Pathogens* **2021**, *10*, 1242. [CrossRef] [PubMed]
43. Nchu, F.; Macuphe, N.; Rhoda, I.; Niekerk, L.-A.; Basson, G.; Keyster, M.; Etsassala, N.G.E.R. Endophytic *Beauveria bassiana* induces oxidative stress and enhances the growth of Fusarium oxysporum-infected tomato plants. *Plants* **2022**, *11*, 3182. [CrossRef]
44. Wang, X.; Yan, G.; Liu, W.; Chen, H.; Yuan, Q.; Wang, Z.; Liu, H. Endophytic Beauveria bassiana of tomato resisted the damage from whitefly *Bemisia tabaci* by mediating theaccumulation of plant-specialized metabolites. *J. Agric. Food Chem.* **2023**, *71*, 13244–13254. [CrossRef]
45. Wang, X.; Yuan, Q.; He, L.Q.; Wang, Z.; Li, G.Y.; Wang, Z.Y.; Liu, H. Biological and physiological effects in *Bemisia tabaci* feeding on tomatoes endophytically colonized by *Beauveria bassiana. Pest Manag. Sci.* **2024**, *80*, 4085–4097. [CrossRef]
46. Vega, F.E. The use of fungal entomopathogens as endophytes in biological control: A review. *Mycologia* **2018**, *110*, 4–30. [CrossRef] [PubMed]

47. Sui, L.; Lu, Y.; Zhou, L.; Li, N.; Li, Q.; Zhang, Z. Endophytic *Beauveria bassiana* promotes plant biomass growth and suppresses pathogen damage by directional recruitment. *Front. Microbiol.* **2023**, *14*, 1227269. [CrossRef] [PubMed]
48. Kurt, B.; Susanne, Z.; Rosalia, C.; Woo, S.L.; Matteo, L.; Kubicek, C.P.; Mach, R.L. Improvement of the fungal biocontrol agent *Trichoderma atroviride* to enhance both antagonism and induction of plant systemic disease resistance. *Appl. Environ. Microbiol.* **2005**, *71*, 3959–3965. [CrossRef]
49. Morris, P.C. MAP kinase signal transduction pathways in plants. *New Phytol.* **2010**, *151*, 67–89. [CrossRef] [PubMed]
50. Claus, W.; Bettrina, H. Jasmonates and octadecanoids: Signals in plant stress responses and development. *Prog. Nucleic Acid Res. Mol. Biol.* **2002**, *72*, 165–221. [CrossRef]
51. Singh, P.; Zimmerli, L. Lectin receptor kinases in plant innate immunity. *Front. Plant Sci.* **2013**, *4*, 124. [CrossRef] [PubMed]
52. Hudspeth, R.L.; Hobbs, S.L.; Anderson, D.M.; Grula, J.W. Characterization and expression of chitinase and 1,3-beta-glucanase genes in cotton. *Plant Mol. Biol.* **1996**, *31*, 911–916. [CrossRef] [PubMed]
53. Zhang, S.; Klessig, D.F. MAPK cascades in plant defense signaling. *Trends Plant Sci.* **2001**, *6*, 520–527. [CrossRef]
54. Yang, Y.; Shah, J.; Klessig, D.F. Signal perception and transduction in plant defense responses. *Genes Dev.* **1997**, *11*, 1621–1639. [CrossRef]
55. Tonnessen, B.W.; Manosalva, P.; Lang, J.M.; Baraoidan, M.; Bordeos, A.; Mauleon, R.; Oard, J.; Hulbert, S.; Leung, H.; Leach, J.E. Rice phenylalanine ammonia-lyase gene OsPAL4 is associated with broad spectrum disease resistance. *Plant Mol. Biol.* **2015**, *87*, 273–286. [CrossRef]
56. Abbate, C.; Toscano, S.; Arcidiacono, R.; Romano, D.; Mazzeo, G. Induced responses of *Bougainvillea glabra* Choisy (Nyctaginaceae) against *Phenacoccus peruvianus* Granara de Willink (Hemiptera: Pseudococcidae) attack: Preliminary results. *J. Arthropod-Plant Interact.* **2017**, *12*, 41–48. [CrossRef]
57. Sinha, M.; Singh, R.P.; Kushwaha, G.S.; Iqbal, N.; Singh, A.; Kaushik, S.; Kaur, P.; Sharma, S.; Singh, T.P. Current overview of allergens of plant pathogenesis related protein families. *Sci. World J.* **2014**, *2014*, 543195. [CrossRef]
58. Shoresh, M.; Harman, G.E.; Mastouri, F. Induced systemic resistance and plant responses to fungal biocontrol agents. *Annu. Rev. Phytopathol.* **2010**, *48*, 21–43. [CrossRef] [PubMed]
59. Bi, J.L.; Felton, G.W. Foliar oxidative stress and insect herbivory: Primary compounds, secondary metabolites, and reactive oxygen species as components of induced resistance. *J. Chem. Ecol.* **1995**, *21*, 1511. [CrossRef]
60. Achakzai, A.K.K.; Achakzai, P.; Masood, A.; Kayani, S.A.; Tareen, R.B. Response of plant parts and age on the distribution of secondary metabolites on plants found in Quetta. *Pak. J. Bot.* **2009**, *41*, 2129–2135. [CrossRef]
61. Verma, N.; Shukla, S. Impact of various factors responsible for fluctuation in plant secondary metabolites. *J. Appl. Res. Med. Aromat. Plants* **2015**, *2*, 105–113. [CrossRef]
62. Johnson, K.S.; Felton, G.W. Plant phenolics as dietary antioxidants for herbivorous insects: A test with genetically modified tobacco. *Chem. Ecol.* **2001**, *27*, 2579–2597. [CrossRef]
63. O'Brien, J.A.; Daudi, A.; Butt, V.S.; Bolwell, G.P. Reactive oxygen species and their role in plant defence and cell wall metabolism. *Planta* **2012**, *236*, 765–779. [CrossRef]
64. Bittner, N.; Trauer-Kizilelma, U.; Hilker, M. Early plant defence against insect attack: Involvement of reactive oxygen species in plant responses to insect egg deposition. *Planta* **2017**, *245*, 993–1007. [CrossRef] [PubMed]
65. Creelman, R.A.; Mulpuri, R. The oxylipin pathway in *Arabidopsis*. *Arab. Book/Am. Soc. Plant Biol.* **2002**, *1*, e0012. [CrossRef] [PubMed]
66. Cheong, J.-J.; Choi, Y.D. Methyl jasmonate as a vital substance in plants. *Trends Genet.* **2003**, *19*, 409–413. [CrossRef] [PubMed]
67. Gathage, J.W.; Lagat, Z.O.; Fiaboe, K.K.M.; Akutse, K.S.; Ekesi, S.; Maniania, N.K. Prospects of fungal endophytes in the control of *Liriomyza* leafminer flies in common bean *Phaseolus vulgaris* under field conditions. *Biocontrol* **2016**, *61*, 741–753. [CrossRef]
68. Parsa, S.; Ortiz, V.; Gomez-Jimenez, M.I.; Kramer, M.; Vega, F.E. Root environment is a key determinant of fungal entomopathogen endophytism following seed treatment in the common bean, *Phaseolus vulgaris*. *Biol. Control* **2018**, *116*, 74–81. [CrossRef] [PubMed]

Journal of
Fungi

MDPI

Article

The Inevitable Fate of *Tetranychus urticae* on Tomato Plants Treated with Entomopathogenic Fungi and Spinosad

Waqas Wakil [1,2,*], Maria C. Boukouvala [3], Nickolas G. Kavallieratos [3,*], Aqsa Naeem [1], Dionysios Ntinokas [3], Muhammad Usman Ghazanfar [4] and Pasco B. Avery [5]

1 Department of Entomology, University of Agriculture, Faisalabad 38040, Pakistan; aqsanaeem231@gmail.com
2 Senckenberg German Entomological Institute, D-15374 Müncheberg, Germany
3 Laboratory of Agricultural Zoology and Entomology, Department of Crop Science, Agricultural University of Athens, 75 Iera Odos Str., 11855 Athens, Greece; mbouk@aua.gr (M.C.B.); dntinokas@aua.gr (D.N.)
4 Department of Plant Pathology, College of Agriculture, Sargodha University, Sargodha 40100, Pakistan; usmanghazanfar@uos.edu.pk
5 Indian River Research and Education Center, Department of Entomology and Nematology, Institute for Agricultural Sciences, University of Florida, Ft. Pierce, FL 34945, USA; pbavery@ufl.edu
* Correspondence: waqaswakeel@hotmail.com (W.W.); nick_kaval@aua.gr (N.G.K.)

Abstract: *Tetranychus urticae* (Acari: Tetranychidae) is a pervasive and damaging mite pest of tomato crops, leading to important economic losses globally. This study evaluated the acaricidal efficacy of spinosad, alone and in combination with *Beauveria bassiana* (*Bb*) WG-21 and *Metarhizium robertsii* (*Mr*) WG-04, in the laboratory (application to tomato leaf discs) and greenhouse (application to tomato plants), considering mortality and establishment, respectively. The combination treatments of *Bb* WG-21 or *Mr* WG-04 with spinosad achieved 100% mortality of *T. urticae* nymphs within 2 days on leaf discs, while individual applications of each control agent resulted in lower mortalities, ranging between 62.91 and 86.25% after 3 days. The paired treatment of *Mr* WG-04 + spinosad killed all exposed adults within 5 d, while that of *Bb* WG-21 + spinosad achieved the same results after 7 d. However, spinosad, *Mr* WG-04, and *Bb* WG-21 alone killed ≥77.08% of adults after 7 d. In the greenhouse, the combination treatment of WG-04 + spinosad deterred the presence of *T. urticae* (adults, immatures, and eggs) on either surface of the tomato leaves, while *Bb* WG-21 + spinosad suppressed the populations only on the adaxial surface. These findings indicate that combined treatments of the tested EPF + spinosad, especially *Mr* WG-04, on tomato plants under greenhouse conditions can provide substantially enhanced control of *T. urticae* life stages compared to each treatment applied alone.

Keywords: two-spotted spider mite; combination treatment applications; *Beauveria bassiana*; *Metarhizium robertsii*; spynosins

1. Introduction

The tetranychid *Tetranychus urticae* Koch (Acari) is widely recognized as a highly adaptable generalist herbivore [1], ranking among the most destructive pests affecting field and greenhouse crops [2–5]. This pest has an extensive host range that includes >1400 plant species from >200 plant families and is found in various geographical regions, such as Australasian, Oriental, Afrotropical, Neotropical, Palaearctic, and Nearctic [1,5–7]. *Tetranychus urticae* also stands out as a highly destructive pest with a significant economic impact on tomato crops worldwide [2,8,9]. Concretely, 10–50% of tomato production losses worldwide are due to *T. urticae* [10]. It feeds on plant sap by puncturing leaves with its mouthparts, causing functional damage to foliage and increasing its vulnerability to viruses and pathogens [4,11]. In addition, due to its short life cycle, *T. urticae* can quickly inflict severe damage on plants resulting in leaf streaking [12–14]. In cases of heavy infestation, this damage can lead to defoliation, stunted plant growth, and complete plant deterioration, and even the death of plants, severely impacting both the quality and yield of affected crops [14,15].

Chemical acaricides have traditionally been the main approach for preventing *T. urticae* from inflicting substantial economic damage [16,17]. However, due to their repeated and intensive use, *T. urticae* has developed resistance to nearly all classes of acaricides [18–21]. *Tetranychus urticae* resistance to chemical acaricides also has developed due to its arrhenotokous parthenogenesis reproduction, which increases genetic diversity and promotes the survival of resistant individuals [18]. In addition, the short life span of *T. urticae*, high offspring production rate, and multiple annual generations facilitate the rapid spread of resistance genes in its populations [18,22].

Due to the ineffectiveness of chemical control methods and environmental impacts, research efforts are increasingly focused on alternative strategies against *T. urticae*, including the use of entomopathogenic fungi (EPF) [10,23–26]. EPF, such as *Metarhizium robertsii* J.F. Bisch., Rehner & Humber (Hypocreales: Clavicipitaceae), and *Beauveria bassiana* (Balsamo-Crivelli) Vuillemin (Hypocreales: Cordycipitaceae) can infect and kill *T. urticae* without affecting the non-target organisms [10,25–28]. EPF are also safe for humans and do not cause any harm to the environment [10,29]. EPF infect and kill insects through a multi-step infection process [30]. Initially, fungal spores (conidia) come into contact with the cuticle of the host and attach, germinate, and produce enzymes to break down the host cuticle [31–34]. Once inside the host, the fungal hyphae proliferate and enter the hemocoel, where they produce blastospores or hyphal bodies that spread throughout the host's internal body, releasing toxins [35,36]. The fungus emerges from the cadaver and sporulates, releasing new conidia to infect other hosts [35,37].

Spinosad is a bioinsecticide, derived from the natural fermentation of *Saccharopolyspora spinosa* Mertz & Yao (Pseudonocardiales: Pseudonocardiaceae), a soil-borne bacterium, consisting of a mixture of two macrocyclic lactones, spynosins A and D [38,39]. Spinosad operates by selectively targeting the nervous system's nicotinic acetylcholine receptors (nAChRs) in pests, functioning as an allosteric modulator. This action rapidly stimulates both the γ-aminobutyric acid (GABA) receptors and nAChRs, causing intense nervous excitation that culminates in paralysis and, eventually, death of the organism [39–41]. Spinosad has widespread use against economically important insects from various orders (i.e., Lepidoptera, Coleoptera, Thysanoptera, and Diptera) across several agricultural crops, stored products, and ornamental plants [42–44]. For instance, spinosad has been successfully applied to control larval stages of pest species within Coleoptera and Lepidoptera in stored products, adult Diptera in certain crops, and even mosquito larvae in disease vector management efforts [39,45–47]. This active ingredient is not intended for the management of spider mites. However, their inadvertent exposure to spinosad does occur, particularly in greenhouse environments due to the widespread use of this pesticide to control caterpillars, leafminers, and thrips [48–52].

It is widely known that the combined use of EPF with pesticides may enhance the effectiveness of pest control (insects and mites) and support integrated pest management (IPM) strategies [26,53–56]. Furthermore, this combined treatment approach helps mitigate the development of pest resistance and promotes environmentally friendly pest management by reducing the amount of pest exposure to broad-spectrum single-mode-of-action synthetic chemicals [26,53,54,56–58]. A thorough investigation of the published literature revealed that the combination of *M. robertsii* or *B. bassiana* with spinosad against *T. urticae* has not been studied in comparison to the single treatments applied alone (i.e., EPF or chemical). Therefore, the aim of this study was to investigate the effectiveness of the combined application of spinosad with *M. robertsii* or *B. bassiana* vs. each treatment alone against different life stages (adults and immatures) of *T. urticae* under laboratory and greenhouse conditions.

2. Materials and Methods

2.1. *Tetranychus urticae*

Tetranychus urticae individuals came from a laboratory colony nurtured at the University of Agriculture, Faisalabad (UAF). The colony originated from mites collected from a natural population that was homogenized [59] and maintained for the past 4 years on tomato plants (*Solanum lycopersicum* L. cv. Moneymaker) inside MIR-254-PE incubators (Panasonic, Kadoma, Japan) set at 26 °C with 65% relative humidity (RH) using daylight lamps under a 16:8 h light (L)/dark (D) photoperiod [60]. Previous studies have reported notable differences in the demography of *T. urticae* feeding on the sap of whole leaves compared to leaf discs [61]. Thus, to minimize this variation and ensure an accurate assessment of *T. urticae* life stages, the rearing method described by Puspitarini et al. [62] was applied for all experiments.

2.2. *Isolates of Metarhizium robertsii and Beauveria bassiana*

The isolates of the EPF *Mr* WG-04 and *Bb* WG-21 corresponding to *M. robertsii* and *B. bassiana*, respectively, were selected for the experiments. *Mr* WG-04 and *Bb* WG-21 were recovered from soil samples collected from Chichawatni, Punjab [63] and Lahore, Punjab [64], respectively. Both EPF isolates were maintained at UAF's microbial culture collection inside Petri dishes that had potato dextrose agar-PDA (Sigma-Aldrich, Taufkirchen, Germany), stored at 4 °C [25,26]. Fungal isolates were cultured separately on Petri dishes (Ø 6 cm × 1.5 cm) that contained Sabouraud dextrose agar (BD-Difco, Franklin Lakes, NJ, USA) + 1% yeast (SDAY). The cultured dishes were closed, sealed with Nescofilm® (Azwell, Osaka, Japan), and incubated in a MIR-254-PE incubator in complete darkness for 14 days at 24 °C [65]. Two weeks after inoculation, these dishes produced a substantial quantity of conidia. A sterilized glass Drigalski spatula (LaborXing, Shenzhen, China) was used to harvest the dry conidia from the SDAY surface, transferring them into a sterile tube (10 mL). Some of the collected conidia were then suspended inside a Falcon tube (50 mL) containing 30 mL of 0.05% v/v of sterile polysorbate (Tween 80) solution (Merck & Co., Inc., Rahway, NJ, USA) to minimize clumping, as noted by Ortucu et al. [66]. The conidial suspension was vortexed for 5 min using a vortexer (Velp Scientifica srl, Usmate, Italy) with eight sterile glass beads. The slurry was passed through a sterilized cheesecloth (double layered) to remove any mycelial clumps into a Falcon tube (50 mL volume) [67,68]. The appropriate concentration was determined using a Neubauer hemocytometer (Marienfeld, Germany) with a microscope at 400× magnification [69].

The germination of the conidia was evaluated in two 6 cm diameter dishes containing SDAY by inoculating 0.1 mL of the conidia suspension (i.e., 1×10^6 conidia/mL). The dishes were sealed with Nescofilm (Azwell, Osaka, Japan) and put in incubator for 18 h set at 25 °C, under a 14:10 h L/D photoperiod. After incubation, each dish was unsealed, the top cover removed, and sterile coverslips were placed on top of the agar inside the dishes. For each fungal isolate, two hundred conidia per EPF isolate and dish were counted [25,26]. If the length of conidia germ tube was twice the diameter, their germination was counted as viable when viewed at 400× magnification using a light microscope (Euromex BB.1152-PLi, Euromex Microscopen bv, Arnhem, The Netherlands) as described by the method of Wakil et al. [25,26]. Prior to the experiments, the conidia viability for all isolates was >92%.

2.3. Insecticide

The spinosad formulation Tracer 240 SC, provided by Dow Agro Sciences (Karachi, Pakistan) containing 240 g/L of the active ingredient (a.i.), was used in the tests. The insecticide was dissolved in Tween 80 (0.05%) solution.

2.4. Effects of Entomopathogenic Fungal Isolates and Spinosad Alone and in Combination Against *T. urticae* Under Laboratory Conditions

The efficacy of *M. robertsii* (*Mr* WG-04), *B. bassiana* (*Bb* WG-21) and spinosad alone and in combination was evaluated against *T. urticae* individuals (nymphs, adult females) under laboratory conditions. The experiments were conducted in plastic Petri dishes (Ø 50 mm × 15 mm) (3 subreplicates/test), holding a tomato (var. "Moneymaker") leaf disc (20 mm diameter) according to the design of Marcossi et al. [70]. This variety was selected for its strong potential and widespread cultivation by Pakistani tomato growers [25,26,71,72]. Prior to spraying, a cork borer was used to prepare the leaf discs. Inside each dish, a damp tissue paper (Ø 3 cm) was placed, then, a leaf disc was placed on top of the tissue paper with the abaxial side facing up and surrounded by a wet cotton swab to maintain moisture [25,26,73]. Groups of twenty *T. urticae* adult females (1 day old) or nymphs (1 day old) [26] were collected and individually placed on each tomato leaf disc [74]. We selected adult females since they are the main vehicle of reproduction. A Master Multipurpose Airbrush (Las Vegas, NM, USA) was then used to spray the leaf discs containing the mites with 1 mL of *Mr* WG-04 or *Bb* WG-21 isolate suspension containing 1×10^8 conidia and allowed to air-dry for 3 h. After preliminary experimentation, examining a range of fungal concentrations from 1×10^4 to 1×10^9, we selected the fungal dose rate that gave approximately 35–55% mortality (nymphs and adult females) on the first day. In the case of spinosad, leaf discs were dipped in a solution at 0.15 mg spinosad/L for 15 sec and allowed to air-dry for 3 h [75]. We mixed spinosad with Tween 80 (0.05%) solution to receive our appropriate concentration. Then, the dry discs were placed in the dishes as described above and groups of 20 adult females or nymphs were released in each disc. Concerning the combination treatment of *Bb* WG-21 or *Mr* WG-04 with spinosad, each leaf disc was first treated with the insecticide alone (leaf dip technique), as described above. The treated dry discs were then sprayed with each fungal isolate at 1×10^8 conidia/mL as described above. Control leaf discs were treated with 1 mL of a Tween 80 (0.05%) solution [26,66,73] with a different airbrush of the same model reserved for treating controls and allowed to air-dry 3 h. Nescofilm was used to seal the dishes, which were then placed in an incubator set at 25 °C with 60% RH, under a 16:8 h L/D photoperiod [26,76,77]. Mortality data were obtained after observing nymphs and adults 1, 2, 3, and 3, 5, 7 days post treatment, respectively, according to Wakil et al. [26]. Per exposure, separate tomato leaf discs containing either nymphs or adult females of *T. urticae* were prepared. Dead individuals were assessed using a Leica stereomicroscope (Wild M3B, Heerbrugg, Switzerland). Adults or nymphs were considered dead if they showed no movement in their appendages after gently prodding with a camel hairbrush [73]. The entire experimental procedure described above was performed 4 times in a completely randomized design, where new dishes, tomato leaf discs, insecticidal solutions, EPF suspensions, and mite individuals were prepared for each test (i.e., 4 replications × 3 sub-replicates × 6 (treatments + control) × 2 life stages).

2.5. Effects of Entomopathogenic Fungal Isolates and Insecticide Against *T. urticae* Under Greenhouse Conditions

Seeds of the tomato variety "Moneymaker" were sown in seedling trays at a rate of two seeds/well and watered daily. Twenty-one days post sowing, each seedling was transplanted individually into 3 L plastic pots (one plant per pot) with sphagnum peat moss. The pots were then transferred to chambers at 25 °C, with 65% RH, under a 16:8 h L/D photoperiod [26,78]. The plants were irrigated every three days. After 15 days, two fertilizers were applied, as recommended by Wakil et al. [26]. After 20 days, plants showing uniform growth traits were selected for the bioassays [26,60]. For each treatment and control, three pots containing three plants in each pot (3 subreplications) were chosen. Each plant had ten to fifteen leaves [26]. A solution (15 mL) of spinosad (30 mg/L H_2O) or a 15 mL conidial suspension (1×10^8 conidia/mL) of either *Mr* WG-04 or *Bb* WG-21 was sprayed until runoff on the plants in their respective groups using a handheld sprayer (Kissan Ghar, Sargodha, Pakistan). Each plant was sprayed thoroughly with each EPF suspension or spinosad solution until runoff to ensure uniform coverage [73]. Control plants were treated until runoff with 15 mL of a 0.05% Tween 80 solution [66,73]. For the combination treatments, *Mr* WG-04 or *Bb* WG-21 with spinosad, two consecutive sprays were applied [26,79]. The plants were initially sprayed with 15 mL of the spinosad solution (30 mg/L H_2O) until runoff and allowed to air-dry for 24 h. This was followed by a second spray with 15 mL of each EPF suspension (1×10^8 conidia/mL) until runoff [26,79]. All plants were allowed to air-dry for 24 h after being sprayed. Using a fine hairbrush, 20 adult females of *T. urticae* were gently placed on the three apical leaflets of the 2nd oldest true leaf on each plant [26]. After 21 days from the initial infestation, all leaves on each plant were cut and examined with a Leica stereomicroscope to assess mite establishment, recording adults (Figure 1), immatures, and eggs of *T. urticae* on both leaf surfaces (adaxial and abaxial) [26,80]. The experiment followed a randomized complete block design (RCBD), which was repeated 4 times, each trial involving new potted plants, mites, insecticidal solutions, and fungal suspension (i.e., 4 replications × 3 subreplicates × 6 (treatments + control) × 2 life stages).

Figure 1. Adult female of *Tetranychus urticae*.

2.6. Statistical Analysis

Prior to analysis, the data set was log (x + 1)-transformed to normalize the variance [81,82]. Abbott's formula was used to correct mortality data [83]. For laboratory assays, the main effects were life stage, treatment, and interval with mortality being the response variable. For data on the number of mites (i.e., eggs, immatures, and adults) per leaf in the greenhouse experiments, the main effects were treatment, life stage, and leaf surface position (abaxial or adaxial). Mite presence on the adaxial or abaxial surface of leaves consisted of the response variable. In all cases, data analysis was performed by a three-way ANOVA, accounting for the main effects and their interactions. Control mortality was below 5%. Mean separation was carried out using the Tukey HSD test at a 5% significance level [84]. The "Minitab" statistical package was used for all statistical analyses [85].

3. Results

3.1. Subsection Mortality of *Tetranychus urticae* Life Stages in the Laboratory Tests

From the ANOVA results, all of the effects alone (interval, life stage, treatment) and in various combinations of interactions were significant; however, only the associated interaction "interval × stage" combination was not significant (Table 1). On the first day of observation, the nymphal mortality ranged from 32.50 to 56.25% after individual applications of either EPF isolates or spinosad alone applied on the tomato leaf discs (Figure 2, Table S1). In contrast, the combined applications of EPF with spinosad resulted in 83.75 and 92.08%, for *Bb* WG-21 + spinosad and *Mr* WG-04 + spinosad, respectively, at 1 day post treatment. Two days after treatment (DAT), complete mortality (100%) was recorded in both fungal–spinosad combinations, whereas mortality ranged between 49.16 and 73.33% for spinosad and fungal treatments alone (Figure 3, Table S1). Three DAT, *Mr* WG-04 (86.25%) killed significantly more immatures than *Bb* WG-21 (74.16%) and spinosad (62.91%) (Figure 4, Table S1).

Table 1. ANOVA parameters for mortality rates of *Tetranychus urticae* exposed to tomato leaf discs treated with *Metarhizium robertsii*, *Beauveria bassiana*, or spinosad, alone and in combination (Total df = 539).

Source	df	F	p
Interval	2	687.77	<0.01
Life stage	1	266.74	<0.01
Treatment	4	1550.48	<0.01
Interval × life stage	2	1.81	0.16
Interval × treatment	8	15.60	<0.01
Life stage × treatment	4	14.13	<0.01
Interval × stage × treatment	8	3.33	<0.01

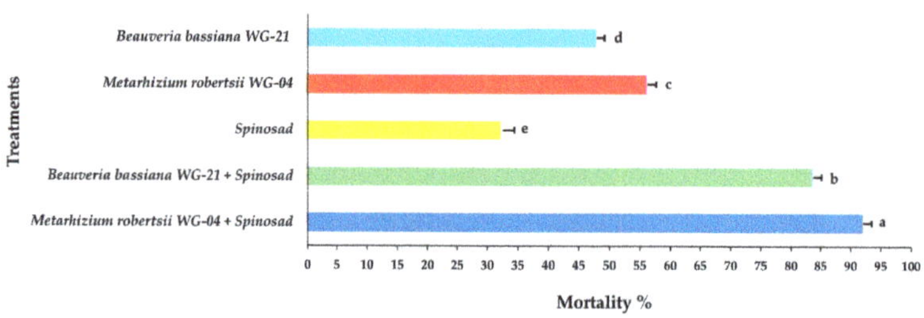

Figure 2. Mean mortality% ± Standard Error (SE) of *Tetranychus urticae* nymphs 1 day after exposure to tomato leaf discs treated with EPF *Beauveria bassiana* WG-21, *Metarhizium robertsii* WG-04 or spinosad, and in combinations (*Beauveria bassiana* WG-21 + Spinosad and *Metarhizium robertsii* WG-04 + Spinosad). Different letters indicate significant differences.

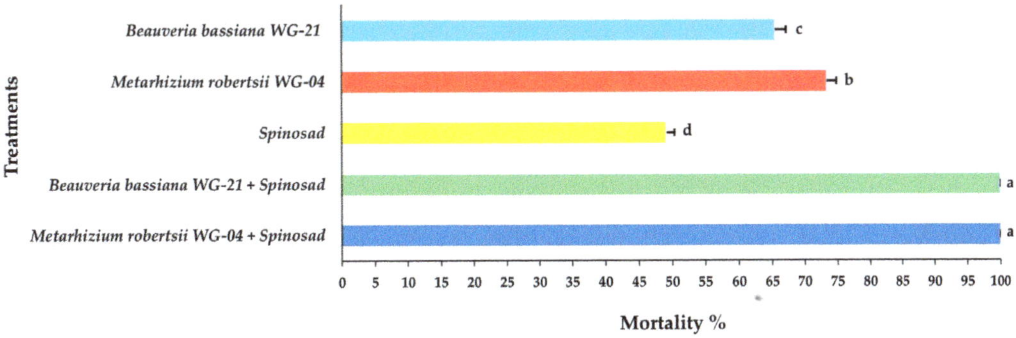

Figure 3. Mean mortality% ± Standard Error (SE) of *Tetranychus urticae* nymphs 2 days after exposure to tomato leaf discs treated with EPF *Beauveria bassiana* WG-21, *Metarhizium robertsii* WG-04 or spinosad, and in combinations (*Beauveria bassiana* WG-21 + Spinosad and *Metarhizium robertsii* WG-04 + Spinosad). Different letters indicate significant differences.

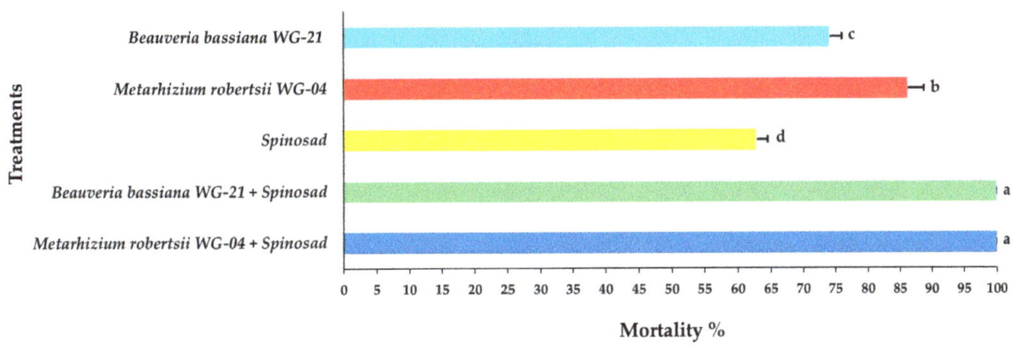

Figure 4. Mean mortality% ± Standard Error (SE) of *Tetranychus urticae* nymphs 3 days after exposure to tomato leaf discs treated with EPF *Beauveria bassiana* WG-21, *Metarhizium robertsii* WG-04 or spinosad, and in combinations (*Beauveria bassiana* WG-21 + Spinosad and *Metarhizium robertsii* WG-04 + Spinosad). Different letters indicate significant differences.

Concerning the mortality of adult females, significant differences were noted among treatments at 3 DAT, with the combined application of *Mr* WG-04 + spinosad showing the highest mortality (85.41%) and the single application of spinosad the lowest (23.75%) (Figure 5, Table S1). After 5 DAT, the combination of *Mr* WG-04 + spinosad killed all (100%) exposed females on the leaf disc. In addition, at 5 DAT, the combination treatment of *Bb* WG-21 + spinosad caused 91.66% mortality of females, while the single application of each fungal isolate *Bb* WG-21 and *Mr* WG-04 alone caused 51.66 and 62.08% mortality, respectively, and spinosad alone killed 38.75% (Figure 6, Table S1). At 7 DAT, 100% mortality of females was recorded on leaf discs exposed to the *Bb* WG-21 + spinosad combination treatment, whereas treatments with females exposed to the fungal isolates or spinosad treatment alone did not surpass 77.08% (Figure 7, Table S1).

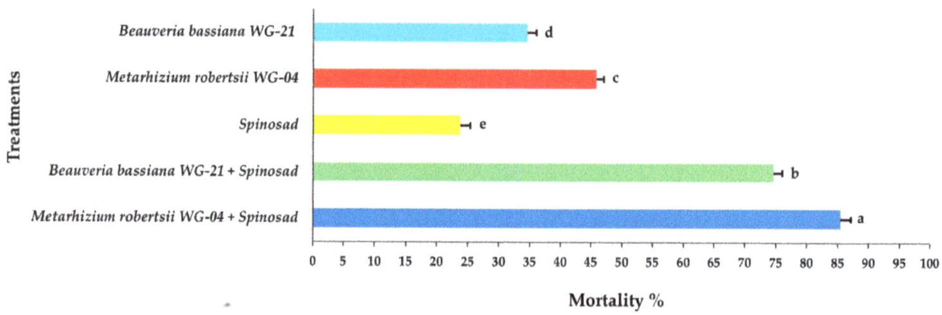

Figure 5. Mean mortality% ± Standard Error (SE) of *Tetranychus urticae* adults 3 days after exposure to tomato leaf discs treated with EPF *Beauveria bassiana* WG-21, *Metarhizium robertsii* WG-04 or spinosad, and in combinations (*Beauveria bassiana* WG-21 + Spinosad and *Metarhizium robertsii* WG-04 + Spinosad). Different letters indicate significant differences.

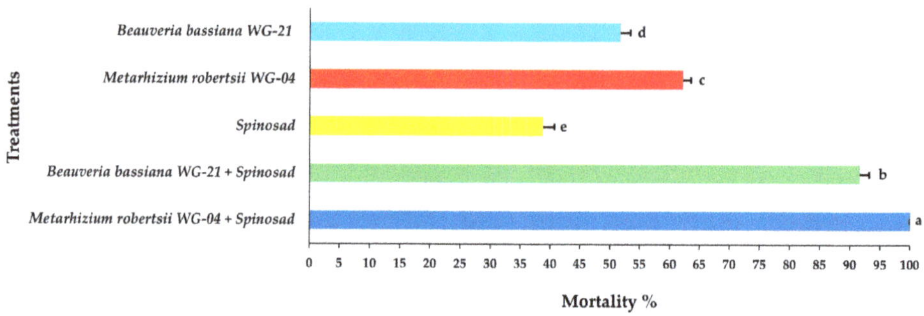

Figure 6. Mean mortality% ± Standard Error (SE) of *Tetranychus urticae* adults 5 days after exposure to tomato leaf discs treated with EPF *Beauveria bassiana* WG-21, *Metarhizium robertsii* WG-04 or spinosad, and in combinations (*Beauveria bassiana* WG-21 + Spinosad and *Metarhizium robertsii* WG-04 + Spinosad). Different letters indicate significant differences.

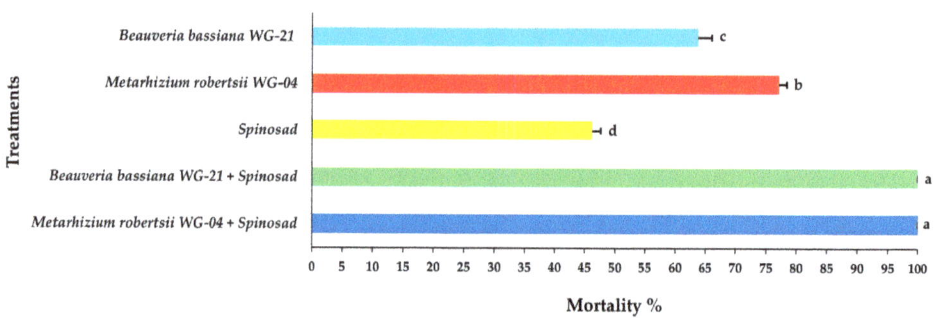

Figure 7. Mean mortality% ± Standard Error (SE) of *Tetranychus urticae* adults 7 days after exposure to tomato leaf discs treated with EPF *Beauveria bassiana* WG-21, *Metarhizium robertsii* WG-04 or spinosad, and in combinations (*Beauveria bassiana* WG-21 + Spinosad and *Metarhizium robertsii* WG-04 + Spinosad). Different letters indicate significant differences.

3.2. *Tetranychus urticae* Populations on Tomato Plants in the Greenhouse

Based on the ANOVA, the main effects and interactions of parameters regarding the presence of *T. urticae* adults, immatures, and eggs on tomato plants following acaricidal treatments in greenhouse trials were significant (Table 2). The number of adults alive on the abaxial side of leaves varied significantly among treated plants compared to control plants

21 DAT (Figure 8, Table S2). No adults were observed on leaves after the application of the *Mr* WG-04 + spinosad combination treatment. A significantly lower number of adults was observed in the combination of *Bb* WG-21 + spinosad (0.87 adults per leaf) and *Mr* WG-04 alone (2.89 adults per leaf) treatments compared to the *Bb* WG-21 (7.26 adults per leaf) and spinosad treatments alone (10.47 adults per leaf). The highest number of adults was found on control plants, with 82.25 adults per leaf compared to the other treatments. Regarding immatures on the bottom (abaxial) surface of the leaves, significantly lower numbers were found in the combination treatments of *Mr* WG-04 + spinosad (0.00 immatures per leaf) and *Bb* WG-21 + spinosad (2.63 immatures per leaf) compared to *Mr* WG-04 (14.87 immatures per leaf), *Bb* WG-21 (23.42 immatures per leaf) and spinosad (39.34 immatures per leaf) applied alone (Figure 9, Table S2). In contrast, 271.67 immatures per leaf were recorded in the control plants, which differed significantly from the treated plants. Concerning the presence of eggs counted on the bottom surface of leaves, control plants had a significantly higher number (186.70 eggs per leaf) compared to spinosad (24.62 eggs per leaf), *Bb* WG-21 (11.29 eggs per leaf), *Mr* WG-04 (7.85 eggs per leaf) alone, and the combination treatments of *Mr* WG-04 + spinosad (0.00 egg per leaf) or *Bb* WG-12 + spinosad (1.15 eggs per leaf) (Figure 10, Table S2).

Table 2. ANOVA parameters for the number of individuals (adults, immatures, eggs) of *Tetranychus urticae* alive on tomato leaves after application with EPF *Beauveria bassiana* (*Bb* WG-21), *Metarhizium robertsii* (*Mr* WG-04) or spinosad alone, and in combination. (Total df = 431).

Source	df	*F*	*p*
Leaf side (abaxial or adaxial)	1	3978.95	<0.01
Life stage	2	852.20	<0.01
Treatment	5	4337.09	<0.01
Leaf side × life stage	2	356.22	<0.01
Leaf side × treatment	5	2439.46	<0.01
Life stage × treatment	10	421.93	<0.01
Leaf side × life stage × treatment	10	207.11	<0.01

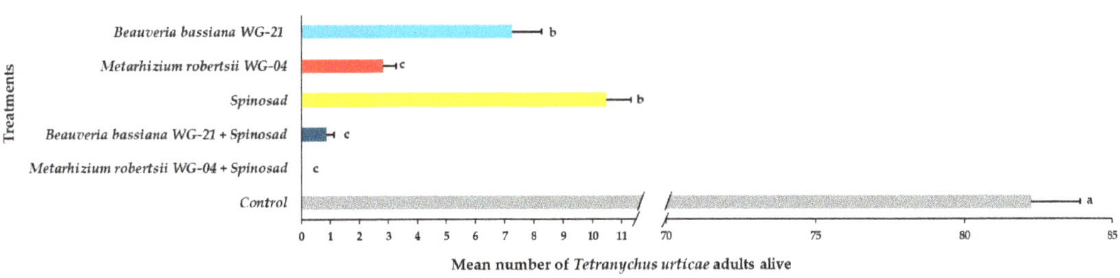

Figure 8. Mean number ± Standard Error (SE) of *Tetranychus urticae* adults alive on the abaxial (bottom) side of tomato leaves after treatment with *Beauveria bassiana* WG-21, *Metarhizium robertsii* WG-04, spinosad, control (water + Tween 80), and in combinations (*Beauveria bassiana* WG-21 + Spinosad and *Metarhizium robertsii* WG-04 + Spinosad) at 21 days after treatment. Different letters indicate significant differences.

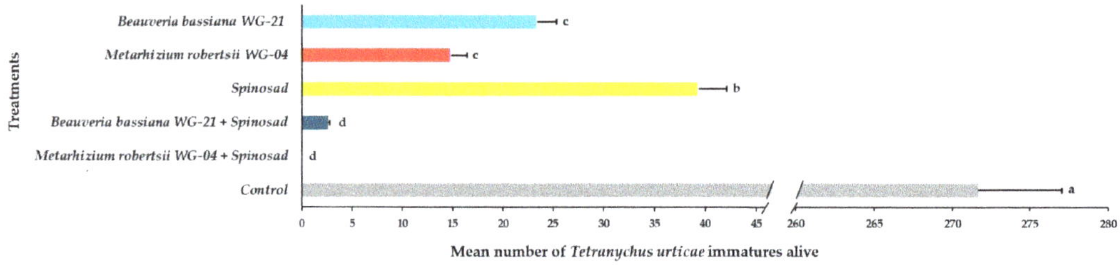

Figure 9. Mean number ± Standard Error (SE) of *Tetranychus urticae* immatures alive on the abaxial (bottom) side of tomato leaves after treatment with *Beauveria bassiana* WG-21, *Metarhizium robertsii* WG-04, spinosad, control (water + Tween 80), and in combinations (*Beauveria bassiana* WG-21 + Spinosad and *Metarhizium robertsii* WG-04 + Spinosad) at 21 days after treatment. Different letters indicate significant differences.

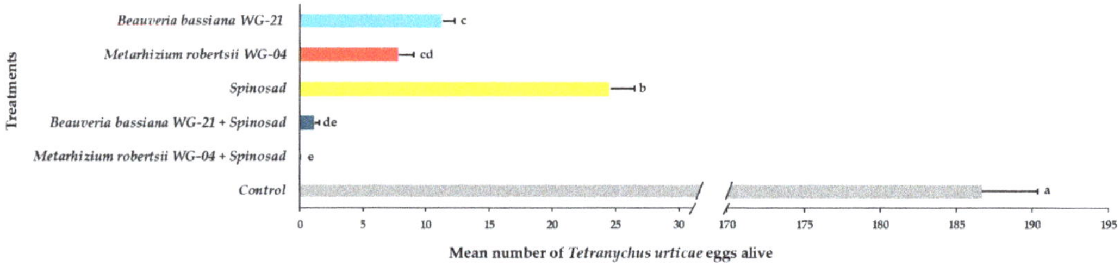

Figure 10. Mean number ± Standard Error (SE) of *Tetranychus urticae* eggs alive on the abaxial (bottom) side of tomato leaves after treatment with *Beauveria bassiana* WG-21, *Metarhizium robertsii* WG-04, spinosad, control (water + Tween 80), and in combinations (*Beauveria bassiana* WG-21 + Spinosad and *Metarhizium robertsii* WG-04 + Spinosad) at 21 days after treatment. Different letters indicate significant differences.

Similar findings were noted regarding the establishment of *T. urticae* on the top (adaxial) surface of the leaves 21 DAT (Figure 11, Table S2). No adults were recorded for either EPF + spinosad combination treatments. Fewer adults were found on plants treated with *Mr* WG-04 (0.57 adults per leaf) and *Bb* WG-21 (1.43 adults per leaf) than on those treated with spinosad (3.45 adults per leaf) alone compared to the control group, which had the highest number of adults recorded (11.64 adults per leaf). The number of adults counted at 21 DAT in all the treated plants was significantly lower compared to that in the control. No immatures were observed on the adaxial surface of the leaves in those plants treated with *Mr* WG-04 or *Bb* WG-21 combined with spinosad (Figure 12, Table S2). In the individual treatments, the number of immatures was significantly higher in the spinosad treatment (17.42 immatures per leaf) compared to either *Bb* WG-21 (5.74 immatures per leaf) or *Mr* WG-04 (0.96 immatures per leaf). In contrast, significantly more immatures were counted in control plants compared to the treated plants. Similarly, no eggs at 21 DAT were found after being exposed to the combination treatment of *Mr* WG-04 or *Bb* WG-21 plus spinosad. The application of each EPF treatment alone (*Mr* WG-04 or *Bb* WG-21) significantly decreased the number of eggs (0.68 and 3.24 eggs per leaf, respectively) on the adaxial surface when compared to spinosad (9.51 eggs per leaf) and control plants (21.28 eggs per leaf) (Figure 13, Table S2). The number of eggs counted in all treated plants was significantly lower compared to that in the control.

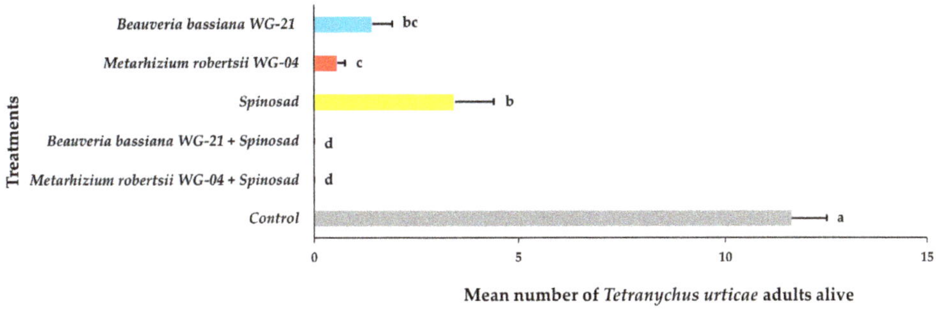

Figure 11. Mean number ± Standard Error (SE) of *Tetranychus urticae* adults alive on the adaxial (top) side of tomato leaves after treatment with *Beauveria bassiana* WG-21, *Metarhizium robertsii* WG-04, spinosad, control (water + Tween 80), and in combinations (*Beauveria bassiana* WG-21 + Spinosad and *Metarhizium robertsii* WG-04 + Spinosad) at 21 days after treatment. Different letters indicate significant differences.

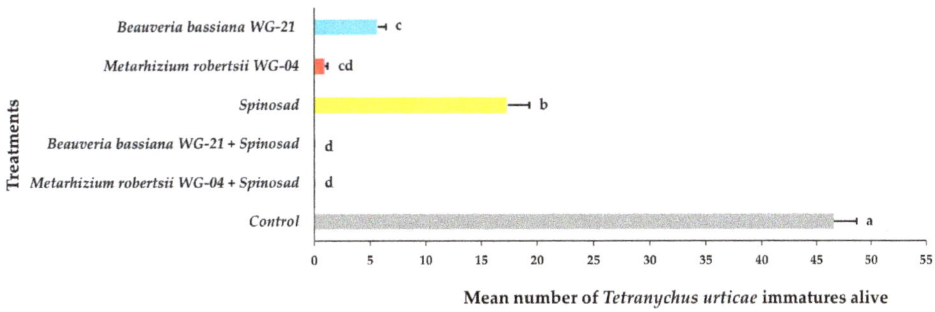

Figure 12. Mean number ± Standard Error (SE) of *Tetranychus urticae* immatures alive on the adaxial (top) side of tomato leaves after treatment with *Beauveria bassiana* WG-21, *Metarhizium robertsii* WG-04, spinosad, control (water + Tween 80), and in combinations (*Beauveria bassiana* WG-21 + Spinosad and *Metarhizium robertsii* WG-04 + Spinosad) at 21 days after treatment. Different letters indicate significant differences.

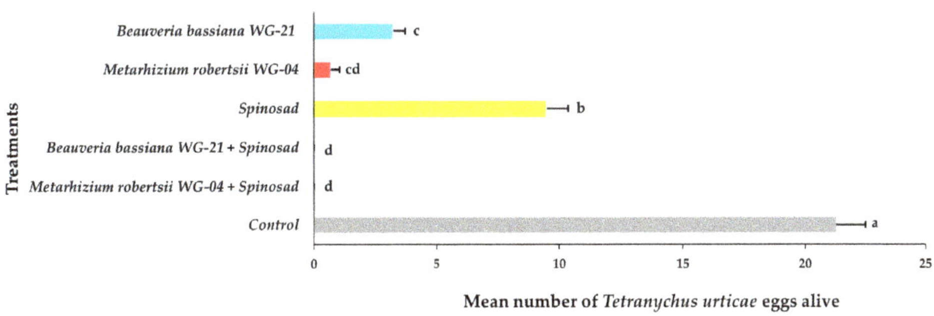

Figure 13. Mean number ± Standard Error (SE) of *Tetranychus urticae* eggs alive on the adaxial (top) side of tomato leaves after treatment with *Beauveria bassiana* WG-21, *Metarhizium robertsii* WG-04, spinosad, control (water + Tween 80), and in combinations (*Beauveria bassiana* WG-21 + Spinosad and *Metarhizium robertsii* WG-04 + Spinosad) at 21 days after treatment. Different letters indicate significant differences.

4. Discussion

Beauveria bassiana and *M. robertsii* isolates have demonstrated considerable potential in the management of *T. urticae* and in providing an effective biological control option as an alternative to the broad-spectrum synthetic acaricides [25–27,86–88]. In the third day of mortality estimation, *B. bassiana* WG-21 killed 34.58% of *T. urticae* females. This was expected since *B. bassiana* normally takes approximately three days to show acaricidal effects. Wu et al. [89] found that germinated conidia successfully penetrated the cuticle of *T. urticae* adults within 2.5 days. For instance, Wakil et al. [25] reported 100.00% mortality of *T. urticae* female adults after exposure to *M. robertsii* WG-02 and *B. bassiana* WG-12 isolates on leaf discs at 10 DAT, whereas the *B. bassiana* WG-19 isolate achieved 94.58% mortality of exposed females over the same duration. Also, in the current work, both fungal isolates demonstrated nymphal and adult mortality rates exceeding 74.00% and 63.00% at 3 DAT and 7 DAT, respectively, significantly outperforming spinosad on leaf discs. In addition, the *M. robertsii* isolate WG-04 consistently caused significantly higher mortality in both nymphs and female adults of *T. urticae* compared to the *B. bassiana* isolate WG-21 throughout the experimental period under laboratory conditions. Similar findings were reported by Wakil et al. [26], with the WG-7 isolate of *M. robertsii* proving to be more virulent against *T. urticae* nymphs and adults (females) than the WG-12 isolate of *B. bassiana* 7 days after application on leaf discs. Recently, Elhakim et al. [27] reported that *M. robertsii* killed more female adults of *T. urticae* (85.00%) compared to *B. bassiana* (70.00%) 7 days after spraying. The authors attributed the difference in the virulence of both EPF isolates against *T. urticae* to the greater protease activity observed in *M. robertsii* compared to *B. bassiana* [27,90] since this activity constitutes one of the primary mechanisms through which EPF exert their pathogenic effects [27,90,91]. Moreover, protease activity is essential in the host penetration process [27,91–94]. Interestingly, in our study, nymphs were more sensitive to EPF than adult females. Similar results have been reported by Wakil et al. [26]. However, previous studies have reported contrary results, with adults having a higher susceptibility to various fungal species, including *M. robertsii* and *B. bassiana*, than nymphs [87,95,96]. This high tolerance is explained due to the ecdysis of nymphs, where the spores of EPF are discarded with the cuticle. For instance, Ranout et al. [96] found mycelial development in adult cadavers vs. no mycelial development in nymphal cadavers, indicating that ecdysis took place before fungal sporulation [95–97]. Therefore, it becomes evident that different EPF species and isolates exhibit variable virulence in adults and nymphs of *T. urticae*.

Previous studies have explored the combined use of EPF with insecticides as a valuable pest control strategy, highlighting their potential to reduce the risk of insecticide resistance [55,98,99]. For instance, Wakil et al. [26] found that the paired treatments of EPF *B. bassiana* WG-12 and *M. robertsii* WG-7 plus abamectin killed all exposed nymphs and adults (females) of *T. urticae* 3 and 7 d post application. In addition, Wakil et al. [26] found that individual treatments of EPF or abamectin caused significantly lower mortality (i.e., <75 and <65% for immatures and adults, respectively) compared to their respective combination treatments (i.e., 100% for nymphs and adults). Similar findings were obtained in the current study using leaf discs, where the combinations of EPF with spinosad caused 100% nymphal and adult female mortality in the same time interval (3 DAT and 7 DAT, respectively). In addition, the treatments of *B. bassiana* WG-21, *M. robertsii* WG-04, and spinosad alone did not exceed 86.50 and 77.50% mortality for nymphs and adults, respectively. Therefore, it is evident that combination treatments of EPF with different active ingredients can have an additive effect upon EPF efficacy and achieve higher mortality rates compared to the application of each control agent alone against *T. urticae* under laboratory conditions [26]. The slow action of EPF on *T. urticae*, when used alone, may be due to the time needed for fungi to attach, penetrate, germinate, and proliferate in a host haemocoel [27,100]. This type of activity is a barrier to using EPF alone because they need considerable time to suppress the mite population, while the infested crop is being damaged [101]. The environmental conditions also slow down EPF activity. For example, temperatures <16 °C delay the germination and growth of EPF, affecting their long-term virulence [29,102]. To overcome these drawbacks, EPF applications should be repeated and used to manage low densities of pests; however, the overall cost of management will be increased [29]. This is also evident from our results where population establishment of *T. urticae* was not avoided on EPF-treated tomato plants, providing the highest immature population in comparison to the combined treatments.

Similarly, spinosad is a comparatively slow-acting toxicant in comparison to other classes of insecticides (e.g., organophosphates or pyrethroids). Intoxication symptoms of spinosad appear very fast and peak cumulative percent mortality increases gradually [103,104]. This progressive mortality is affected by the slow penetration of the cuticle by spinosad [104]. Sparks et al. [103] reported that spynosin A penetrated into *Heliothis virescens* (F.) (Lepidoptera: Noctuidae) larvae slowly, but when it entered the insect, spynosin A was not easily metabolized by the host either. Sparks et al. [103] noted that this limited metabolism in larvae of *H. virescens* contributed to the observed increased activity against this pest, thereby compensating slower penetration rate of spynosin A. Presumably, the protease activity of EPF [90] accelerated the penetration of spinosad into mites, providing increased rates of mortality faster than the single applications of EPF or spinosad alone. This combined activity may explain the highest and most rapid efficacy of the paired treatment of *M. robertsii* WG-04 + spinosad against *T. urticae*; however, this hypothesis needs to be further investigated. There is no precise report available describing the mechanisms of rapid mortality in combinations of EPF and spinosad. We hypothesized that one agent might stress the pest while enhancing its vulnerability to infection or susceptibility to insecticides by giving an additive effect [34,105,106]. The possible mechanism of EPF–spinosad interaction is that the pesticide weakens the immune system of the target organism [107–110]. In addition, Rivero-Borja et al. [110] reported that when spinosad application preceded the application of EPF (*M. robertsii* isolate ETL, or *B. bassiana* isolate Bb88), fungal performance improved in comparison to the case where EPF preceded the application of the pesticide. In our study, spinosad was applied first, followed by the application of EPF.

Concerning the establishment of *T. urticae* on tomato plants, the mites predominantly prefer to inhabit and feed on the abaxial surface of leaves [26,111–113]. The findings from this study revealed a greater number of *T. urticae* on the abaxial surface of the tomato leaves compared to the adaxial side, where the application of each treatment significantly suppressed the presence of mites compared to the control. However, most importantly, the combination treatment of *M. robertsii* WG-04 + spinosad completely prevented the establishment of *T. urticae* on either side of the leaves. In plants treated with *B. bassiana* WG-21 + spinosad, only a few mites were found in the abaxial surface of leaves (i.e., <2.70 individuals/leaf). However, in contrast to the combination treatments, the least effective treatment was spinosad on either side of the leaf surface. In corroboration, Wakil et al. [26] observed similar results, where the combination treatments of *M. robertsii* WG-7 or *B. bassiana* WG-12 plus abamectin significantly reduced *T. urticae* presence on both leaf surfaces, whereas abamectin alone showed lower efficacy than either of the combination treatments. Therefore, plant protection and potentially increased yield can be achieved after the combined application of EPF with a compatible acaricide, especially the combination treatments of *M. robertsii* + spinosad (current study) and abamectin combinations [26], which can cause a significant reduction of *T. urticae* populations.

The outcome of this study indicates that *B. bassiana* WG-21 + spinosad or *M. robertsii* WG-04 + spinosad combination treatments are effective tools against various life stages of *T. urticae*. The combination treatments were more efficacious compared to the application of each component alone, which could be due to their rapid activity, increased mortality rates, and resultant low pest populations. It is widely recognized that when two agents with different modes of action are combined, their activity may be independent (additive effect) or one agent may induce sufficient stress on the pest, boosting the effectiveness of the second agent (synergistic effect) [99,107,114]. Furthermore, since different isolates of *M. robertsii* and *B. bassiana* have shown varying mortality rates against *T. urticae* [25,26], the selection of appropriate fungal isolates is essential for the efficacy of EPF against pests [10,27], especially when used in combination with a compatible pesticide. Thus, based on the results of this study, the use of new entomopathogenic fungal isolates from different geographical areas in combination with spinosad may be a promising alternative strategy for managing this mite pest successfully.

Supplementary Materials: The following supporting information can be downloaded at: https://www.mdpi.com/article/10.3390/jof11020138/s1, Table S1: Mean mortality % ± Standard Error (SE) of nymphs and adult females of *Tetranychus urticae* exposed to tomato leaf discs after the treatment with the entomopathogenic fungi *Beauveria bassiana* (*Bb* WG-21), *Metarhizium robertsii* (*Mr* WG-04) or spinosad, and in combinations (*Bb* WG-21 + Spinosad and *Mr* WG-04 + Spinosad) at different days after treatment. Within each row, means with the same uppercase letter are not significantly different (df = 2.35, Tukey HSD test at $p = 0.05$). Within each column, means followed by the same lowercase letter are not significantly different (df = 4.59, Tukey HSD test at $p = 0.05$); Table S2: Mean number ± Standard Error (SE) of individuals of *Tetranychus urticae* alive at different life stages (adults, immatures, eggs) on the abaxial and adaxial side of tomato leaves after treatment with *Beauveria bassiana* (*Bb* WG-21), *Metarhizium robertsii* (*Mr* WG-04), spinosad, control (water + Tween 80), and the respective paired combinations of the entomopathogenic fungi with spinosad (*Bb* WG-21 + Spinosad and *Mr* WG-04 + Spinosad) per leaf side at 21 days after treatment. Within each row means with the same uppercase letter(s) are not significantly different (df = 2.35, Tukey HSD test at $p = 0.05$). Within each column, means followed by the same lowercase letter are not significantly different (df = 5.71, Tukey HSD test at $p = 0.05$). Where dashes exist, no statistics were applied.

Author Contributions: Conceptualization, W.W. and N.G.K.; methodology, W.W., N.G.K. and P.B.A.; software, W.W., M.C.B., N.G.K., A.N. and D.N.; validation, W.W., M.C.B., N.G.K., A.N., D.N., M.U.G. and P.B.A.; formal analysis, W.W., M.C.B., N.G.K., A.N. and D.N.; investigation, W.W., M.C.B., N.G.K., A.N., D.N. and M.U.G.; resources, W.W. and N.G.K.; data curation, W.W. and A.N.; writing—original draft preparation, W.W., M.C.B., N.G.K., A.N., D.N., M.U.G. and P.B.A.; writing—review and editing, W.W., M.C.B., N.G.K., A.N., D.N., M.U.G. and P.B.A.; visualization, W.W., M.C.B., N.G.K., A.N., D.N., M.U.G. and P.B.A.; supervision, W.W.; project administration, W.W. and M.U.G.; funding acquisition, W.W. All authors have read and agreed to the published version of the manuscript.

Funding: Project 3244 of the Higher Education Commission, Islamabad, Pakistan, partly funded this research.

Institutional Review Board Statement: Not applicable.

Informed Consent Statement: Not applicable.

Data Availability Statement: The data are contained within this article.

Conflicts of Interest: The authors declare no conflicts of interest.

References

1. Adesanya, A.W.; Lavine, M.D.; Moural, T.W.; Lavine, L.C.; Zhu, F.; Walsh, D.B. Mechanisms and management of acaricide resistance for *Tetranychus urticae* in agroecosystems. *J. Pest Sci.* **2021**, *94*, 639–663. [CrossRef]
2. Leppla, N.C.; Johnson, M.W.; Merritt, J.L.; Zalom, F.G. Applications and trends in commercial biological control for arthropod pests of tomato. In *Sustainable Management of Arthropod Pests of Tomato*; Wakil, W., Brust, G.E., Perring, T.M., Eds.; Academic Press: London, UK, 2018; pp. 283–303.
3. Pekár, S.; Raspotnig, G. Defences of Arachnids: Diversified arsenal used against range of enemies. *Entomol. Gen.* **2022**, *42*, 663–679. [CrossRef]
4. Zhao, S.; Zhao, Q.; Dai, X.; Lv, B.; Wang, R.; Yin, Z.; Zhang, F.; Liu, Y.; Su, L.; Chen, H.; et al. Control of two-spotted spider mite, *Tetranychus urticae*, on strawberry by integrating with cyetpyrafen and *Phytoseiulus persimilis*. *CABI Agric Biosci.* **2023**, *4*, 54. [CrossRef]
5. Migeon, A.; Dorkled, F. Spider Mites Web: A Comprehensive Database for the Tetranychidae. Available online: https://www1.montpellier.inrae.fr/CBGP/spmweb (accessed on 14 November 2024).
6. Rioja, C.; Zhurov, V.; Bruinsma, K.; Grbic, M.; Grcic, V. Plant-herbivore interactions: A case of an extreme generalist, the two-spotted spider mite *Tetranychus urticae*. *Mol. Plant Microbe Interact* **2017**, *30*, 935–945. [CrossRef] [PubMed]
7. Namin, H.H.; Zhurov, V.; Spenler, J.; Grbic, M.; Grbic, V.; Scott, I.M. Resistance to pyridaben in Canadian greenhouse populations of two-spotted spider mites, *Tetranychus urticae* (Koch). *Pestic. Biochem. Physiol.* **2020**, *170*, 104677. [CrossRef]
8. Assouguem, A.; Kara, M.; Mechchate, H.; Korkmaz, Y.B.; Benmessaoud, S.; Ramzi, A.; Abdullah, K.R.; Noman, O.M.; Farah, A.; Lazraq, A. Current situation of Tetranychus urticae (Acari: Tetranychidae) in Northern Africa: The sustainable control methods and priorities for future research. *Sustainability* **2022**, *14*, 2395. [CrossRef]
9. Tabary, L.; Navia, D.; Auger, P.; Migeon, A.; Navajas, M.; Tixier, M.S. Plant, pest and predator interplay: Tomato trichomes effects on *Tetranychus urticae* (Koch) and the predatory mite *Typhlodromus* (*Anthoseius*) *recki* Wainstein. *Exp. Appl. Acarol.* **2024**, *93*, 169–195. [CrossRef]
10. Al-Zahrani, J.K.; Al-Abdalall, A.H.; Osman, M.A.; Aldakheel, L.A.; AlAhmady, N.F.; Aldakeel, S.A.; AbdulAzeez, S.; Borgio, J.F.; ElNaggar, M.A.; Alabdallah, N.M.; et al. Entomopathogenic fungi and their biological control of *Tetranychus urticae*: Two-spotted spider mites. *J. King Saud Univ. Sci.* **2023**, *35*, 102910. [CrossRef]
11. Nieberding, C.M.; Kaiser, A.; Visser, B. Inbreeding and learning affect fitness and colonization of new host plants, a behavioral innovation in the spider mite *Tetranychus urticae*. *Entomol. Gen.* **2022**, *42*, 531–538. [CrossRef]
12. Suekane, R.; Degrande, P.E.; De Melo, E.P.; Bertoncello, T.F.; Junior, I.D.S.L.; Kodama, C. Damage level of the two-spotted spider mite *Tetranychus urticae* Koch (acari: Tetranychidae) in soybeans. *Rev. Ceres* **2012**, *59*, 77–81. [CrossRef]
13. Attia, S.; Grissa, K.L.; Lognay, G.; Bitume, E.; Hance, T.; Mailleux, A.C. A review of the major biological approaches to control the worldwide pest *Tetranychus urticae* (Acari: Tetranychidae) with special reference to natural pesticides. *J. Pest Sci.* **2013**, *86*, 361–386. [CrossRef]
14. Estrella Santamaria, M.; Arnaiz, A.; Rosa-Diaz, I.; González-Melendi, P.; Romero-Hernandez, G.; Ojeda-Martinez, D.A.; Garcia, A.; Contreras, E.; Martinez, M.; Diaz, I. Plant Defenses against *Tetranychus urticae*: Mind the Gaps. *Plants* **2020**, *9*, 464. [CrossRef] [PubMed]
15. Bensoussan, N.; Santamaria, M.E.; Zhurov, V.; Diaz, I.; Grbic, M.; Grbic, V. Plant-herbivore interaction: Dissection of the cellular pattern of *Tetranychus urticae* feeding on the host plant. *Front. Plant Sci.* **2016**, *7*, 1105. [CrossRef]
16. Van Leeuwen, T.; Tirry, L.; Yamamoto, A.; Nauen, R.; Dermauw, W. The economic importance of acaricides in the control of phytophagous mites and an update on recent acaricide mode of action research. *Pestic. Biochem. Physiol.* **2015**, *121*, 12–21. [CrossRef]
17. İnak, E.; Demirci, B.; Vandenhole, M.; Söylemezoğlu, G.; Van Leeuwen, T.; Toprak, U. Molecular mechanisms of resistance to spirodiclofen and spiromesifen in *Tetranychus urticae*. *Crop Prot.* **2023**, *172*, 106343. [CrossRef]
18. Van Leeuwen, T.; Vontas, J.; Tsagkarakou, A.; Tirry, L. Acaricide resistance mechanism in the two-spotted spider mite *Tetranychus urticae* and other important Acari: A review. *Insect Biochem. Mol. Biol.* **2010**, *40*, 563–572. [CrossRef] [PubMed]
19. Tian, T.; Wu, M.M.; Zhang, Y.; Xu, D.D.; Wu, M.Y.; Xie, W.; Su, Q.; Wang, S.L. Pesticide Resistance and Related Mutation Frequencies of *Tetranychus urticae* in Hainan, China. *Horticulturae* **2022**, *8*, 590. [CrossRef]
20. De Rouck, S.; İnak, E.; Dermauw, W.; Van Leeuwen, T. A Review of the Molecular Mechanisms of Acaricide Resistance in Mites and Ticks. *Insect Biochem. Mol. Biol.* **2023**, *159*, 103981. [CrossRef] [PubMed]
21. Zhu, Y.; Wu, T.; Hu, Q.; He, W.; Zheng, Y.; Xie, Y.; Rao, Q.; Liu, X. Plant essential oils: Dual action of toxicity and egg-laying inhibition on *Tetranychus urticae* (Acari: Tetranychidae), unveiling their potential as botanical pesticides. *Plants* **2024**, *13*, 763. [CrossRef]
22. Riga, M.; Tsakireli, D.; Ilias, A.; Morou, E.; Myridakis, A.; Stephanou, E.G.; Nauen, R.; Dermauw, W.; Van Leeuwen, T.; Paine, M.; et al. Abamectin is metabolized by CYP392A16, a cytochrome P450 associated with high levels of acaricide resistance in *Tetranychus urticae*. *Insect Biochem. Mol. Biol.* **2014**, *46*, 43–53. [CrossRef]

23. Tehri, K.; Gulati, R.; Geroh, M.; Dhankhar, S.K. Dry weather: A crucial constraint in the field efficacy of entomopathogenic fungus *Beauveria bassiana* against *Tetranychus urticae* Koch (Acari: Tetranychidae). *J. Entomol. Zool. Stud.* **2015**, *3*, 287–291.
24. Gámez-Guzmán, A.; Torres-Rojas, E.; Gaigl, A. Potential of a *Cladosporium cladosporioides* strain for the control of *Tetranychus urticae* Koch (Acari: Tetranychidae) under laboratory conditions. *Agron. Colomb.* **2019**, *37*, 84–89. [CrossRef]
25. Wakil, W.; Boukouvala, M.C.; Kavallieratos, N.G.; Naeem, A.; Ghazanfar, M.U.; Alhewairini, S.S. Impact of three entomopathogenic fungal isolates on the growth of tomato plants—Ectoapplication to explore their effect on Tetranychus urticae. *Agronomy* **2024**, *14*, 665. [CrossRef]
26. Wakil, W.; Boukouvala, M.C.; Kavallieratos, N.G.; Riasat, T.; Ghazanfar, M.U.; Avery, P.B. Acaricidal efficacy of abamectin against *Tetranychus urticae* populations when combined with entomopathogenic fungi. *Horticulturae* **2024**, *10*, 1019. [CrossRef]
27. Elhakim, E.; Mohamed, O.; Elazouni, I. Virulence and proteolytic activity of entomopathogenic fungi against the two-spotted spider mite, *Tetranychus urticae* Koch (Acari: Tetranychidae). *Egypt. J. Biol. Pest Control* **2020**, *30*, 30. [CrossRef]
28. Yucel, C. Effects of local isolates of *Beauveria bassiana* (Balsamo) Vuillemin on the two-spotted spider mite, *Tetranychus urticae* (Koch) (Acari: Tetranychidae). *Egypt. J. Biol. Pest Control* **2021**, *31*, 63. [CrossRef]
29. Jaronski, S.T. Ecological factors in the inundative use of fungal entomopathogens. *BioControl* **2020**, *55*, 159–185. [CrossRef]
30. Leao, M.P.; Tiago, P.V.; Andreote, F.D.; de Araujo, W.L.; de Oliveira, N.T. Differential expression of the pr1A gene in *Metarhizium anisopliae* and *Metarhizium acridum* across different culture conditions and during pathogenesis. *Genet. Mol. Biol.* **2015**, *38*, 86–92. [CrossRef]
31. Ortiz-Urquiza, A.; Keyhani, N.O. Action on the surface: Entomopathogenic fungi versus the insect cuticle. *Insects* **2013**, *4*, 357–374. [CrossRef] [PubMed]
32. Santi, L.; Beys da Silva, W.O.; Berger, M.; Guimarães, J.A.; Schrank, A.; Vainstein, M.H. Conidial surface proteins of *Metarhizium anisopliae*: Source of activities related with toxic effects, host penetration and pathogenesis. *Toxicon* **2010**, *55*, 874–880. [CrossRef] [PubMed]
33. Gabarty, A.; Salem, H.; Fouda, M.; Abas, A.; Ibrahim, A. Pathogencity induced by the entomopathogenic fungi *Beauveria bassiana* and *Metarhizium anisopliae* in *Agrotis ipsilon* (Hufn.). *J. Radiat. Res. Appl. Sci.* **2014**, *7*, 95–100. [CrossRef]
34. Butt, T.M.; Coates, C.J.; Dubovskiy, I.M.; Ratcliffe, N.A. Entomopathogenic Fungi: New Insights into Host-Pathogen Interactions. In *Advances in Genetics*, 1st ed.; Lovett, B., St. Leger, R., Eds.; Elsevier: Cambridge, MA, USA, 2016; Volume 94, pp. 307–364.
35. Islam, W.; Adnan, M.; Shabbir, A.; Naveed, H.; Abubakar, Y.S.; Qasim, M.; Tayyab, M.; Noman, A.; Nisar, M.S.; Khan, K.A.; et al. Insect-fungal-interactions: A detailed review on entomopathogenic fungi pathogenicity to combat insect pests. *Microb. Pathog.* **2021**, *159*, 105122. [CrossRef]
36. Francis, F.; Fingu-Mabola, J.C.; Ben Fekih, I. Direct and endophytic effects of fungal entomopathogens for sustainable aphid control: A review. *Agriculture* **2022**, *12*, 2081. [CrossRef]
37. Jaber, L.R.; Ownley, B.H. Can we use entomopathogenic fungi as endophytes for dual biological control of insect pests and plant pathogens? *Biol. Control* **2018**, *116*, 36–45. [CrossRef]
38. Copping, G.L.; Menn, J.J. Biopesticides: A review of their action, applications and efficacy. *Pest Manag. Sci.* **2000**, *56*, 651–676. [CrossRef]
39. Santos, V.S.V.; Pereira, B.B. Properties, toxicity and current applications of the biolarvicide spinosad. *J. Toxicol. Environ. Health—B Crit.* **2020**, *23*, 13–26. [CrossRef]
40. Salgado, V.L. Studies on the Mode of Action of Spinosad: Insect Symptoms and Physiological Correlates. *Pestic. Biochem. Physiol.* **1998**, *60*, 91–102. [CrossRef]
41. Biondi, A.; Mommaerts, V.; Smagghe, G.; Vinuela, E.; Zappala, L.; Desneux, N. The non-target impact of spinosyns on beneficial arthropods. *Pest Manag. Sci.* **2012**, *68*, 1523–1536. [CrossRef]
42. Abd El-Samei, E.M.; Hamama, H.M.; El-Enien, M.G.; Awad, H.H. Interaction of spinosad and *Bacillus thuringiensis* on certain toxicological, biochemical and molecular aspects in the Egyptian cotton leafworm, *Spodoptera littoralis* (Boisduval) (Lepidoptera: Noctuidae). *Afr. Entomol.* **2019**, *27*, 508–522. [CrossRef]
43. Khan, H.A.A. Variation in susceptibility to insecticides and synergistic effect of enzyme inhibitors in Pakistani strains of *Trogoderma granarium*. *J. Stored Prod. Res.* **2021**, *91*, 101775. [CrossRef]
44. Ishadi, N.A.M.; Norida, M.; Oma, D.; Hong, L.W. Resistance against spinosad in a lab-rearing *Plutella xylostella* population and its impact on fitness cost. *J. Anim. Plant Sci.* **2022**, *32*, 479–488.
45. Darriet, F.; Duchon, S.; Hougard, J.M. Spinosad: A New Larvicide against Insecticide-Resistant Mosquito Larvae. *J. Am. Mosq. Control Assoc.* **2005**, *21*, 495–496. [CrossRef] [PubMed]
46. Yasir, M.; Hasan, M.; Sagheer, M.; Rasul, A.; Ali, R.A.; Rehman, H. Evaluation of spinosad applied to grain commodities for the control of stored product insect pests. *Sarhad J. Agric.* **2021**, *37*, 24–31. [CrossRef]
47. Ganjisaffar, F.; Gress, B.E.; Demkovich, M.R.; Nicola, N.L.; Chiu, J.C.; Zalom, F.G. Spatio-temporal variation of spinosad susceptibility in *Drosophila suzukii* (Diptera: Drosophilidae), a three-year study in California's Monterey Bay Region. *J. Econ. Entomol.* **2022**, *115*, 972–980. [CrossRef] [PubMed]

48. Jones, T.; Scott-Dupree, C.; Harris, R.; Shipp, L.; Harris, B. The efficacy of spinosad against the western flower thrips, *Frankliniella occidentalis*, and its impact on associated biological control agents on greenhouse cucumbers in southern Ontario. *Pest Manag. Sci.* **2005**, *61*, 179–185. [CrossRef] [PubMed]
49. Dripps, J.E.; Boucher, R.E.; Chloridis, A.; Cleveland, C.B.; DeAmicis, C.V.; Gomez, L.E.; Paroonagian, D.L.; Pavan, L.A.; Sparks, T.C.; Watson, G.B. The spinosyn insecticides. In *Green Trends in Insect Control*; López, O., Fernández-Bolaños, J.G., Eds.; Royal Society of Chemistry: London, UK, 2011; pp. 163–212.
50. van der Linden, A.; van der Staaij, M. Effectiviness of pesticides and potential for biological control of the tomato leaf miner *Tuta absoluta* (Meyrick) (Lepidoptera: Gelechiidae) in Europe. *IOBC/wprs Bull.* **2011**, *68*, 97–100.
51. Santis, E.L.; Hernandez, L.A.; Martinez, A.M.; Campos, J.; Figueroa, J.I.; Lobit, P.; Chavarrieta, J.M.; Vinuela, E.; Smagghe, G.; Pineda, S. Long-term foliar persistence and efficacy of spinosad against beet armyworm under greenhouse conditions. *Pest Manag. Sci.* **2012**, *68*, 914–921. [CrossRef] [PubMed]
52. Međo, I.; Stojnić, B.; Marčić, D. Acaricidal activity and sublethal effects of the microbial pesticide spinosad on *Tetranychus urticae* (Acari: Tetranychidae). *Syst. Appl. Acarol.* **2017**, *22*, 1748–1762. [CrossRef]
53. Hernández, M.M.; Martínez-Villar, E.; Peace, C.; Pérez-Moreno, I.; Marco, V. Compatibility of the entomopathogenic fungus *Beauveria bassiana* with flufenoxuron and azadirachtin against *Tetranychus urticae*. *Exp. Appl. Acarol.* **2012**, *58*, 395–405. [CrossRef]
54. Seyed-Talebi, F.-S.; Kheradmand, K.; Talaei-Hassanloui, R.; Talebi-Jahromi, K. Synergistic effect of *Beauveria bassiana* and spirodiclofen on the two-spotted spider mite (*Tetranychus urticae*). *Phytoparasitica* **2014**, *3*, 405–412. [CrossRef]
55. Nawaz, A.; Razzaq, F.; Razzaq, A.; Gogi, M.D.; Fernández-Grandon, G.M.; Tayib, M.; Ayub, M.A.; Sufyan, M.; Shahid, M.R.; Qayyum, M.A.; et al. Compatibility and synergistic interactions of fungi, *Metarhizium anisopliae*, and insecticide combinations against the cotton aphid, *Aphis gossypii* Glover (Hemiptera: Aphididae). *Sci. Rep.* **2022**, *12*, 4843. [CrossRef] [PubMed]
56. Dearlove, E.L.; Chandler, D.; Edgington, S.; Berry, S.D.; Martin, G.; Svendsen, C.; Hesketh, H. Improved control of *Trialeurodes vaporariorum* using mixture combinations of entomopathogenic fungi and the chemical insecticide spiromesifen. *Sci. Rep.* **2024**, *14*, 15259. [CrossRef]
57. Abidin, A.F.; Ekowati, N.; Ratnaningtyas, N.I. Insecticide compatibility to the entomopathogenic fungi *Beauveria bassiana* and *Metarhizium anisopliae*. *Scr. Biol.* **2017**, *4*, 273–279. [CrossRef]
58. Al-Ani, L.K.T.; Aguilar-Marcelino, L.; Salazar-Vidal, V.E.; Becerra, A.G.; Raza, W. Role of Useful Fungi in Agriculture Sustainability. In *Recent Trends in Mycological Research*; Yadav, A.N., Ed.; Springer: Berlin/Heidelberg, Germany, 2021; pp. 1–44.
59. Golizadeh, A.; Ghavidel, S.; Razmjou, J.; Fathi, S.A.A.; Hassanpour, M. Comparative life table analysis of *Tetranychus urticae* Koch (Acari: Tetranychidae) on ten rose cultivars. *Acarologia* **2017**, *57*, 607–616. [CrossRef]
60. Keskin, N.; Kumral, N.A. Screening tomato varietal resistance against the two-spotted spider mite [*Tetranychus urticae* (Koch)]. *Int. J. Acarol.* **2015**, *41*, 300–309. [CrossRef]
61. Kavousi, A.; Chi, H.; Talebi, K.; Bandani, A.; Ashouri, A.; Naveh, V.H. Demographic traits of *Tetranychus urticae* (Acari: Tetranychidae) on leaf discs and whole leaves. *J. Econ. Entomol.* **2009**, *102*, 595–601. [CrossRef]
62. Puspitarini, R.D.; Fernando, I.; Rachmawati, R.; Hadi, M.S.; Rizali, A. Host plant variability affects the development and reproduction of Tetranychus urticae. *Int. J. Acarol.* **2021**, *47*, 381–386. [CrossRef]
63. Tahir, T.; Wakil, W.; Ali, A.; Sahi, S.T. Pathogenicity of *Beauveria bassiana* and *Metarhizium anisopliae* isolates against larvae of the polyphagous pest *Helicoverpa armigera*. *Entomol. Gen.* **2019**, *38*, 225–242. [CrossRef]
64. Alwaneen, W.S.; Wakil, W.; Kavallieratos, N.G.; Qayyum, M.A.; Tahir, M.; Rasool, K.G.; Husain, M.; Aldawood, A.S.; Shapiro-Ilan, D. Efficacy and persistence of entomopathogenic fungi against *Rhynchophorus ferrugineus* on date palm: Host to host transmission. *Agronomy* **2024**, *14*, 642. [CrossRef]
65. Rasool, S.; Markou, A.; Hannula, S.E.; Biere, A. Effects of tomato inoculation with the entomopathogenic fungus Metarhizium brunneum on spider mite resistance and the rhizosphere microbial community. *Front. Microbiol.* **2023**, *14*, 1197770. [CrossRef]
66. Ortucu, S.; Algur, O.F. A laboratory assessment of two local strains of the Beauveria bassiana (Bals.) Vuill. against the Tetranychus urticae (Acari: Tetranychidae) and their potential as a mycopesticide. *J. Pathog.* **2017**, *2017*, 7628175.
67. Wakil, W.; Tahir, M.; Al-Sadi, A.M.; Shapiro-Ilan, D. Interactions between two invertebrate pathogens: An endophytic fungus and an externally applied bacterium. *Front. Microbiol.* **2020**, *11*, 2624. [CrossRef] [PubMed]
68. Rasool, S.; Cárdenas, P.D.; Pattison, D.I.; Jensen, B.; Meyling, N.V. Isolate-specific effect of entomopathogenic endophytic fungi on population growth of two-spotted spider mite (*Tetranychus Urticae* Koch) and levels of steroidal glycoalkaloids in tomato. *J. Chem. Ecol.* **2021**, *47*, 476–488. [CrossRef] [PubMed]
69. Wakil, W.; Kavallieratos, N.G.; Eleftheriadou, N.; Haider, S.A.; Qayyum, M.A.; Tahir, M.; Rasool, K.G.; Husain, M.; Aldawood, A.S. A winning formula: Sustainable control of three stored-product insects through paired combinations of entomopathogenic fungus, diatomaceous earth, and lambda-cyhalothrin. *Environ. Sci. Pollut. Res.* **2024**, *31*, 1–15. [CrossRef]
70. Marcossi, Í.; Francesco, L.S.; Fonseca, M.M.; Pallini, A.; Groot, T.; De Vis, R.; Janssen, A. Predatory mites as potential biological control agents for tomato russet mite and powdery mildew on tomato. *J. Pest Sci.* **2024**. [CrossRef]

71. Chaudhry, Z.; Abbas, S.; Yasmin, A.; Rashid, H.; Ahmed, H.; Anjum, M.A. Tissue culture studies in tomato (*Lycopersicon esculentum*) var. Moneymaker. *Pak. J. Bot.* **2010**, *42*, 155–163.
72. Ijaz, A.; Khan, I.; Zareen, S.; Khan, M.I.; Khan, R.; Haroon, M. Yield and yield attributes of tomato (*Lycopersicon esculentum* Mill) cultivars influenced by weed management techniques. *Pak. J. Weed Sci. Res.* **2017**, *23*, 431–438.
73. Wu, S.; Sarkar, S.C.; Lv, J.; Xu, X.; Lei, Z. Poor infectivity of *Beauveria bassiana* to eggs and immatures causes the failure of suppression on *Tetranychus urticae* population. *BioControl* **2020**, *65*, 81–90. [CrossRef]
74. Seiedy, M.; Moezipour, M. The entomopathogenic fungus Beauveria bassiana and its compatibility with Phytoseiulus persimilis (Acari: Phytoseiidae): Effects on Tetranychus urticae (Acari: Tetranychidae). *Persian J. Acarol.* **2017**, *6*, 329–338.
75. Sharma, P.; Sharma, P.L.; Verma, S.C.; Sharma, N.; Sharma, P.; Thakur, S.; Sharma, S. Effect of fungicides on the functional response of *Neoseiulus longispinosus* (Phytoseiidae) to *Tetranychus urticae* (Tetranychidae) eggs. *Int. J. Acarol.* **2024**, *50*, 151–158. [CrossRef]
76. Castillo-Ramírez, O.; Guzmán-Franco, A.W.; Santillán-Galicia, M.; Tamayo-Mejía, F. Interaction between predatory mites (Acari: Phytoseiidae) and entomopathogenic fungi in *Tetranychus urticae* populations. *BioControl* **2020**, *65*, 433–445. [CrossRef]
77. Golec, J.R.; Hoge, B.; Walgenbach, J.F. Effect of biopesticides on different *Tetranychus urticae* Koch (Acari: Tetranychidae) life stages. *Crop Prot.* **2020**, *128*, 105015. [CrossRef]
78. Nishi, O.; Sushida, H.; Higashi, Y.; Iida, Y. Epiphytic and endophytic colonisation of tomato plants by the entomopathogenic fungus Beauveria bassiana strain GHA. *Mycology* **2021**, *12*, 39–47. [CrossRef] [PubMed]
79. Amjad, M.; Bashir, M.H.; Afzal, M.; Sabri, M.A.; Javed, N. Synergistic effect of some entomopathogenic fungi and synthetic pesticides, against two spotted spider mite, Tetranychus urticae Koch (Acari: Tetranychidae). *Pak. J. Zool.* **2012**, *44*, 977–984.
80. Maniania, N.K.; Ekesi, S.; Kungu, M.M.; Salifu, D.; Srinivasan, R. The effect of combined application of the entomopathogenic fungus *Metarhizium anisopliae* and the release of predatory mite *Phytoseiulus longipes* for the control of the spider mite *Tetranychus evansi* on tomato. *Crop Prot.* **2016**, *90*, 49–53. [CrossRef]
81. Zar, J.H. *Biostatistical Analysis*; Pearson Education Limited: Essex, UK, 2014.
82. Scheff, D.S.; Arthur, F.H. Fecundity of *Tribolium castaneum* and *Tribolium confusum* adults after exposure to deltamethrin packaging. *J. Pest Sci.* **2018**, *91*, 717–725. [CrossRef]
83. Abbott, W.S. A method of computing the effectiveness of an insecticide. *J. Econ. Entomol.* **1925**, *18*, 265–267. [CrossRef]
84. Sokal, R.R.; Rohlf, F.J. *Biometry*; Freeman & Company: New York, NY, USA, 1995.
85. Minitab, LLC. Available online: https://www.minitab.com (accessed on 4 November 2024).
86. Souza, R.K.; Azevedo, R.F.; Lobo, A.O.; Rangel, D.E. Conidial water affinity is an important characteristic for thermotolerance in entomopathogenic fungi. *Biocontrol Sci. Technol.* **2014**, *24*, 448–461. [CrossRef]
87. Dogan, Y.O.; Hazir, S.; Yildiz, A.; Butt, T.M.; Cakmak, I. Evaluation of entomopathogenic fungi for the control of *Tetranychus urticae* (Acari: Tetranychidae) and the effect of *Metarhizium brunneum* on the predatory mites (Acari: Phytoseiidae). *Biol. Control* **2017**, *111*, 6–12. [CrossRef]
88. Usman, M.; Gulzar, S.; Wakil, W.; Wu, S.; Piñero, J.C.; Leskey, T.C.; Nixon, L.J.; Oliveira-Hofman, C.; Toews, M.D.; Shapiro-Ilan, D. Virulence of entomopathogenic fungi to *Rhagoletis pomonella* (Diptera: Tephritidae) and interactions with entomopathogenic nematodes. *J. Econ. Entomol.* **2020**, *113*, 2627–2633. [CrossRef] [PubMed]
89. Wu, S.; Xie, H.; Li, M.; Xu, X.; Lei, Z. Highly virulent *Beauveria bassiana* strains against the two-spotted spider mite, *Tetranychus urticae*, show no pathogenicity against five phytoseiid mite species. *Exp. Appl. Acarol.* **2016**, *70*, 421–435. [CrossRef] [PubMed]
90. Saleem, A.R.; Ibrahim, R.A. Assessment of the virulence proteolytic activity of three native entomopathogenic fungi against the larvae of *Oryctes agamemnon* (Burmeister) (Coleoptera: Scarabaeidae) Egypt. *J. Biol. Pest Control* **2019**, *29*, 1–8. [CrossRef]
91. Zare, M.; Talaei-Hassanloui, R.; Fotouhifar, K. Relatedness of proteolytic potency and virulence in entomopathogenic fungus *Beauveria bassiana* isolates. *J. Crop Prot.* **2014**, *3*, 425–434.
92. Mustafa, U.; Kaur, G. Extracellular enzyme production in *Metarhizium anisopliae* isolates. *Folia Microbiol.* **2009**, *54*, 499–504. [CrossRef] [PubMed]
93. Mondal, S.; Baksi, S.; Koris, A.; Vatai, G. Journey of enzymes in entomopathogenic fungi. *Pac. Sci. Rev. A Nat. Sci. Eng.* **2016**, *18*, 85–99. [CrossRef]
94. Petrisor, C.; Stoian, G. The role of hydrolytic enzymes produced by entomopathogenic fungi in pathogenesis of insects mini review. *Rom. J. Plant Prot.* **2017**, *10*, 66–72.
95. Bugeme, D.M.; Knapp, M.; Boga, H.I.; Ekesi, S.; Maniania, N.K. Susceptibility of developmental stages of *Tetranychus urticae* (Acari: Tetranychidae) to infection by *Beauveria bassiana* and *Metarhizium anisopliae* (Hypocreales: Clavicipitaceae). *Int. J. Trop. Insect Sci.* **2014**, *34*, 190–196.
96. Ranout, A.S.; Kaur, R.; Kumar, R.; Nadda, G. Pathogenicity and compatibility studies of native *Tolypocladium inflatum* and *Clonostachys krabiensis* against *Tetranychus urticae*. *J. Appl. Entomol.* **2024**, *148*, 1210–1222. [CrossRef]

97. Kim, J.J.; Roberts, D.W. The relationship between conidial dose, moulting and insect developmental stage on the susceptibility of cotton aphid, *Aphis gossypii*, to conidia of *Lecanicillium attenuatum*, an entomopathogenic fungus. *Biocontr. Sci. Technol.* **2012**, *22*, 319–331. [CrossRef]
98. López-Manzanares, B.; Martínez-Villar, E.; Marco-Mancebón, V.S.; Pérez-Moreno, I. Compatibility of the entomopathogenic fungus *Beauveria bassiana* with etoxazole, spirodiclofen and spiromesifen against *Tetranychus urticae*. *Biol. Control* **2022**, *169*, 104892. [CrossRef]
99. Salem, H.H.A.; Mohammed, S.H.; Eltaly, R.I.; Moustafa, M.A.M.; Fónagy, A.; Farag, S.M. Co-application of entomopathogenic fungi with chemical insecticides against *Culex pipiens*. *J. Invertebr. Pathol.* **2023**, *198*, 107916. [CrossRef]
100. Ekesi, S. Pathogenicity and antifeedant activity of entomopathogenic hyphomycetes to the cowpea leaf beetle, *Ootheca mutabilis* Shalberg. *Int. J. Trop. Insect Sci.* **2001**, *21*, 55–60. [CrossRef]
101. Vivekanandhan, P.; Swathy, K.; Sarayut, P.; Patcharin, K. Classification, biology and entomopathogenic fungi-based management and their mode of action against *Drosophila* species (Diptera: Drosophilidae): A review. *Front. Microbiol.* **2024**, *15*, 1443651. [CrossRef]
102. Ihara, F.; Toyama, M.; Mishiro, K.; Yaginuma, K. Laboratory studies on the infection of stink bugs with *Metarhizium anisopliae* strain FRM515. *Appl. Entomol. Zool.* **2008**, *43*, 503–509. [CrossRef]
103. Sparks, T.C.; Crouse, G.D.; Durst, G. Natural products as insecticides: The biology, biochemistry and quantitative structure—Activity relationships of spinosyns and spinosoids. *Pest. Manag. Sci.* **2001**, *57*, 896–905. [CrossRef] [PubMed]
104. Salgado, V.L.; Sparks, T.C.; Gilbert, L.I.; Gill, S.S. The Spinosyns: Chemistry, Biochemistry, Mode of Action, and Resistance. In *Insect Control: Biological and Synthetic Agents*; Academic Press: Cambridge, MA, USA, 2010; pp. 137–169.
105. Wakil, W.; Gulzar, S.; Prager, S.M.; Ghazanfar, M.U.; Shapiro-Ilan, D.I. Efficacy of entomopathogenic fungi, nematodes and spinetoram combinations for integrated management of *Thrips tabaci*: A two-year onion field study. *Pest Manag. Sci.* **2023**, *79*, 3227–3238. [CrossRef]
106. El Aalaoui, M.; Rammali, S.; Bencharki, B.; Sbaghi, M. Efficacy of biorational insecticides and entomopathogenic fungi for controlling *Cassida vittata* Vill. (Coleoptera: Chrysomelidae) in sugar beet crops. *Neotrop. Entomol.* **2025**, *54*, 2. [CrossRef]
107. Hiromori, H.; Nishigaki, J. Joint action of an entomopathogenic fungus (*Metarhizium anisopliae*) with synthetic insecticides against the scarab beetle, *Anomala cuprea* (Coleoptera: Scarabaeidae) larvae. *Appl. Entomol. Zool.* **1998**, *33*, 77–84. [CrossRef]
108. Hiromori, H.; Nishigaki, J. Factor analysis of synergistic effect between the entomopathogenic fungus *Metarhizium anisopliae* and synthetic insecticides. *Appl. Entomol. Zool.* **2001**, *36*, 231–236. [CrossRef]
109. Ericsson, J.D.; Kabaluk, J.T.; Goettel, M.S.; Myers, J.H. Spinosad interacts synergistically with the insect pathogen *Metarhizium anisopliae* against the exotic wireworms *Agriotes Lineatus* and *Agriotes Obscurus* (Coleoptera: Elateridae). *J. Econ. Entomol.* **2007**, *100*, 31–38. [CrossRef] [PubMed]
110. Rivero-Borja, M.; Guzmán-Franco, A.W.; Rodríguez-Leyva, E.; Santillán-Ortega, C.; Pérez-Panduro, A. Interaction of *Beauveria bassiana* and *Metarhizium anisopliae* with chlorpyrifos ethyl and spinosad in *Spodoptera frugiperda* larvae. *Pest Manag. Sci.* **2018**, *74*, 2047–2052. [CrossRef] [PubMed]
111. Numa, S.; Rodríguez, L.; Rodríguez, D.; Coy-Barrera, E. Susceptibility of *Tetranychus urticae* Koch to an ethanol extract of *Cnidoscolus aconitifolius* leaves under laboratory conditions. *Springerplus* **2015**, *4*, 338. [CrossRef]
112. Tsolakis, H.; Ragusa, E.; Sinacori, M.; Lombardo, A. On the perception of leaf morphology and visible light by *Tetranychus urticae* Koch (Acariformes, Tetranychidae). *Acarologia* **2022**, *62*, 404–417. [CrossRef]
113. Al-Shammery, K.A.; Al-Khalaf, A.A. Effect of host preference and micro habitats on the survival of *Tetranychus urticae* Koch (Acari: Tetranychidae) in Saudi Arabia. *J. King Saud Univ. Sci.* **2022**, *34*, 102030. [CrossRef]
114. Sanyang, S.; Van Emden, H.F. The combined effects of the fungus *Metarhizium flavoviride* Gams & Rozsypal and the insecticide cypermethrin on *Locusta migratoria migratorioides* (Reiche & Fairmaire) in the laboratory. *Inter. J. Pest Manag.* **1996**, *42*, 183–187.

Journal of *Fungi*

MDPI

Article

The Susceptibility of Two *Beauveria bassiana* Strains on Rice Pests *Nilaparvata lugens* and *Sogatella furcifera*

Zhongwei Chen [1], Hanqing Mu [1], Yifan Peng [2], Rui Huo [2] and Jiaqin Xie [1,3,4,*]

1 School of Life Sciences, Genetic Engineering Research Center, Chongqing University, Chongqing 400044, China; 202326131053@stu.cqu.edu.cn (Z.C.); mhq19900828@163.com (H.M.)
2 Wuhan Kernel Bio-tech Co., Ltd., Guannanyuan Road No.17, Guannan Industrial Park, Wuhan 430074, China; pyifan@yeah.net (Y.P.); 597568519@163.com (R.H.)
3 National Engineering Research Center of Microbial Pesticides (Joint Institute-Chongqing University) and Chongqing Engineering Research Center for Fungal Insecticides, Chongqing 400044, China
4 Key Laboratory of Gene Function and Regulation Technology Under Chongqing Municipal Education Commission, Chongqing 400044, China
* Correspondence: xiejqin@cqu.edu.cn

Abstract: Entomopathogenic fungi represent a valuable natural resource with significant potential as biological agents for pest management. However, different species or strains of fungi demonstrate varying effectiveness against specific targets. In this study, we assessed the impact of two fungal strains, *Beauveria bassiana* KN801 and KN802, on the rice planthoppers *Ninaparvata lugens* and *Sogatella furcifera*, in combination with insecticides. Our findings indicate that both *B. bassiana* strains can effectively infect the nymphs and adults of *N. lugens* and *S. furcifera*, resulting in a significantly higher mortality rate compared to the control groups. Notably, the *B. bassiana* strain KN801 demonstrated greater virulence than *B. bassiana* KN802 against these pests. However, no significant differences were observed when using different concentrations of the same fungal strain (*B. bassiana* KN801 or *B. bassiana* KN802) against these targets. Additionally, both fungi showed a germination rate of over 90% after treatment when combined with several common insecticides like chlorfenapyr and dinotefuran. The combined application of *B. bassiana* with chlorfenapyr or dinotefuran could improve pest control efficacy for these two pests. This study suggests that the two *B. bassiana* strains have the potential to infect rice planthoppers *N. lugens* and *S. furcifera*, indicating their promise as agents for the control of these pests.

Keywords: entomopathogenic fungi; biological agent; rice planthopper; pest control

1. Introduction

The rice planthopper is a key pest for rice plants, causing significant damage by directly feeding on the sap of rice plants [1], and also by transmitting diseases such as *the rice ragged stunt virus*, *the rice grassy stunt virus*, and *the south rice black-streaked dwarf virus* as vectors [2,3]. Traditional control methods, primarily chemical insecticides, have proven unsustainable for crop development [4]. Misuse of chemical insecticides has led to numerous environmental and nontarget organism side effects, including the decline of pollinator populations and disruption of natural enemy behavior [5,6]. Additionally, these insect pests have developed high resistance to commonly used chemicals such as imidacloprid and pymetrozine [7]. Therefore, the search for more effective and environmentally friendly approaches or alternatives for controlling these pests is imperative.

Entomopathogenic fungi (EF) are abundant in nature and play crucial roles in regulating insect populations. Certain EF strains have been reported to be effective in infecting insect pests, such as *Diaphorina citri* [8] and *Busoniomimus manjunathi* [9]. However, it is important to note that different fungal species or strains exhibit varying effects on targeting insect hosts, suggesting the need for specific fungal strains to effectively infect and control particular insect pests. In addition to their impact on insects, EF have also been reported to have beneficial effects on plants, yielding positive ecological outcomes [10]. Some EF can enhance host plants' nutrient intake, resistance to pathogens, and promote growth [11]. Several studies have demonstrated that EF improve host plant utilization of nitrogen and carbon via the endosphere [12,13] while also enhancing resistance to insect pests [14]. These findings underscore the potential of EF as a valuable resource for sustainable pest management and plant health.

EF possess a unique ability to directly penetrate the cuticle of insect hosts through complex processes [15]. Once inside the host's body, they release toxins that harm and weaken the host's immune response, facilitating their infection [16]. Subsequently, the EF acquire nutrients from the host's body for their own growth, ultimately leading to the death of the insect host. The penetration strategy of entomopathogenic fungi makes it challenging for insect hosts to develop resistance to fungal infection [17]. As a result, certain fungal strains have been employed to control resistant insects, and engineered fungal strains have the potential to enhance their performance against such pests [18,19]. In contrast to chemical approaches, EF present fewer risks to the environment and may have sustained effects on insect populations, due to their survival traits in soil or water [11,20]. Therefore, the utilization of EF holds significant promise as an alternative to insecticides for the control of insect pests.

The outbreak of pest populations in a specific area, due to climate change or human activity, can cause significant damage to crops within a short period. While EF can be utilized to target and control these pests, it often requires some time to effectively reduce their numbers and suppress the population to a low level. Certain fungal strains may show limited control potential when used alone [10]. In such scenarios, integrated pest management (IPM) or the combined use of EF and other control approaches becomes a favorable option [11,21]. In practice, some successful applications have involved the combined use of fungi and insecticides for controlling insect pests, yielding positive results [22]. When EF poses a risk (albeit lower than that associated with chemical pesticides) to beneficial insects or natural enemies of the target pest, significant biological control effects can still be achieved by staggering the application times of the two agents [23]. Consequently, the careful and strategic evaluation of EF in conjunction with insecticides can provide a more effective approach for the control of insect pests.

In this study, we examined the impact of two different strains of *B. bassiana* on different stages of significant rice pests, specifically *N. lugens* and *S. furcifera*. Furthermore, we evaluated the compatibility of these two strains with 10 commonly used insecticides for rice pest management. Our investigation focused on the synergistic application of these fungi with low dosages of insecticides to combat rice planthoppers. The outcomes of this research may introduce novel *B. bassiana* strains for effectively managing rice planthoppers, providing alternative approaches to rice pest control.

2. Materials and Methods

2.1. Insects and B. bassiana Strains

The rice planthoppers *N. lugens* and *S. furcifera* utilized in this study were sourced from laboratory populations in the insectary room at Chongqing University, Chongqing, China. All insects were maintained on rice seedlings within plastic cages (30 cm × 30 cm × 40 cm) under ambient conditions, with a temperature of 26 ± 1 °C and a light cycle of 14:10 h (Dark: Light). Prior to the commencement of the experiments, female and male rice planthoppers

N. lugens or *S. furcifera* were conducted into new cages with fresh rice seedlings. After 48 h, all individuals were removed from the cages following egg laying. Subsequently, five-instar nymphs and newly emerged *N. lugens* and *S. furcifera* were collected for use in this experiment. The conditions of the rice plants were consistently monitored throughout the experiments, ensuring they were watered daily to maintain their freshness. This meticulous care and maintenance of the experimental setup were crucial to ensuring the health and viability of the rice planthoppers and the accuracy of the study's results.

The entomopathogenic fungal strains *B. bassiana* KN801 and KN802 were obtained from Wuhan Kernel Bio-tech Co. in Wuhan, China. Strain *B. bassiana* KN801 (Identification No. 20231211-1) and KN802 (Identification No. 20231211-2) were recently isolated from aphids in Wuhan and *Spodoptera frugiperda* in Yunnan, respectively. These strains have been identified and characterized by the China Microbial Culture Preservation Management Committee (Agricultural Microbiology Center, Beijin, China). To conduct bioassay evaluations, *B. bassiana* KN801 and KN802 were initially inoculated onto SDAY medium (containing 10 g glucose, 5 g yeast extract, 2.5 g peptone, and 18 g agar per liter of sterilized water) by streaking on a Petri dish. These plates were then placed in a constant temperature incubator set at 27 °C for a period of two weeks. Once the spores reached maturity, an inoculation loop was used to collect an appropriate quantity of spores, which were subsequently transferred to a 0.05% Tween-80 solution.

The spore-Tween-80 mixture was then vigorously shaken and mixed for three to five minutes using a vortex oscillator. Following this, the mixed solution was filtered through pre-sterilized four-layer lens paper to eliminate impurities such as mycelium and medium, resulting in the original spore suspension. This original suspension was then gradient-diluted with 0.05% Tween-80 solution to obtain spore suspensions at concentrations of 1×10^5, 1×10^6, 1×10^7, and 1×10^8 conidia per milliliter for use in further experiments.

2.2. Bioassay Analysis of *B. bassiana* to *N. lugens* and *S. furcifera*

To investigate the potential impacts of *B. bassiana* strains KN801 and KN802 on rice pests *N. lugens* and *S. furcifera*, the survival of these pests was evaluated after being exposed to various concentrations of *B. bassiana*. A plastic cup measuring with dimensions of 9 cm in diameter and 15 cm in height, containing fresh rice seedlings, was used for this experiment. Twenty individuals of *N. lugens* or *S. furcifera*, including five-instar nymphs and adults, were collected and transferred into each cup. The cups were then covered with gauze to facilitate air circulation and to prevent the planthoppers from escaping.

Following the transfer, a Potter Precision Spray Tower manufactured by Burkard, Hertfordshire, UK, was utilized to spray 500 μL of each concentration (i.e., 1×10^5, 1×10^6, 1×10^7, and 1×10^8 conidia/mL) of *B. bassiana* KN801 or KN802 suspension into the cups. The control group was sprayed with an equal volume of 0.05% Tween-80 solution. The spraying operation pressure was set at 0.5 Bar. Subsequently, the plastic cups were transferred to a biotest chamber for cultivation, maintaining constant environmental conditions including a temperature of 26 ± 1 °C, a light cycle of 14 h of light and 10 h of dark, and humidity within the range of $70 \pm 5\%$. Each treatment was conducted with three replicates. Throughout the experiment, the survival of *N. lugens* and *S. furcifera* in each treatment cup was monitored, and fresh water was provided daily up to the 8th days. The LT_{50} of each treatment was determined through Probit analysis using GraphPad Prism 9 and further analyzed with Tukey's test with the control treatment. This approach allowed us to observe and assess the impact of *B. bassiana* strains on the survival of the rice pests in a controlled environment.

2.3. Compatibility Analysis of B. bassiana with Insecticides

In our further assessment, we aimed to determine the compatibility of *B. bassiana* strains KN801 and KN802 with ten commonly used insecticides for the control of rice pests (refer to Table 1). These insecticides were prepared at twice the recommended dosages as specified by the producers. Subsequently, a spore suspension of *B. bassiana* KN801 and KN802 at a concentration of 2×10^7 conidia/mL was prepared utilizing the method previously described.

Table 1. The ten common insecticides were used for the compatibility test with *B. bassiana*.

Insecticides	Concentrations/Dosages	Producers
Dinotefuran	865 μg/mL	Tongzhou Zhongda Agrochemical Co., Ltd., Nantong, China
Thiazinone	0.337 5 μL/mL	Shenzhen Noposion Agrochemicals Co., Ltd., Shenzhen, China
Bifenthrin	0.165 μL/mL	Yangnong Chemical, Yangzhou, China
Pyrazon	5.325 mg/mL	Syngenta Nantong Crop Protection Co., Ltd., Nantong, China
Chlorobenzamide (Eufoban)	1.1725 mg/mL	Bayer AG, Leverkusen, Germany
Ethyl-multicidin	120 μg/mL	Dow AgroSciences China Limited, Nantong, China
Bromofluramide	0.1 μL/mL	Mitsui Chemicals, Inc., Shanghai, China
Broclofenamide	0.2 μL/mL	Hebei Longping Agricultural Science and Technology Co., Ltd., Shijiazhuang, China
Chlorobenzamide (Kangkuan)	133.4 μg/mL	DuPont, Wilmington, DE, USA
Chlorfenpropyl	0.265 μL/mL	BASF Plant Protection Limited Company, Nantong, China

The spore suspension was then combined with each insecticide in a 1:1 ratio to create a compound liquid. This compound liquid was thoroughly mixed using a vortex oscillator for 3 to 5 min after the components were added together. Following this initial mixing, the mixture was allowed to stand at room temperature for 4 h, with the compound liquid being shaken again using the vortex oscillator every hour. Finally, 100 μL of the compound liquid was spread on SDAY medium and placed in a constant temperature incubator at 27 °C. We observed the samples at 6 and 12 h to determine the germination rate under an optical microscope (germination rate = number of germinated spores/total number of spores, with spore germination being determined by the emergence of germ tubes). For each sample, 100 individual conidia were counted, and three replicates were conducted for each treatment. Subsequently, a t-test was performed to analyze the compatibility of chemical insecticides with the two fungus strains. This rigorous approach allowed us to assess the impact of the insecticides on the germination of *B. bassiana* spores, providing valuable insights into their compatibility.

2.4. Effects of the Combined Use of B. bassiana with Insecticides on Rice Planthopper

In our investigation of the combined effects of fungi and insecticides, we have selected chlorfenapyr and dinotefuran as the target insecticides for testing with *B. bassiana* strains KN801 and KN802, respectively, based on their compatibility with the fungi. To begin, 26.5 μL of the chlorfenapyr formulation was added to 10 mL of 0.05% Tween-80 solution to create a 2× solution. Subsequently, this was further diluted to 20% and 40%, and then mixed with a 2×10^7 conidia/mL *B. bassiana* KN801 spore suspension at a 1:1 ratio. The solution was thoroughly mixed using a vortex oscillator for 3 to 5 min, and then allowed to stand at room temperature for 4 h, with intermittent shaking using the vortex oscillator every hour.

After the preparation, *N. lugens* and *S. furcifera* adults were treated according to the bioassay method outlined previously. Each treatment was conducted with three replicates, and the daily number of surviving *N. lugens* and *S. furcifera* individuals was recorded. The approach allowed us to assess the impact of the combined application of the selected insecticides and *B. bassiana* strains on the survival of these target insects, providing insights into the potential synergistic or antagonistic effects of the combined treatments.

In the investigation of the combined effects of dinotefuran and *B. bassiana* KN802, a total of 0.173 g of the dinotefuran formulation was dissolved in 10 mL of 0.05% Tween-80 solution to create a 2× solution. This solution was then further diluted to 6% and 10% and mixed with a 2×10^7 conidia/mL *B. bassiana* KN802 spore suspension at a 1:1 ratio. The resulting mixture was thoroughly agitated using a vortex oscillator for three to five minutes and then allowed to stand at room temperature for four hours, with intermittent shaking using the vortex oscillator every hour. Following this preparation, a group of 20 *N. lugens* and *S. furcifera* adults were treated according to the bioassay method outlined previously. Each treatment was conducted with three replicates, and the daily number of surviving *N. lugens* and *S. furcifera* individuals was recorded up to the eighth day. The LT_{50} and LT_{90} of each treatment were determined through Probit analysis with χ^2 using GraphPad Prism 9 and further analyzed using Tukey's test with the control treatment. The approach enabled us to assess the potential combined impact of dinotefuran and *B. bassiana* KN802 on the survival of these target insects, providing insights into the interactions between the insecticide and the fungal strain.

2.5. Statistical Analysis

To analyze the data, the normality of the data and the homogeneity of variances were assessed using the Shapiro–Wilk and Levene tests, respectively. The effects of different concentrations of KN801 and KN802 on *N. lugens* and *S. furcifera* individuals were analyzed using one-way analysis of variance (ANOVA) followed by Tukey's test. The compatibility was examined by *t*-test, while combined use were also analyzed by ANOVA post with Tukey's test. Subsequently, the results of the LT_{50} or LT_{90} were then subjected to Probit analysis with χ^2 using GraphPad Prism 9, followed by further analysis using Tukey's test. A significance level of $p < 0.05$ was applied to determine statistical significance, ensuring that any observed differences were not due to random variation.

3. Results

3.1. Susceptibility of B. bassiana to S. furcifera

This study focused on assessing the efficacy of the fungal *B. bassiana* against the rice pests *N. lugens* and *S. furcifera*, specifically targeting both the fifth instar nymphs and adults of *S. furcifera*. It was observed that *B. bassiana* strains KN801 and KN802 demonstrated effective control against the targeted pests. Following treatment with *B. bassiana* KN801 at a concentration of 1×10^5 conidia/mL, the LT_{50} values were 7.08 ± 0.27 days ($F_{4,10} = 266.6$, $p < 0.001$; Figure 1A,B) for nymphs and 2.96 ± 0.34 days ($F_{4,10} = 24.86$, $p < 0.001$; Figure 1C,D) for adults. Similarly, treatment with *B. bassiana* KN802 at the same concentration resulted in LT_{50} values of 5.02 ± 0.27 days ($F_{4,10} = 16.43$, $p = 0.0002$; Figure 1E,F) for nymphs and 4.08 ± 1.42 days ($F_{4,10} = 9.113$, $p = 0.0023$; Figure 1G,H) for adults. These values were significantly lower than those of the control group, indicating the substantial impact of the fungal strains on pest mortality. Furthermore, it was noted that as the concentration of the spore suspension increased, there was no significant difference in LT_{50} values between the nymphs and adults of *S. furcifera* for both *B. bassiana* KN801 and KN802.

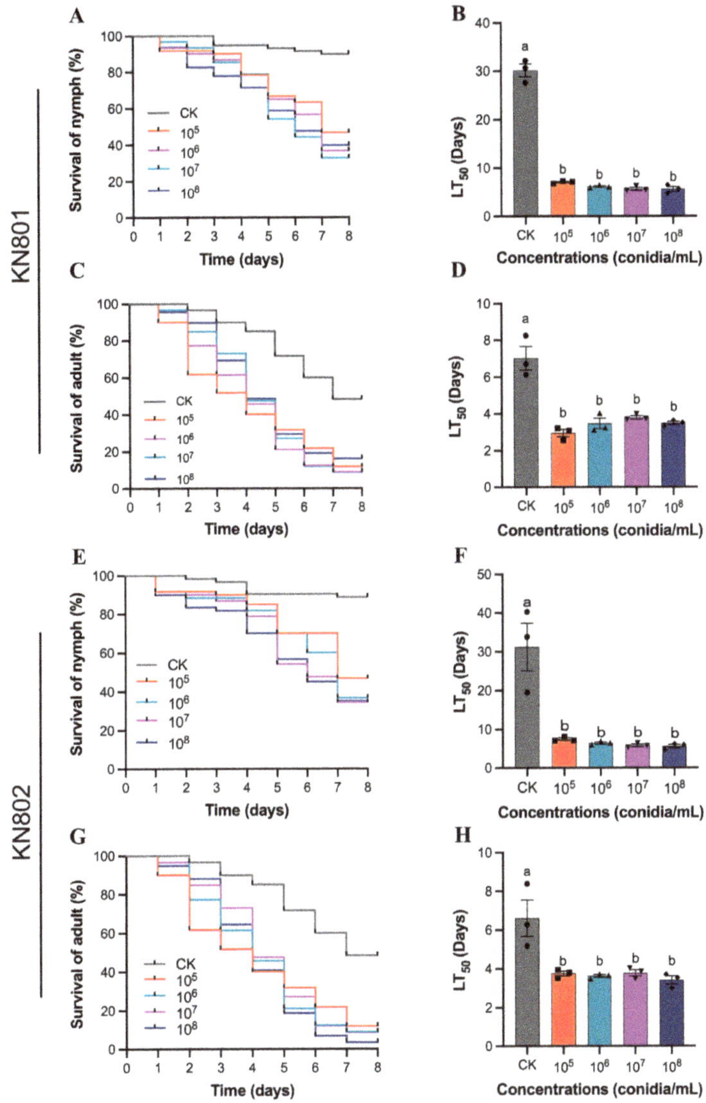

Figure 1. The susceptibility evaluation of *B. bassiana* strains KN801 and KN802 to *S. furcifera*. (**A**,**B**) the survival rate and LT_{50} of *B. bassiana* strain KN801 to *S. furcifera* nymph. (**C**,**D**) the survival rate and LT_{50} of *B. bassiana* strain KN801 to *S. furcifera* adult. (**E**,**F**) the survival rate and LT_{50} of *B. bassiana* KN802 to *S. furcifera* nymph. (**G**,**H**) the survival rate and LT_{50} of *B. bassiana* KN802 to *S. furcifera* adult. The error bars indicate Standard error (SE) in the graphs. Different letters indicate a significant difference with $p < 0.05$.

3.2. Susceptibility of B. bassiana to N. lugens

The study also demonstrated the effectiveness of *B. bassiana* strains KN801 and KN802 in controlling the fifth instar nymphs and adults of the rice planthopper *N. lugens*. After treatment with *B. bassiana* KN801 at a concentration of 1×10^5 conidia/mL, the LT_{50} values were 7.92 ± 2.2 days ($F_{4,10} = 4.496$, $p = 0.0245$; Figure 2A,B) for nymphs and 3.91 ± 0.82 days ($F_{4,10} = 4.149$, $p = 0.0310$; Figure 2C,D) for adults. Similarly, treatment with *B. bassiana* KN802 at the same concentration resulted in LT_{50} values of 5.02 ± 0.27 days ($F_{4,10}$ = 4.205, p = 0.0298; Figure 2E,F) for nymphs and 4.08 ± 1.42 days ($F_{4,10}$ = 32.16, p < 0.0001; Figure 2G,H) for adults. These values were lower than those of the control group, indicating the significant impact of the fungal strains on pest mortality.

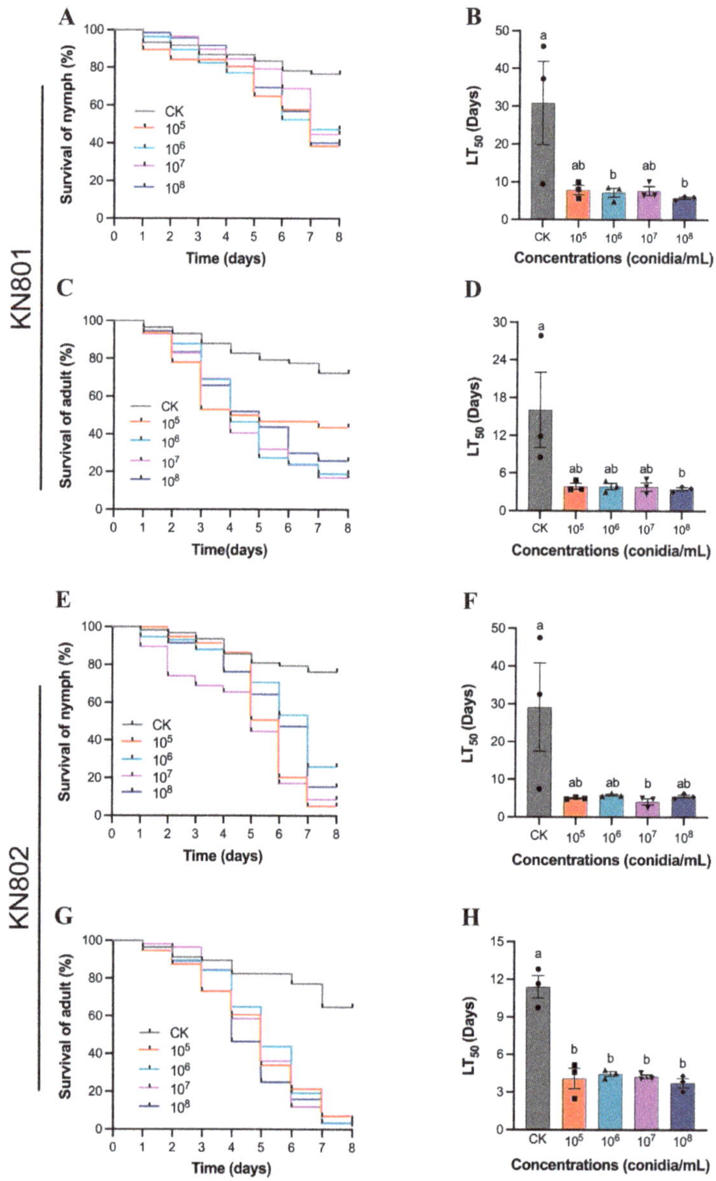

Figure 2. The susceptibility evaluation of *B. bassiana* strains KN801 and KN802 to *N. lugens*. (**A**,**B**) the survival rate and LT_{50} of *B. bassiana* KN801 to *N. lugens* nymph. (**C**,**D**) the survival rate and LT_{50} of *B. bassiana* KN801 to *N. lugens* adult. (**E**,**F**) the survival rate and LT_{50} of *B. bassiana* KN802 to *N. lugens* nymph. (**G**,**H**) the survival rate and LT_{50} of *B. bassiana* KN802 to *N. lugens* adult. The error bars indicate the Standard error (SE) in the graphs. Different letters indicate a significant difference with p < 0.05.

In addition, for *B. bassiana* KN801, spore suspensions at 1×10^6 and 1×10^8 conidia/mL showed significant lethal effects on *N. lugens* nymphs. The 1×10^8 conidia/mL suspension exhibited a significant lethal effect on brown planthopper adults ($p = 0.0495$, Tukey's test). Similarly, for *B. bassiana* KN802, the 1×10^7 conidia/mL spore suspension showed a significant lethal effect on nymphs ($p = 0.0438$, Tukey's test), while all concentrations showed a significant lethal effect on adults (all $p < 0.0001$, Tukey's test). These results highlight the potential of *B. bassiana* strains KN801 and KN802 as effective agents for controlling *N. lugens* populations, particularly in targeting both nymphs and adults.

3.3. Compatibility of B. bassiana Strains with Insecticides

The study also evaluated the compatibility of *B. bassiana* strains KN801 and KN802 with ten commonly used insecticides for rice pests. The results revealed that the germination of *B. bassiana* KN801 was significantly inhibited by the insecticide bromochloronicotinyl at 6 h ($p = 0.0145$, Figure 3A; *t*-test), with only half the germination rate of the control group. However, chlorfenapyr and cyflumetofen did not have a significant effect on spore germination for KN801 (chlorfenapyr, $p = 0.0420$; cyflumetofen, $p = 0.0003$; Figure 3A). At 12 h, only thiacloprid exhibited significant inhibition, with a germination rate only one-third of the control group ($p < 0.0001$, Figure 3B).

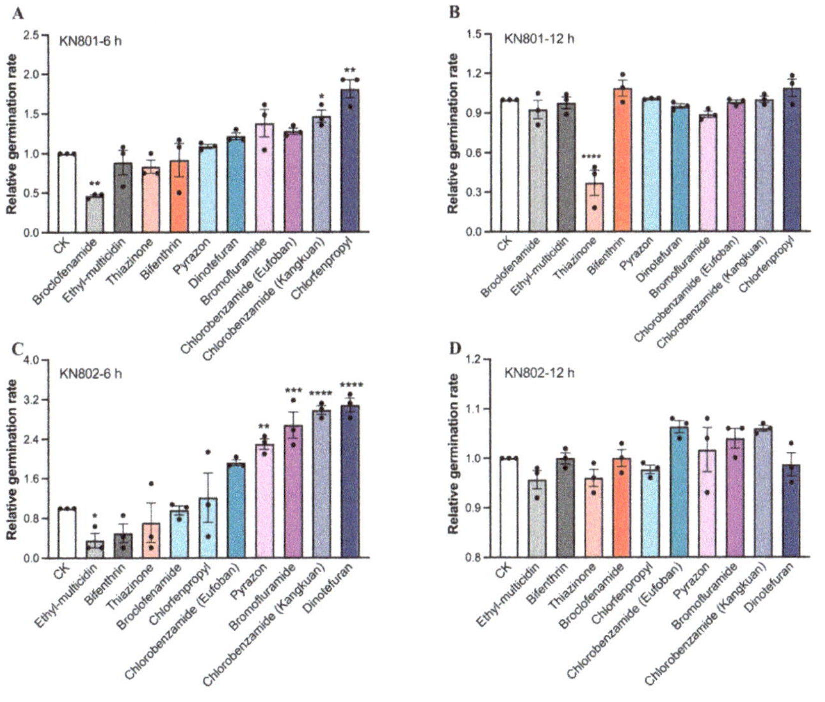

Figure 3. The compatibility of *B. bassiana* strains KN801 and KN802 with insecticides. (**A**,**B**) the relative germination of *B. bassiana* KN801 mixed with insecticides at 6 and 12 h. (**C**,**D**) the relative germination of *B. bassiana* KN802 mixed with insecticides at 6 and 12 h. The error bars indicate the Standard error (SE) in the graphs. An asterisk "*" indicates a significant difference between two treatments with $p < 0.05$, "**" indicates a significant difference between two treatments with $p < 0.01$, "***" indicates a significant difference between two treatments with $p < 0.001$, "****" indicates a significant difference between two treatments with $p < 0.0001$.

For *B. bassiana* strain KN802, the spore germination rates in the groups treated with ethaboxam, deltamethrin, and thiacloprid were lower than the control group but had no significant effect at 6 h (Figure 3C). Conversely, acetamiprid, flonicamid, chlorfenapyr, and flubendiamide promoted spore germination of *B. bassiana* KN802, with germination rates about three times that of the control group (acetamiprid, $p = 0.0050$; flonicamid, $p = 0.0003$; chlorfenapyr_convict, $p < 0.0001$; flubendiamide, $p < 0.0001$; Figure 3C). However, the other tested insecticides had no significant effect on *B. bassiana* germination. Furthermore, all insecticides had no significant effect on KN802 germination after 12 h (Figure 3D). These findings underscore the differential effects of various insecticides on the germination of *B. bassiana* strains KN801 and KN802.

3.4. Effects of the Combined Used for *N. lugens* and *S. furcifera*

Based on the results provided, it appears that the effects of *B. bassiana* and the insecticides chlorfenapyr and dinotefuran were investigated on *N. lugens* and *S. furcifera*. When *S. furcifera* was treated with low concentrations of chlorfenapyr (10% and 20% of the recommended dosage) in combination with *B. bassiana*, it led to significantly shorter LT_{50} compared to using *B. bassiana* KN801 alone, regardless of whether the concentration was at 10% or 20%. However, the LT_{90} for the combined use at 20% chlorfenapyr was significantly lower than the use of *B. bassiana* alone, whereas 10% chlorfenapyr showed no significant difference in LT_{90} ($\chi^2 = 165.2$, $p < 0.001$; Table 2 and Figure 4A).

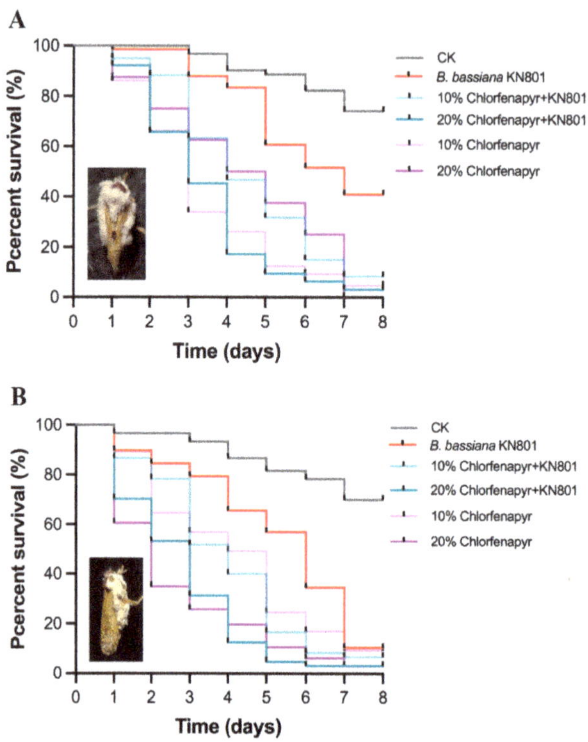

Figure 4. Effects of combined use of *B. bassiana* KN801 and insecticide chorfenapyr for *S. furcifera* and *N. lugens* adults. (**A**) the survival of *S. furcifera* after treating with combined use of *B. bassiana* KN801 and chlorfenapyr. (**B**) the survival of *N. lugens* after treating with combined use of *B. bassiana* KN801 and chlorfenapyr.

Table 2. Effects of the combined use of *B. bassiana* strain KN801 and insecticide chlorfenapyr for *S. furcifera* and *N. lugens*.

Lethal Time	Targets	*B. bassiana*	10% Chlorfenapyr + *B. bassiana*	20% Chlorfenapyr + *B. bassiana*	10% Chlorfenapyr	20% Chlorfenapyr	CK
LT_{50} with 95% CIs (days)	*N. lugens*	4.84 (4.46~5.22)	3.4 (2.6~4.2)	2.28 (2.11~2.45)	3.99 (3.75~4.23)	1.76 (1~2.52)	8.87 (8.37~9.38)
	S. furcifera	5.87 (5.34~6.4)	4.02 (3.97~4.06)	2.89 (2.31~3.47)	2.9 (2.79~3.01)	2.63 (2.34~2.92)	8.89 (7.56~10.21)
LT_{90} with 95% CIs (days)	*N. lugens*	7.88 (6.87~8.89)	5.7 (3.94~7.45)	4.48 (4.14~4.83)	6.75 (5.84~7.65)	3.9 (2.33~5.47)	13.92 (12.9~14.93)
	S. furcifera	8.83 (8.47~9.19)	6.44 (5.63~7.26)	4.76 (3.54~5.97)	5.38 (5.14~5.61)	4.56 (4.19~4.94)	12.73 (10.85~14.61)

When testing on *N. lugens*, using 10% chlorfenapyr showed no significant difference compared to using *B. bassiana* KN801 alone. However, when using 20% chlorfenapyr, the LT_{50} and LT_{90} for the combined use were significantly shorter than using abamectin alone ($\chi^2 = 154.6$, $p < 0.001$; Table 2 and Figure 4B). Similarly, when using *B. bassiana* KN802 in combination with 3% or 5% of dinotefuran ($\chi^2 = 148.9$, $p <$ 001; Table 3 and Figure 5A), it resulted in significantly shorter LT_{50} and LT_{90} compared to using *B. bassiana* KN802 alone for *S. furcifera*. However, when considering *N. lugens* (Table 3 and Figure 5B), using 3% dinotefuran alone showed no significant difference compared to using *B. bassiana* KN802 alone. Nevertheless, when using 5% dinotefuran, the LT_{50} for the combined use was significantly shorter than using dinotefuran alone, while the LT_{90} showed no significant difference ($\chi^2 = 85.82$, $p < 0.001$; Table 3 and Figure 5B).

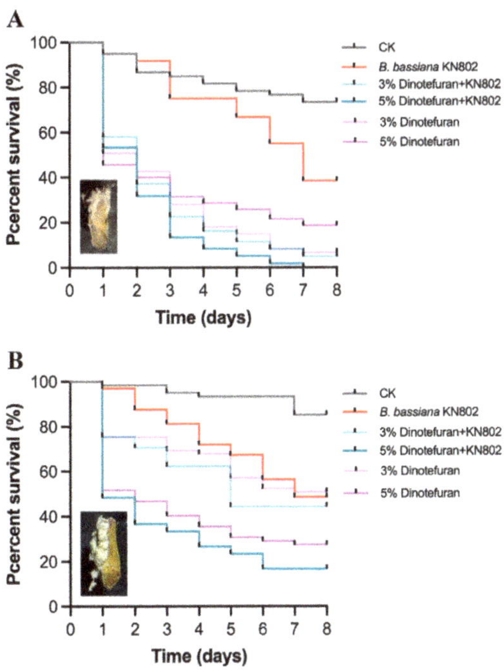

Figure 5. Effects of combined use of *B. bassiana* KN802 and insecticide dinotefuran for *S. furcifera* and *N. lugens* adults. (**A**) the survival of *S. furcifera* after treating with combined use of *B. bassiana* KN802 and dinotefuran. (**B**) the survival of *N. lugens* after treating with combined use of *B. bassiana* KN802 and dinotefuran.

Table 3. Effects of the combined use of *B. bassiana* strain KN802 and insecticide dinotefuran for *S. furcifera* and *N. lugens*.

Lethal Time	Targets	*B. bassiana*	3% Dinotefuran + *B. bassiana*	5% Dinotefuran + *B. bassiana*	3% Dinotefuran	5% Dinotefuran	CK
LT_{50} with 95% Cis (days)	*N. lugens*	6.72 (7.66~5.78)	5.23 (6.05~4.41)	2.18 (2.91~1.44)	6.42 (7.74~5.11)	3.01 (4.3~1.73)	10.83 (13~8.66)
	S. furcifera	6.15 (6.43~5.86)	2.07 (2.96~1.18)	1.61 (1.78~1.43)	2.02 (2.1~1.95)	1.61 (2.48~0.74)	9.77 (10.35~9.2)
LT_{90} with 95% Cis (days)	*N. lugens*	11.4 (14.16~8.64)	11.09 (11.78~10.41)	6.74 (8.13~5.34)	13.28 (15.09~11.47)	8.71 (11.24~6.19)	15.68 (20.58~10.78)
	S. furcifera	10.18 (11.02~9.34)	4.63 (6.82~2.45)	3.57 (4.27~2.88)	5.46 (6.02~4.9)	5.46 (6.19~4.72)	17.07 (17.68~16.46)

4. Discussion

Entomopathogenic fungi offer a safer and more effective alternative to commonly used insecticides for managing pests and pathogens [14,24]. However, it is important to note that the efficacy and specificity of EF (i.e., whether they target a wide range of insects or specific species) can vary, even within the same insect species [25,26]. Therefore, careful consideration of the specific fungi that target particular insect pests is crucial before implementing them for pest control. In this study, the focus was on evaluating the effects of two new strains of the fungal species *B. bassiana* (strains KN801 and KN802) on a major rice pest. Additionally, the study examined the susceptibility of these fungi to several insecticides. The results indicated that *B. bassiana* can infect and cause the death of the rice planthopper, and the combined use of these fungi and certain insecticides may enhance their efficiency, thereby presenting a promising approach for rice pest control. Their unique mode of action and potential for sustained effectiveness make EF a compelling option for pest management, particularly in light of the challenges posed by insecticide resistance and environmental concerns associated with chemical insecticides.

The rice planthopper is a persistent pest for rice plants, and has been known to cause substantial damage. Numerous studies have documented that populations of rice planthoppers have developed high resistance to many insecticides, including pyrethroids and imidacloprid [7,27]. Insects have evolved to resist these chemicals by modifying the channels or targets that the insecticides act upon [28]. This resistance has led to a situation where new or more potent insecticides are being used to combat this rice pest, often with high environmental risks and increased costs associated with their application. Importantly, there is a growing need to reduce the reliance on chemical insecticides and seek safer alternatives [29]. This demand has prompted the search for effective approaches or products for pest control, and entomopathogenic fungi emerge as a promising choice due to their inherent traits.

While some entomopathogenic fungi, such as *Cladosporium* sp., *M. indicum*, and *B. bassiana*, have been utilized for the control of various pests, there remains a wealth of untapped resources in the form of more effective fungi [30–32]. Only a few entomopathogenic fungi have been specifically studied and reported as useful for controlling rice pests in field applications [26,33]. This study contributes to the evaluation of two new *B. bassiana* strains for the control of the rice planthopper. The findings of this study revealed that the two *B. bassiana* strains exhibited differing pathogenic effects, with different lethal efficiencies (i.e., LT_{50} = 4.84 vs. 6.72 for *N. lugens* and 5.87 vs. 6.15 for *S. furcifera*) on the rice planthopper. This indicates the variable functionality of entomopathogenic fungi with respect to specific insect hosts. Similarly, other studies have reported that different fungal strains, such as *M. anisopliae* Ma41 and *M. anisopliae* Ma22, demonstrated varying functions [34]. As a result, it is crucial to thoroughly assess the potential of fungal strains for specific hosts before considering their further use or commercialized development. However, it was noted that the different concentrations of *B. bassiana* showed no significant difference from two tested rice planthoppers. This lack of differentiation could potentially be attributed to a loading threshold on the insect body achieved by spraying methods with a similar volume. Further examination is required for a more precise assessment in this regard.

The complex nature of the process by which fungi act on insect hosts implies that it may take a longer time to induce the death of insect pests compared to the immediate action of insecticides [35–37]. This prolonged process may limit the efficiency of entomopathogenic fungi, especially when facing rapid pest outbreaks. Additionally, the cost associated with using entomopathogenic fungi for pest control can be relatively high due to the extended timeframe [38]. Several studies have demonstrated that the combined use of fungi and insecticides can be an effective approach for pest control [39,40]. However, the success of employing a combined approach hinges on the susceptibility of different fungi strains to various insecticides, a factor that can vary even among the same type of insecticide produced by different manufacturers [41]. Different chemical insecticides contain distinct components that target specific areas within the insect body. Therefore, it is essential to evaluate the susceptibility of specific fungi strains, such as *B. bassiana* KN801 and KN802, to a range of commonly used insecticides for rice pest control. In our evaluation, we noted variability in the susceptibility of *B. bassiana* strains to these insecticides, highlighting the importance of thoughtful selection when combining insecticides with entomopathogenic fungi. Furthermore, studies to investigate the underlying mechanisms behind the diverse outcomes observed with different insecticide-fungus combinations would be an intriguing avenue for future research. Understanding these mechanisms can provide valuable insights for optimizing pest management strategies and enhancing the efficacy of integrated pest control approaches.

In the combined use of EF and insecticides, it is important to note that using a lower dosage of insecticides, as per the recommended guidelines, may potentially reduce the side effects on the environment. Furthermore, this approach can weaken the target pests and even enhance the effectiveness of the entomopathogenic fungi. Our study has specifically evaluated the combined effects of *B. bassiana* strains KN801 and KN802 with the insecticides chlorfenapyr and dinotefuran, demonstrating that this combination indeed improves the efficiency of *B. bassiana* in controlling rice planthoppers. The combined use of these fungi with specific insecticides has been shown to significantly shorten the time required to induce the death of the target rice planthopper. In practical applications, this combination approach offers the dual benefits of pest prevention and reduction in the workload required for pest control.

5. Conclusions

This study focused on evaluating two identified *B. bassiana* strains KN801 and KN802, for their effectiveness in controlling major rice planthopper pests. The results revealed that *B. bassiana* KN801 exhibited a superior performance compared to KN802 in terms of different LT_{50} values for the targeted pests. Moreover, the susceptibility of both strains to various insecticides suggested that *B. bassiana* KN801 combined with chlorfenapyr and *B. bassiana* KN802 combined with dinotefuran could serve as promising approaches for practical use. These findings suggest that the newly identified *B. bassiana* strains hold promise for integrated pest management strategies and could contribute to more effective and sustainable control of rice pests in agricultural settings.

Author Contributions: J.X. conceptualized the study. Z.C. and J.X. designed the experiments. Y.P. and R.H. Collected the strains. Z.C. and H.M. conducted the experiments and data collection. J.X. acquired the fundings. J.X. and Z.C. wrote the draft. H.M., Z.C., Y.P. and J.X. reviewed and edited the manuscript. All authors have read and agreed to the published version of the manuscript.

Funding: The project acquired the fundings from the Natural Science Foundation of Chongqing (Project No. CSTB2024TIAD-GPX0024), the Fundamental Research Funds for the Central Universities with project No. 2023CDJKYJH043 and 2024CDJXY016.

Institutional Review Board Statement: Not applicable.

Informed Consent Statement: Not applicable.

Data Availability Statement: Data will be made available on request.

Acknowledgments: We appreciate Chunfang Huang for assisting maintaining insect population.

Conflicts of Interest: Authors Yifan Peng and Rui Huo were employed by the Wuhan Kernel Bio-tech Co., Ltd. The remaining authors declare that the research was conducted in the absence of any commercial or financial relationships that could be construed as a potential conflict of interest.

References

1. Lou, Y.G.; Zhang, G.R.; Zhang, W.Q.; Hu, Y.; Zhang, J. Biological control of rice insect pests in China. *Biol. Control* **2013**, *67*, 8–20. [CrossRef]
2. Tu, Z.; Ling, B.; Xu, D.; Zhang, M.; Zhou, G. Effects of southern rice black-streaked dwarf virus on the development and fecundity of its vector, Sogatella furcifera. *Virol. J.* **2013**, *10*, 145. [CrossRef] [PubMed]
3. Wei, T.Y.; Li, Y. Rice reoviruses in insect vectors. *Ann.Rev. Phytopathol.* **2016**, *54*, 99–120. [CrossRef]
4. Gandara, L.; Jacoby, R.; Laurent, F.; Spatuzzi, M.; Vlachopoulos, N.; Borst, N.O.; Ekmen, G.; Potel, C.M.; Garrido-Rodriguez, M.; Bohmert, A.L.; et al. Pervasive sublethal effects of agrochemicals on insects at environmentally relevant concentrations. *Science* **2024**, *386*, 446–453. [CrossRef]
5. Whitehorn, P.R.; O'Connor, S.; Wackers, F.L.; Goulson, D. Neonicotinoid pesticide reduces bumble bee colony growth and queen production. *Science* **2012**, *336*, 351–352. [CrossRef]
6. Gill, R.J.; Raine, N.E. Chronic impairment of bumblebee natural foraging behaviour induced by sublethal pesticide exposure. *Funct. Ecol.* **2015**, *28*, 1459–1471. [CrossRef]
7. Wu, S.F.; Zeng, B.; Zheng, C.; Mu, X.C.; Zhang, Y.; Hu, J.; Zhang, S.; Gao, C.F.; Shen, J.L. The evolution of insecticide resistance in the brown planthopper (*Nilaparvata lugens* Stål) of China in the period 2012–2016. *Sci. Rep.* **2018**, *8*, 4586. [CrossRef] [PubMed]
8. Cisneros, J.; Wendel, J.; Jaronski, S.; Vitek, C.; Ciomperlik, M.; Flores, D. Assessment of two novel host-derived *Beauveria bassiana* (Hypocreales: Cordycipitaceae) isolates against the citrus pest, *Diaphorina citri* (Hemiptera: Liviidae). *J. Econ. Entomol.* **2022**, *115*, 56–64. [CrossRef]
9. Senthil Kumar, C.M.; Jacob, T.K.; Devasahayam, S.; Rajeshkumar, K.C.; Lad, S.S.; D'Silva, S.; Geethu, C. *Metarhizium indicum*, a new species of entomopathogenic fungus infecting leafhopper, *Busoniomimus manjunathi* from India. *J. Invertebr. Pathol.* **2023**, *198*, 107919. [CrossRef] [PubMed]
10. Wang, L.; Keyhani, N.O.; Xia, Y.; Xie, J. The potential and limitations of entomopathogenic fungi as biocontrol agents for insect pest management. *Entomol. Gen.* **2024**, *44*, 797–811. [CrossRef]
11. Quesada-Moraga, E.; Garrido-Jurado, I.; Yousef-Yousef, M.; González-Mas, N. Multitrophic interactions of entomopathogenic fungi in BioControl. *Biocontrol* **2022**, *67*, 457–472. [CrossRef]
12. Behie, S.W.; Bidochka, M.J. Endophytic insect-parasitic fungi translocate nitrogen directly from insects to plants. *Science* **2012**, *336*, 1576. [CrossRef] [PubMed]
13. Behie, S.W.; Moreira, C.C.; Sementchoukova, I.; Barelli, L.; Zelisko, P.M.; Bidochka, M.J. Carbon translocation from a plant to an insect-pathogenic endophytic fungus. *Nat. Commun.* **2017**, *8*, 14245. [CrossRef] [PubMed]
14. Panwar, N.; Szczepaniec, A. Endophytic entomopathogenic fungi as biological control agents of insect pests. *Pest Manag. Sci.* **2024**, *80*, 6033–6040. [CrossRef] [PubMed]
15. Ortiz-Urquiza, A.; Keyhani, N. Action on the surface: Entomopathogenic fungi versus the insect cuticle. *Insects* **2013**, *4*, 357–374. [CrossRef] [PubMed]
16. Li, J.; Xia, Y.X. Host-Pathogen Interactions between spp. and Locusts. *J. Fungi* **2022**, *8*, 602. [CrossRef]
17. Gao, T.N.; Wang, Z.L.; Huang, Y.; Keyhani, N.O.; Huang, Z. Lack of resistance development in to after multiple generations of selection. *Sci. Rep.* **2017**, *7*, 42727.

18. Peng, G.; Xia, Y. Integration of an insecticidal scorpion toxin (BjαIT) gene into *Metarhizium acridum* enhances fungal virulence towards *Locusta migratoria manilensis*. *Pest Manag. Sci.* **2015**, *71*, 58–64. [CrossRef]
19. Fan, Y.H.B.D.; Hawkings, C.; Ortiz-Urquiza, A.; Keyhani, N.O. Exploiting host molecules to augment mycoinsecticide virulence. *Nat. Biotechnol.* **2012**, *30*, 35–37. [CrossRef] [PubMed]
20. Quesada-Moraga, E.; Garrido-Jurado, I.; González-Mas, N.; Yousef-Yousef, M. Ecosystem services of entomopathogenic ascomycetes. *J. Invertebr. Pathol.* **2023**, *201*, 108015. [CrossRef]
21. García-Espinoza, F.; Yousef-Yousef, M.; del Rosal, M.J.G.; Cuenca-Medina, M.; Quesada-Moraga, E. Greenhouse melon crop protection and production through the compatible use of a parasitoid with endophytic entomopathogenic ascomycetes. *J. Pest Sci.* **2024**, *97*, 1899–1912. [CrossRef]
22. Tang, J.; Liu, X.; Ding, Y.; Jiang, W.; Xie, J. Evaluation of *Metarhizium anisopliae* for rice planthopper control and its synergy with selected insecticides. *Crop Prot.* **2019**, *121*, 132–138. [CrossRef]
23. Wakil, W.; Yasin, M.; Shapiro-Ilan, D. Effects of single and combined applications of entomopathogenic fungi and nematodes against *Rhynchophorus ferrugineus* Olivier. *Sci. Rep.* **2017**, *7*, 5971. [CrossRef] [PubMed]
24. Senthil-Nathan, S. A review of biopesticides and their mode of action against insect pests. In *Environmental Sustainability*; KoganPage: London, UK, 2015; pp. 49–63.
25. Peng, G.X.; Xie, J.Q.; Guo, R.; Keyhani, N.O.; Zeng, D.Y.; Yang, P.Y.; Xia, Y.X. Long-term field evaluation and large-scale application of a *Metarhizium anisopliae* strain for controlling major rice pests. *J. Pest Sci.* **2021**, *94*, 969–980. [CrossRef]
26. Peng, Y.; Tang, J.; Hong, M.; Xie, J. Suppression of rice planthopper populations by the entomopathogenic fungus *Metarhizium anisopliae* without affecting the rice microbiota. *Appl. Environ. Microb.* **2020**, *86*, e01337-20. [CrossRef]
27. Su, J.; Wang, Z.; Zhang, K.; Tian, X.; Yin, Y.; Zhao, X.; Shen, A.; Gao, C.F. Status of insecticide resistance of the whitebacked planthopper, *Sogatella furcifera* (Hemiptera: Delphacidae). *Fla. Entomol.* **2013**, *96*, 948–956. [CrossRef]
28. Guo, J.; Xu, C.; Wu, D.; Zhao, Y.; Qiu, Y.; Wang, X.; Ouyang, Y.; Cai, B.; Liu, X.; Jing, S.; et al. Bph6 encodes an exocyst-localized protein and confers broad resistance to planthoppers in rice. *Nat. Genet.* **2018**, *50*, 297–306. [CrossRef]
29. Wing, R.A.; Purugganan, M.D.; Zhang, Q. The rice genome revolution: From an ancient grain to Green Super Rice. *Nat. Rev. Genet.* **2018**, *19*, 505–517. [CrossRef] [PubMed]
30. Farrokhzadeh, H.; Sharifi, S.; Eroglu, G.B.; Karimi, J. A new fungal entomopathogen has potency as a biocontrol agent of longhorn beetle larva, *Osphranteria coerulescencs*. *Int. J. Trop. Insect Sci.* **2024**, *44*, 1185–1193. [CrossRef]
31. Rice, S.J.; Baker, D.K.; Leemon, D.M. Development of mycoinsecticide formulations with *Beauveria bassiana* and *Metarhizium anisopliae* for the control of lesser mealworm, *Alphitobius diaperinus*, in chicken broiler houses. *Biocontrol* **2019**, *64*, 489–500. [CrossRef]
32. Daza, F.F.F.; Roman, G.R.; Rodriguez, M.V.; Vargas, I.A.G.; Heano, H.C.; Cereda, M.P.; Mulet, R.A.C. Spores of *Beauveria bassiana* and Trichoderma lignorum as a bioinsecticide for the control of *Atta cephalotes*. *Biol. Res.* **2019**, *52*, 51. [CrossRef] [PubMed]
33. Jin, S.F.; Feng, M.G.; Ying, S.H.; Mu, W.J.; Chen, J.Q. Evaluation of alternative rice planthopper control by the combined action of oil-formulated *Metarhizium anisopliae* and low-rate buprofezin. *Pest Manag. Sci.* **2011**, *67*, 36–43. [CrossRef] [PubMed]
34. Cruz-Avalos, A.M.; Bivián-Hernández, M.D.; Ibarra, J.E.; Del Rincón-Castro, M.C. High virulence of mexican entomopathogenic fungi against fall armyworm, (Lepidoptera: Noctuidae). *J. Econ. Entomol.* **2019**, *112*, 99–107. [CrossRef]
35. Wang, C.; Wang, S. Insect pathogenic fungi: Genomics, molecular interactions, and genetic improvements. *Ann. Rev. Entomol.* **2017**, *62*, 73–90. [CrossRef] [PubMed]
36. Ortiz-Urquiza, A.; Keyhani, N.O. Molecular genetics of infection of insects. *Adv. Genet.* **2016**, *94*, 165–249.
37. Tang, C.; Hu, X.; Tang, J.; Wang, L.; Liu, X.; Peng, Y.; Xia, Y.; Xie, J. The symbiont *Acinetobacter baumannii* enhances the insect host resistance to entomopathogenic fungus *Metarhizium anisopliae*. *Commun. Biol.* **2024**, *7*, 1184. [CrossRef] [PubMed]
38. de Santana, A.S.; Cavalcante, U.M.T.; Sampaio, E.V.D.B.; Maia, L.C. Production, storage and costs of inoculum of arbuscular mycorrhizal fungi (AMF). *Braz. J. Bot.* **2014**, *37*, 159–165. [CrossRef]
39. Ajvad, F.T.; Madadi, H.; Michaud, J.P.; Zafari, D.; Khanjani, M. Combined applications of an entomopathogenic fungus and a predatory mite to control fungus gnats (Diptera: Sciaridae) in mushroom production. *Biol. Control* **2020**, *141*, 104101.
40. Mohammed, A.A.; Hatcher, P.E. Combining entomopathogenic fungi and parasitoids to control the green peach aphid. *Biol. Control* **2017**, *110*, 44–55. [CrossRef]
41. Gu, M.Q.; Tian, J.X.; Lou, Y.L.; Ran, J.; Mohamed, A.; Keyhani, N.O.; Jaronski, S.; Wang, G.M.; Chen, X.Y.; Zang, L.S.; et al. Efficacy of granules for the control of and its synergistic effects with chemical pesticide, sex pheromone and parasitoid. *Entomol. Gen.* **2023**, *43*, 1211–1219. [CrossRef]

Journal of *Fungi*

MDPI

Article

Temperature-Dependent Modeling and Spatial Predictions for Identifying Geographical Areas in Brazil Suitable for the Use of *Cordyceps javanica* in Whitefly Control

Heloiza A. Boaventura [1,2,*], Lidiane A. Queirós [1], José Francisco A. Silva [2], Tarryn A. Goble [3], Kelly Pazolini [3], Allan F. Marciano [3] and Eliane D. Quintela [1,2,*]

[1] Escola de Agronomia, Universidade Federal de Goiás, Goiânia 74690-900, Brazil; lidianekeiroz@gmail.com
[2] Embrapa Arroz e Feijão, Santo Antônio de Goiás 75375-000, Brazil; jose.arruda-silva@embrapa.br
[3] Lallemand Plant Care, Patos de Minas 38706-420, Brazil; tgoble@lallemand.com (T.A.G.); kpazolini@lallemand.com (K.P.); amarciano@lallemand.com (A.F.M.)
* Correspondence: boaventuraheloiza@gmail.com (H.A.B.); eliane.quintela@embrapa.br (E.D.Q.)

Abstract: Lalguard C99 WP, based on the *Cordyceps javanica* BRM 27666 strain, is registered in Brazil for whitefly control. Spatial prediction is crucial for optimizing its field use and efficacy. In this study, the optimal temperature for mycelial growth and conidial production of *C. javanica is* 25–30 °C, with no growth at 33–35 °C. The highest nymphal mortality occurred at 25 and 30 °C, showing lower LT_{50} values at 30 °C. Mycelial growth was similar at 15, 20, 25, 30, and 35 °C when the fungus was exposed for 6 h and then transferred to a 27.4 °C environment; however, growth was slower at 35 °C with daily 6 h exposure alternating over 18 h at room temperature (mean of 28.5 °C). When the second instar whitefly nymphs were exposed for 6 h or 6 h daily at 15, 20, 25, 30, and 35 °C, followed by 7 days at fluctuating temperatures (mean of 28.4–30.2 °C), nymphal mortality was similar across temperatures. Although other abiotic factors (solar radiation, humidity, rainfall, etc.) must be considered for fungal efficacy, spatial predictions based on fluctuating temperatures indicated that *C. javanica* is suitable for use throughout Brazil, though its performance varied at constant temperatures in different locations.

Keywords: *Bemisia tabaci*; entomopathogenic fungi; mathematical models; control; IPM

1. Introduction

Cordyceps javanica (Frieder. & Bally) (Hypocreales: Cordycipitaceae) is an entomopathogenic fungus capable of infecting a wide range of agricultural pests [1]. This species is the most prevalent *Cordyceps* species in Brazil, originally isolated from both soil and various insect orders, and is predominant within populations of the whitefly, *Bemisia tabaci* (Hemiptera: Aleyrodidae) [2,3]. Many strains previously referred to as *C. fumosorosea* were shown to be *C. javanica*, such as the strains ESALQ 1296 and PFR-97, active ingredients of commercial mycoinsecticides registered in Brazil and USA, respectively, used for whitefly control [3].

In Central Brazil, several *C. javanica* strains were isolated from whiteflies under epizootic conditions [2]. Since 2012, the Embrapa Arroz e Feijão research team has conducted studies in the laboratory, screenhouse, and field to select the best *C. javanica* strain for a bioproduct development based on mass production characteristics, tolerance to detrimental environmental factors (UV radiation, high temperatures, oxidative, and osmotic stress), and virulence to all stages of *B. tabaci* [4–6]. Then, after seven years of a collaborative research project between Embrapa and Lallemand Plant Care (Patos de Minas, MG, Brazil), a product named Lalguard C99 based on aerial conidia of the *C. javanica* strain BRM 27666 formulated as a wettable powder was registered in August 2022 in Brazil. This was the first biopesticide based on this fungus species for the control of *B. tabaci* in Brazil [7], and in 2023, it was also registered in Paraguay.

The whitefly, *B. tabaci,* is an economically significant pest in Brazil and damages more than 40 agricultural crops [2,8]. Prior to the introduction of biotype B throughout the American Tropics, *B. tabaci* was an important pest and virus vector in bean and tomato crops only [2,9]. Following its introduction in 1989, the spread and increase in population size of biotype B was facilitated by the agricultural farming system of Brazil (three growing seasons), combined with favorable climatic conditions [2,10]. This pest established itself and continues to persist in cotton, dry bean, soybean, tomato, and several other crops and wild plant species.

The increased area planted with soybean is credited as one of the main factors driving the development of *B. tabaci* in Brazil, often reaching extremely high population levels and resulting in enormous 'clouds' of whiteflies [2,10,11]. In 30 years, the increase in soybean production area jumped by 306%, from 10.7 million hectares to 45.2 million hectares in the 2022/23 season [12]. In soybeans, damage may be caused by direct feeding and indirectly by deposition of honeydew, on which sooty mold fungi can grow and impede photosynthesis, and by transmission of the stem necrosis CpMMV (Cowpea mild mottle virus) [13–15].

In general, soybeans are cultivated from October to March in Brazil when high infestation levels of *B. tabaci* can be observed, particularly in the 'veranico' period, characterized by cycles of hot and dry days (lasting for 2–4 weeks) in the central–west (Mato Grosso, Mato Grosso do Sul, Goiás, and Federal District), north (Tocantins), and northeast (Maranhão, Piauí, and Bahia) [11,16,17]. In these regions, soybean production can suffer from the high incidence of whiteflies, intensified by high temperatures that reduce their life cycle and increase the number of generations [18].

Fungal efficacy is intrinsically mediated by abiotic factors, especially humidity, temperature, rainfall, and solar radiation [19–23]. For humidity, an increasing number of studies indicated that sufficient moisture exists within the microhabitat of many insect hosts or within the microenvironment of the host's body surface to support infection independent of ambient moisture conditions [24–29]. A study by Wraight et al. [30] demonstrated that a moisture-saturated environment was not required for *C. fumosorosea* infection of *B. tabaci* nymphs on excised hibiscus leaves, and it could infect third instars incubated at 25–30% RH. Boaventura et al. [6] also showed the ability of *C. javanica* to exploit the moist conditions of the leaf or insect boundary layer for germination and host infection, and the low temperature seemed to be more detrimental for *Cordyceps* virulence to *Bemisia* nymphs than RH. Furthermore, in the field, the soybean canopy microclimate is also a factor that favors fungal infection in the host (dew formation that creates a humid environment near the vicinity of the leaf) [31].

Furthermore, temperature influences all fungal development stages, including viability, morphogenesis, germination, host interaction, virulence, and conidiogenesis [32,33]. *Cordyceps* spp. are known to be mesophilic since they grow over a range of 8 to 32 °C, with thermal optima ranging from 20 to 30 °C and limits at 35 °C. Their thermal tolerance is related to their history, including their geo-climatic origin [34–37]. Boaventura et al. [6] observed that temperatures $\geq$ 35 °C for 4 to 6 h daily did not affect the efficacy of the three strains of *C. javanica*, including the strain BRM 27666, against nymphs.

In Brazil, climate varies considerably depending on factors such as latitude, sea conditions, and air masses. Thus, various climate types can be identified in Brazil: equatorial (hot and humid, with year-round rainfall and temperatures between 26 and 28 °C, reaching 35 °C in summer); semi-arid (hot and dry, with low rainfall and temperatures above 25 °C, often exceeding 28 °C); tropical (dry winters and rainy summers, with temperatures ranging from 18 to 27 °C, occasionally exceeding 30 °C in the hottest months); high-altitude tropical (distinct winter and summer seasons, with temperatures ranging from 15 to 22 °C and frost occurring in some areas); tropical Atlantic (encompassing the coastal region, with temperatures ranging from mild to hot, between 18 and 26 °C); and subtropical (mild climate, with temperatures ranging from 18 °C to below 0 °C in winter, with high temperatures in summer) [38]. Planting areas in Brazil are predominantly located in a tropical climate.

To further elucidate how this fungus, specifically the strain BRM 27666, will perform in different temperature environments, this study was conducted to explore the environmental factors associated with the virulence of *C. javanica* to the whitefly, *B. tabaci*. Thus, the objectives of this work were to develop temperature-dependent modeling and spatial prediction areas suited for the use of *C. javanica* BRM 27666 for whitefly control in Brazil, according to studies of mycelial growth, conidial production, and virulence under different temperature regimes.

2. Materials and Methods

This study resulted from a collaboration between Embrapa and Lallemand Plant Care (Patos de Minas, Minas Gerais, Brazil) under a collaborative research and development agreement. The experiments were conducted in the Insect Pathology Laboratory and Screenhouse at the Brazilian Agricultural Research Corporation (Embrapa Rice and Beans) located in Santo Antônio de Goiás, Goiás state (Central Brazil) (16°30′24.57″ S, 49°17′06.53″ W).

2.1. Insect Colony

The whitefly, *B. tabaci*, used in all experiments was identified as biotype B by molecular gene sequence markers from mtDNA cytochrome oxidase I (mtCOI) [2]. The whiteflies originated from a colony reared on common bean plants (*Phaseolus vulgaris* L., cv. Pérola) maintained under screenhouse with no environmental control at the Embrapa Rice and Beans Research Station.

2.2. Fungal Strain and Preparations

The *C. javanica* strain BRM 27666 was originally isolated from infected *B. tabaci* adults collected from soybeans in Porangatu, Goiás state, Central Brazil (13°27′ S, 49°10′ W, 600 m altitude). The region has a tropical savanna climate (Aw, megathermal) per Köppen classification, with two distinct seasons: a dry season (May to September) and a rainy season (October to April). The average annual temperature is 26 °C, reaching up to 39 °C in September and October [39]. This strain was preserved in liquid nitrogen and deposited at the Invertebrate Fungal Collection at Embrapa Genetic Resources and Biotechnology (Brasília, DF, Brazil) and in the Invertebrate Fungal Collection at Embrapa Rice and Beans. The *C. javanica* was identified using multi-loci-based phylogenetic methods, according to Lopes et al. [3].

For the mycelial growth and conidiogenesis experiments, the fungus was cultured on potato dextrose agar (PDA) in 9 cm Petri plates in the dark at 26 ± 1 °C for 7–10 days. After this period, the conidia were suspended in 10 mL of sterile aqueous solution of 0.01% (*v*/*v*) Tween 80 in 50 mL plastic centrifuge tubes. The suspension was vigorously agitated on a vortex mixer for 1 min and filtered through two layers of 30 µm pore-sized nylon cheesecloth. The filtered suspension (10 mL) was vortexed again for 1 min before application, and conidial concentrations were enumerated by a hemocytometer (Brightline Improved Neubauer, New Optik®, Brazil) at 400× magnification.

For the virulence experiments, the commercial product Lalguard C99 containing conidia of the strain BRM 27666 formulated as a wettable powder was provided by Lallemand Plant Care (Patos de Minas, MG, Brazil). For all experiments, conidial germination exceeded 90% on PDA after 16 h in darkness at 26 ± 1 °C. Only conidia with germ tubes greater than the conidial diameter were considered germinated.

2.3. Climate Conditions at the Main Cities of Soybean Production

The records of temperature (°C), relative humidity (%) (maximum and minimum), and rainfall (mm) for the largest soybean-producing municipalities in each region of Brazil were obtained from the website of the National Meteorological Institute in 2017 [39]. The cities were selected according to the Systematic Survey of Agricultural Production conducted by the Brazilian Institute of Geography and Statistics for soybean production [40]. The meteorological data were used to build graphics with the number of hours per day that the temperatures were in the ranges of <15 °C, 16–20 °C, 21–25 °C, 26–30 °C, 31–35 °C, and >35 °C for all months of 2017. The average daily relative humidity (%) for each month was added to the graph (Figure S1A–L). The number of hours per day that the temperatures were in the ranges described above never exceeded a 6 h period for the main municipal production regions.

2.4. Experimental Procedure at Different Temperature Ranges

Based on the graphics of the temperatures (Figure S1A–L) observed in the different regions of Brazil, temperatures ranging from 10 to 35 °C were chosen to conduct the experiments on fungal mycelial growth, conidial production, and virulence to nymphs. All experiments were carried out in incubators. Temperature and relative humidity were monitored at 1 h intervals by two dataloggers (Hobo® U12-012, Onset Computer Corp. Ltd., Bourne, MA, USA) per incubator (Figures S2–S7). To maintain the temperature of each incubator with the lowest possible variation and ensure humidity between 60% and 70%, daily monitoring was conducted for five days prior to the experiments. To increase the humidity, when necessary, particularly at temperatures exceeding 30 °C, water containers were placed at various locations within the incubator until the desired standard was achieved.

2.5. Fungal Growth and Conidial Production at Different Temperatures and Hourly Intervals

The methodological information for the experiments on mycelial growth and conidial production of *C. javanica* on PDA is provided in Table 1.

Table 1. Temperatures and assessment days for mycelial growth and conidial production of *C. javanica* on PDA medium. A 6 h period was chosen since the temperature ranges of <15 °C, 16–20 °C, 21–25 °C, 26–30 °C, 31–35 °C, and >35 °C never exceeded 6 h daily during the soybean cultivation season (October through March) [39].

Experiment	Temperature (°C)	Exposure Time	Assessment Days	
			Radial Growth	Conidial Production
1	10, 15, 20, 25, 30, 35	24 h for 10 days	3, 4, 5, 6, 7, 10	10
2	25, 27, 29, 31, 33	24 h for 10 days	3, 4, 5, 6, 7, 10	Not determined
3	26, 28, 30	24 h for 17 days	3, 4, 5, 6, 7, 10, 11, 12, 13, 14, 17	Not determined
4	15, 20, 25, 30, 35	6 h at each temperature and then 27.4 °C for 10 days [1]	1, 3, 6, 8	10
5	15, 20, 25, 30, 35, 28.5 °C	6 h alternating with 18 h at 28.5 °C for 10 days [2]	1, 3, 6, 8	10

[1] Mean: 27.4 °C; min: 27.0 °C; max: 29.9 °C; RH: between 60 to 70%. [2] Room temperature: mean 28.5 °C; min: 26.4 °C; max: 31.6 °C; RH: between 60 to 70%.

For the experiments listed in Table 1, two microliters of a suspension at 1×10^7 conidia mL^{-1} were placed in the center of Petri dishes (60 × 15 mm) containing 5 mL of PDA medium. Inoculated plates were incubated at temperatures described in Table 1. For all experiments, each treatment was replicated 10 times (1 Petri dish per treatment).

Fungal growth colonies (length and width of colonies in mm) were determined at days described in Table 1 using a digital caliper (150 mm/6" Digimess-100.174BL). Conidial production was determined after 10 days by cutting the fungal colony from the culture plate with a sterile scalpel. The colony disk was added in 30 mL of 0.01% Tween 80 and vortexed for five min at 700 rpm to dislodge conidia from the PDA. Successive dilutions were carried out until a desirable suspension was obtained for conidia counting using a Neubauer chamber (400× magnification). Conidial production was expressed as the number of conidia per colony.

2.6. Virulence of *C. javanica* to 2nd Instar Nymphs at Different Temperatures and Hourly Intervals

Three experiments were conducted to determine the effect of different temperatures on the virulence of *C. javanica* to *B. tabaci* nymphs according to the methods described by Boaventura et al. [6] (Table 2).

Table 2. Temperatures and assessment days for virulence of *C. javanica* against 2nd instar nymphs of *B. tabaci*.

Experiment	Temperature (°C)	Exposure Time	Mortality (Assessment Days)
1	15, 20, 25, 30, 35, 28.4 °C [1]	24 h	
2	15, 20, 30, 35	6 h at each temperature and then 28.4 °C	3, 4, 5, 6, and 7
3	15, 20, 25, 30, 35, 30.2 °C [2]	6 h alternating with 18 h at 30.2 °C	

[1] Room temperature: mean: 28.4 °C; min: 25.2 °C; max: 32.4 °C; RH: between 60 to 70%. [2] Room temperature: mean: 30.2 °C; min: 26.5 °C; max: 34.4 °C; RH: between 60 to 70%.

Experimental units contained two 10-day-old common bean plants (*Phaseolus vulgaris*, cv. Pérola) grown in plastic pots filled with oxisol soil (2 L) and kept in a screenhouse (9 × 8 m) covered with a fine screen fabric (50 mesh). The experiments were carried out on common bean plants since soybean plants were highly sensitive to the elevated temperatures in the incubator. Adult-infested plants were shaken and placed near pest-free bean plants (10 days old with two primary leaves) for 6 to 8 h to allow oviposition. This procedure provided more than 100 eggs per leaf. The adult whiteflies were removed, and newly infested plants moved to another screenhouse until nymphs reached the 2nd instar (0.30– 0.44 mm in length and 0.18–0.36 mm in width) [41].

For all experiments, treatments were applied to the abaxial side of primary leaves containing 2nd instar nymphs with a microsprayer (0.3 mm needle, Paasche® airbrush type H-set) connected to a vacuum pump and calibrated to 500 µL per leaf in an even coverage. Fungal treatments were sprayed with 2×10^7 conidia mL^{-1}. Controls consisted of nymphs sprayed with distilled water.

All experiments were conducted in a completely randomized design with four replicates, each consisting of three seedlings per pot (six primary leaves). One leaf from each pot was collected for nymphal mortality assessment at 3, 4, 5, 6, and 7 days after spraying and evaluated under a dissecting stereomicroscope at 40× magnification. A small red dot was marked with a permanent marker pen on the leaf next to the dead nymphs to confirm fungal infection. Nymphs that became desiccated or developed yellowish signs with mycelial or conidial growth on the insect cadaver were considered dead by *C. javanica*. The leaves were incubated inside gerbox-type boxes (11 × 11 × 3.5 cm) with wet cotton added to the leaf petiole for 4 days at room temperature (mean 26.1–30.2 °C, 53.0–73.7% RH). The dead marked nymphs presenting conidiogenesis (i.e., mycosed insects) were also considered infected by the fungus.

2.7. Modeling the Performance of C. javanica for Whitefly Control in Different Regions of Brazil

Temperature is considered the main climatic factor in the context of this work. To determine the performance of *C. javanica* at different temperature environments, several parameters were analyzed, including mycelial growth, conidial production, speed of insect mortality (LT_{50} value), and virulence of *C. javanica* to *B. tabaci* nymphs under temperatures ranging from 10 to 35 °C. Temperature data were obtained from the WorldClim database (http://www.worldclim.org/, accessed on 14 March 2023). These datasets are made of layers containing monthly mean, minimum, and maximum records in raster files in "Flat" format (.flt and.hdr file), with a spatial resolution varying from 10 arc-minutes (~20 km) to 30 arc-seconds (~1 km). The 30 arc-seconds (~1 km) were used in the context of this study.

An open-source computer-aided tool built on R codes and Java interface, the entomopathogenic fungi application (EPFA) software version 1.045, was used for modeling the virulence of *C. javanica* against *B. tabaci*. The recorded mortality was plotted against the corresponding temperature values from which nonlinear models were fitted to the observed data [42]. The model parameters were estimated by fitting equations to the recorded mortality and the corresponding temperature values. First, 82 models were fitted to the data; then, the best-fitted models were selected based on their coefficient of determination (R^2), adjusted R^2, Akaike's information criteria, the root mean square error, and residual sum of squares (RSS) [42].

The temperature-dependent mathematical expression obtained during the modeling step was run at each grid of the raster files of Brazil using the monthly minimum and maximum temperature datasets obtained from WorldClim. The gridded temperature datasets were loaded into EPFA software, simultaneously extracted from the database, and then organized in matrix format using longitude as the columns and latitude as the rows [42]. The monthly temperature data from Brazil, in raster format as required by the EPFA software, are from 2018, which were the most recent data available. A point object picks the temperature-dependent mathematical expression of the fungal virulence, which was consecutively applied in each geographical coordinate of the grid. The results were converted into ASCII file format (.asc) and transferred into an open-source software Q-GIS60 for visualization [42]. The virulence map was produced for Brazil after completing the fitting process. Details of procedures and steps to estimate the best model parameters as per the defined input data can be found in Guimapi et al. [42].

2.8. Data Analysis

For mycelial growth and virulence experiments, curves were adjusted according to nonlinear models (Logistic, Log Logistic, Log-normal, Weibull, and Gompertz) and compared using the Chi-square test ($p < 0.05$). This nonparametric statistical method was used for the comparison of two unpaired groups to verify whether or not they belong to the same population and when the requirements for the application of the Student's *t*-test were not met.

To estimate the median lethal time (LT_{50}), nonlinear models were fitted, and values were compared by the overlap of their 95% confidence intervals (95% CI) using the package "drc 3.0-1" [43]. The LT_{50} was not estimated for treatments where mortality did not reach 50%. For conidial production, data were submitted to analysis of variance, and a multiple means comparison was made using Tukey's HSD test ($p < 0.05$).

Nonlinear models were used to determine the optimal temperature for mycelial growth and virulence of *C. javanica* to *B. tabaci*. The best-fitted model was selected based on the goodness-of-fit of the model using the coefficient for nonlinear regression (R^2) and the RSS. Higher R^2 values and lower RSS values suggest a better fit [44]. The models and equations used to describe optimal temperatures are shown in Table 3.

Table 3. Mathematical equations describing the relationship between temperature and colony growth (Experiments 1, 2, and 3) and virulence (Experiment 1) of *C. javanica* conducted at constant temperatures. For models, r_T represents the fungal development rate, T is the observed temperature (°C), T_1 and T_2 are the minimum and maximum temperatures, $\sqrt{c_c}k_1$ and c_c correspond to the regression slope, and k and k_2 are constants.

Experiments	Model	Mathematical Expressions	Model Parameters
Colony Growth	Ratkowsky 1 [1]	$r_T = \sqrt{c_c}k_1(T - T_1)(1 - \exp[k_2(T - T_2)]))^2$	$\sqrt{c_c}k_1,\ k_2,\ T_1,\ T_2$
	Ratkowsky 2 [2,3]	$r_T = (c_c(T - T_1)(1 - \exp[k(T - T_2)]))^2$	$c_c,\ k,\ T_1,\ T_2$
Virulence	Ratkowsky 1 [4]	$r_T = \sqrt{c_c}k_1(T - T_1)(1 - \exp[k_2(T - T_2)]))^2$	$\sqrt{c_c}k_1, k_2,\ T_1,\ T_2$

[1] Experiment 1: 10, 15, 20, 25, 30, and 35 °C; [2] Experiment 2: 27, 29, 30, and 31 °C; [3] Experiment 3: 26, 28, and 30 °C). [4] Experiment 1: 15, 20, 25, 30, and 35 °C.

In all the model equations described in Table 3, r_T represents the fungal development rate and T (°C) is the observed temperature. T_1 and T_2 are the minimum and maximum temperatures, respectively, at which the growth rate is zero. The $\sqrt{c_c}\ k_1$ and c_c correspond to the slope of the regression as described in the rootsq_82 equation, and k and k_2 are constants [45–48]. These expressions are included in the "devRate" package [available at https://cran.r-project.org/web/packages/devRate/devRate.pdf, accessed on 27 February 2023].

3. Results

3.1. Fungal Mycelial Growth

Mycelial growth of *C. javanica* occurred at all temperatures except 35 °C (Experiment 1) (Figure 1A). However, the mean diameter of colonies was significantly slower and reduced at 10, 15, and 20 °C (38.9 to 196.4 mm^2) compared to 25 and 30 °C (425.7 and 554.4 mm^2, respectively) (Figure 1A, Table 4).

Table 4. *p*-values ($p \leq$ value) of the comparisons of mycelial growth curves of *C. javanica* at different temperatures. Curves were considered significantly different at $p \leq 0.05$. Experiment 1, 2, and 3 = constant temperatures for 10 days.

Mycelial Growth					
Experiment 1					
Temperature (°C)	10	15	20	25	30
15	0.220	-	-	-	-
20	0.068	0.527	-	-	-
25	<0.001	0.004	0.022	-	-
30	<0.001	<0.001	0.001	0.281	-
35	0.735	0.118	0.031	<0.001	<0.001
Experiment 2					
Temperature (°C)	27	29	31		
29	0.034	-	-		
31	0.845	0.049	-		
33	<0.001	<0.001	<0.001		
Experiment 3					
Temperature (°C)	26	28			
28	0.338	-			
30	0.836	0.298			

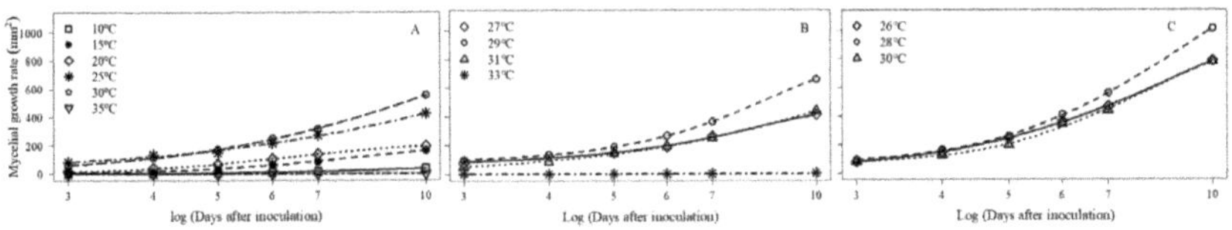

Figure 1. The mean diameter of *Cordyceps javanica* colonies at different constant temperatures for 10 days. Curves were adjusted according to nonlinear models: Log logistic (**A**), Logistic (**B**), and Log-normal (**C**).

In the second experiment, *C. javanica* colony growth was significantly higher at 29 °C (655.6 mm^2) than at 27 °C (411.8 mm^2) and 31 °C (439.1 mm^2) for 10 days (Figure 1B, Table 4). No fungal growth was observed at 33 °C (Figure 1B). In Experiment 3, the mean diameter of colonies was similar at 26, 28, and 30 °C (779.2, 1013.4, and 787.3 mm^2, respectively) for 10 days (Figure 1C, Table 4).

No difference in colony growth was observed when the fungus was held for 6 h at 15, 20, 25, 30, or 35 °C and then transferred to 27.4 °C for 8 days (Figure 2A, Table 5). When the fungus was held for 6 h at 15, 20, 25, 30, or 35 °C, alternating with 18 h at 28.5 °C for 8 days, mycelial growth occurred at all temperatures. However, the growth of colonies was slower and reduced at 35 °C (152.9 mm^2) compared to 20, 25, and 30 °C and 28.5 °C (476.6 to 565.9 mm^2) for 8 days (Figure 2B, Table 5).

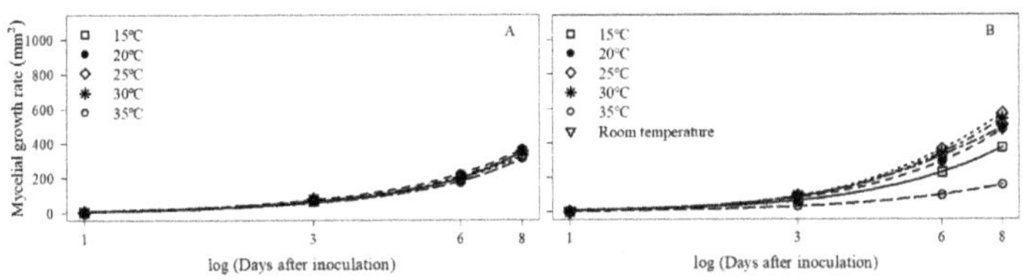

Figure 2. The mean diameter of *Cordyceps javanica* colonies after exposure to different temperatures, followed by mean temperatures of 27.4 and 28.5 °C for 10 days. (**A**) = 6 h at each temperature before being transferred to the mean temperature of 27.4 °C (Experiment 4). (**B**) = 6 h alternating with 18 h at fluctuating temperatures (mean of 28.5 °C) (Experiment 5). Curves were adjusted according to nonlinear models: Logistic (**A**) and Log Logistic (**B**).

Table 5. *p*-values ($p \leq$ value) of the comparisons of mycelial growth curves of *C. javanica* at different temperatures. Experiment 4 = 6 h at each temperature before being transferred to an incubator (mean of 27.4 °C). Experiment 5 = 6 h alternating with 18 h at fluctuating temperatures (mean of 28.5 °C). Curves were considered significantly different at $p \leq 0.05$.

Mycelial Growth				
Experiment 4				
Temperature (°C)	15	20	25	30
20	0.852	-	-	-
25	0.941	0.832	-	-
30	0.942	0.901	0.924	-
35	0.868	0.739	0.902	0.829

Experiment 5					
Temperature (°C)	15	20	25	30	35
20	0.378	-	-	-	-
25	0.077	0.374	-	-	-
30	0.134	0.536	0.786	-	-
35	0.072	0.007	<0.001	<0.001	-
28.5	0.227	0.726	0.566	0.755	0.003

The optimal temperature for *C. javanica* mycelial growth was estimated at 28.0 to 29.1 °C based on parameters from the nonlinear Ratkowsky 1 and Ratkowsky 2 models fitted to the mean diameter of colonies versus temperature (Figure 3). Statistics of the goodness-of-fit of models describing the relationship of temperature with the mean diameter of *C. javanica* colonies are presented in Table 6.

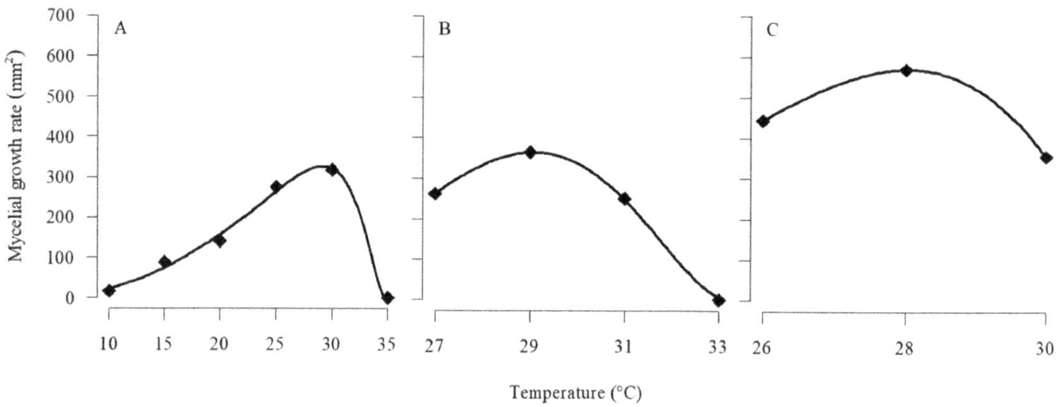

Figure 3. Relationship between temperature and mycelial growth rate of *C. javanica*. Nonlinear models fitted to observed values of the mycelial growth rate of *C. javanica* at constant temperatures of (**A**) 10, 15, 20, 25, 30, and 35 °C (Experiment 1); (**B**) 27, 29, 31, and 33 °C (Experiment 2); and (**C**) 26, 28, and 30 °C (Experiment 3).

Table 6. Estimated parameters for nonlinear Ratkowsky models 1 and 2 describing the relationship between temperature and mean diameter of *C. javanica* colonies for Experiments 1 to 3 conducted at constant temperatures. R^2 and RSS were used to select the best-fitted model.

Model	Parameters	Experiments [1]		
		1	2	3
Ratkowsky 1	c_c	0.089	-	-
	k_1	2.620	-	-
	k_2	0.435	-	-
	T_1	3.926	-	-
	T_2	34.845	-	-
Ratkowsky 2	c_c	-	5.919	3.670
	k	-	0.177	0.410
	T_1	-	22.907	19.634
	T_2	-	33.250	31.671
R^2		0.972	0.911	0.873
RSS		2473	7009	1424

[1] Experiment 1: 10, 15, 20, 25, 30, and 35 °C; Experiment 2: 27, 29, 30, and 31 °C; Experiment 3: 26, 28, and 30 °C.

3.2. Conidial Production on PDA Medium

At constant temperatures, fungal conidial production at 15 and 20 °C was 0.8 and 2.7×10^6 conidia colony^{-1}, respectively, and significantly lower than 25 °C (100×10^6 conidia colony^{-1}) and 30 °C (220×10^6 conidia colony^{-1}) after 10 days (Table 7). There was no conidial production at 10 and 35 °C because there was no mycelial growth (Table 7).

Table 7. Conidial production of *C. javanica* after exposure for 10 days to constant temperatures (Experiment 1), 6 h at 15, 20, 25, 30, or 35 °C and then at the incubator (mean of 27.4 °C) for 8 days (Experiment 4), and 6 h at 15, 20, 25, 30, or 35 °C, alternating with 18 h at fluctuating temperatures (mean of 28.5 °C) for 8 days (Experiment 5).

	Number of Conidia × 10^6/Colony		
Temperature (°C)	**Experiment 1**	**Experiment 4**	**Experiment 5**
10	0.0 ± 0.0 c	- *	-
15	0.8 ± 0.7 b	118.9 ± 79.4 ab	32.5 ± 7.7 c
20	2.7 ± 2.0 b	78.2 ± 41.5 b	110.0 ± 41.4 ab
25	100.0 ± 180.0 a	162.8 ± 99.2 ab	146.2 ± 54.2 a
30	220.0 ± 90.2 a	120.8 ± 63.8 ab	160.8 ± 57.1 a
35	0.0 ± 0.0 c	193.9 ± 67.5 a	48.7 ± 57.4 bc

* This temperature was not tested. Each value is expressed as mean ± SD. Different lowercase letters show significant differences (Tukey's HSD test) in sporulation among temperature regimes. Experiment 1 ($F = 143.60$; $df = 4{,}45$; $p < 0.001$); Experiment 4 ($F = 2.67$; $df = 4{,}32$; $p = 0.049$); Experiment 5 ($F = 13.23$; $df = 4{,}41$; $p < 0.001$).

When the fungus was held for 6 h at 15, 20, 25, 30, or 35 °C before being transferred to 27.4 °C for 10 days, conidial production at 35 °C (193.9 × 10^6 conidia colony^{-1}) was significantly higher than at 20 °C (78.2 × 10^6 conidia colony^{-1}). However, no differences were observed for 15, 25, and 30 °C compared to 20 and 35 °C (Table 7).

On the other hand, when *C. javanica* was incubated for 6 h at 15, 20, 25, 30, or 35 °C, alternating with 18 h at 28.5 °C for 10 days, conidial production at 25 and 30 °C (146.2 and 160.8 × 10^6 conidia colony^{-1}, respectively) were significantly higher than at 15 °C (32.5 × 10^6 conidia colony^{-1}) and 35 °C (48.7 × 10^6 conidia colony^{-1}). No differences in sporulation were observed at 20 °C compared to other temperatures tested, except at 15 °C (Table 7).

3.3. Virulence of *C. javanica* to 2nd Instar Nymphs at Different Temperatures and Hourly Intervals

At constant temperatures for 7 days, nymphal mortality at 15, 20, and 25 °C was very low (5.3–14.9%) and significantly slower than for 25 and 30 °C (60.9% and 68.5%, respectively) (Figure 4A, Table 8). At 25 and 30 °C, mortality was significantly higher than the untreated control ($p < 0.001$). The fungus killed faster at 30 °C (LT_{50} 5.4 days) compared to 25 °C (6.3 days) (Table 9).

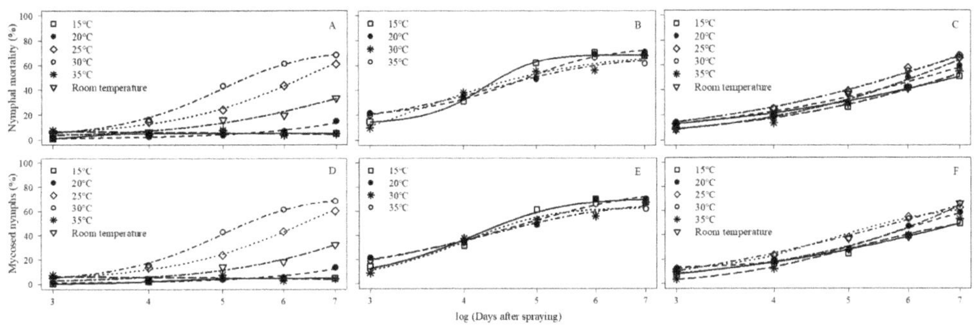

Figure 4. Nymphal mortality (**A–C**) and mycosed nymphs (**D–F**) on different days post-inoculation of 2nd instar *B. tabaci* with *C. javanica* after exposure to different temperatures for 7 days. (**A,D**) = Constant temperatures for 7 days (Experiment 1). (**B,E**) = 6 h at each temperature before being transferred to fluctuating temperatures (mean of 28.4 °C) for 7 days (Experiment 2). (**C,F**) = 6 h at 15, 20, 25, 30, or 35 °C, alternating with 18 h at fluctuating temperatures (mean of 30.2 °C) for 7 days (Experiment 3). Curves were adjusted according to nonlinear Logistic (**A,C,F**), Weibull (**B,D**), and Gompertz (**E**) models.

Table 8. *p*-values ($p \leq$ value) of the comparisons of nymphal mortality curves for *B. tabaci* after treatment with *C. javanica* at different temperatures. Curves were considered significantly different at $p \leq 0.05$. Experiment 1 = Constant temperatures for 7 days. Experiment 2 = 6 h at each temperature before being transferred to fluctuating temperatures (mean of 28.4 °C) for 7 days. Experiment 3 = 6 h at 15, 20, 25, 30, or 35 °C alternating with 18 h at fluctuating temperatures (mean of 30.2 °C) for 7 days. Curves were adjusted according to nonlinear Logistic (A, C) and Weibull models (B).

Nymphal Mortality					
		Experiment 1			
Temperature (°C)	15	20	25	30	35
20	0.752	-	-	-	-
25	0.004	0.007	-	-	-
30	<0.001	<0.001	0.210	-	-
35	0.869	0.677	0.004	<0.001	-
28.4 *	0.181	0.255	0.119	0.006	0.172
		Experiment 2			
Temperature (°C)	15	20	30		
20	0.552	-	-		
30	0.524	0.637	-		
35	0.414	0.628	0.781		
		Experiment 3			
Temperature (°C)	15	20	25	30	35
20	0.685	-	-	-	-
25	0.307	0.535	-	-	-
30	0.318	0.550	0.980	-	-
35	0.770	0.588	0.253	0.262	-
30.2 *	0.541	0.728	0.500	0.510	0.568

* Average room temperature after seven days.

Table 9. LT_{50} values 7 days post-inoculation of whitefly nymphs treated with *C. javanica*. Experiment 1 = Constant temperatures for 7 days. Experiment 2 = 6 h at each temperature before being transferred to fluctuating temperatures (mean of 28.4 °C) for 7 days. Experiment 3 = 6 h at 15, 20, 25, 30, and 35 °C, alternating with 18 h at fluctuating temperatures (mean of 30.2 °C) for 7 days. Dash (-) means the LT_{50} value was not estimated for cumulative mortality less than 50% at 7 days post-treatment.

Temperature	**LT_{50} (d) (CI95%)**		
	Experiment 1	**Experiment 2**	**Experiment 3**
15	-	4.5 (4.4–4.5)	6.9 (6.8–7.0)
20	-	4.9 (4.7–5.0)	6.3 (6.3–6.4)
25	6.3 (6.1–6.6)	Not determined *	5.7 (5.6–5.7)
30	5.4 (5.3–5.4)	4.7 (3.8–5.7)	5.7 (5.6–5.7)
35	-	4.9 (4.6–5.3)	6.7 (6.7–6.8)
Room temperature	-	Not determined *	6.1 (6.1–6.2)

* Treatment not evaluated due to material loss.

No difference in nymphal mortality was observed when the fungus was held for 6 h at 15, 20, 30, and 35 °C before being transferred to 26.1 °C (min. 24.3 and max. 28.9 °C) for 7 days (Figure 4B, Table 8). Fungal treatments killed significantly more nymphs than the untreated control at all temperatures ($p < 0.001$). The LT_{50} values at 15 °C (4.5 days) were significantly different from those at 20 °C and 35 °C (4.9 days for both), but the mortality at all temperatures was similar to 30 °C (4.7 days) (Table 9).

There was no difference in nymphal mortality among the fungus treatments (Figure 4C, Table 8) when the treated nymphs were held for 6 h at temperatures of 15, 20, 25, 30, or 35 °C, alternating with 18 h at 30.2 °C (min. 26.5 and max. 34.4 °C) for 7 days. The nymphal mortality was significantly higher when compared to the untreated control ($p < 0.001$). The speed of infection of *C. javanica* was faster at 25 and 30 °C than at 15, 20, 35, and 28.4 °C.

The LT_{50} was 5.7 days at both 25 and 30 °C, while it ranged from 6.11 to 6.94 days under the other temperature conditions (Table 9).

For all three experiments, nymphal mortalities were very similar to the percentage of cadavers with fungal sporulation (i.e., mycosed nymphs) at all temperatures tested (Figure 4D–F). The *p*-values of the comparisons of confirmed mortality curves for *B. tabaci* nymphs are also reported as a Supplementary File (Table S1).

The Ratkowsky 1 model had a good fit to the mortality data, and the model adjusted for the mortality versus temperature is shown in Figure 5. The nonlinear model predicted an optimal temperature for higher mortality of *B. tabaci* nymphs by *C. javanica* at 28.1 °C.

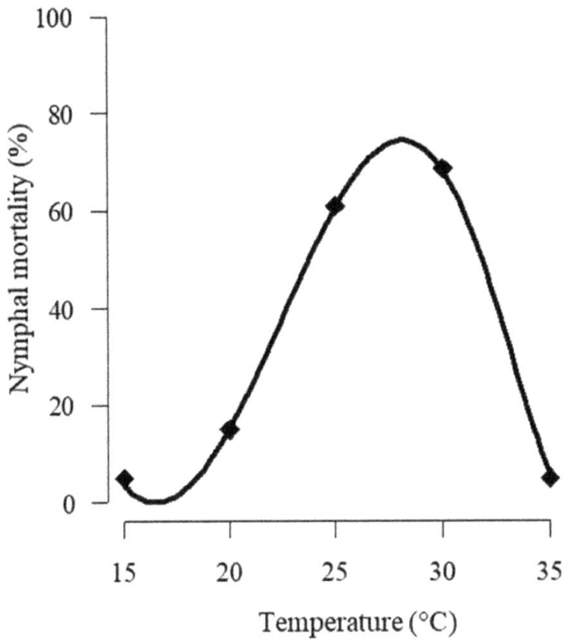

Figure 5. Graphical comparison of a temperature-dependent model describing nymphal mortality of *C. javanica* after constant temperature. The curve represents the Ratkowsky 1 nonlinear model that best fitted the experimental data points and was used to predict the virulence of the fungus against the whitefly. (Parameters: c_c = 2.999, k_1 = 0.084, k_2 = 0.090, T_1 = 16.502, T_2 = 36.051; R^2: 0.959, RSS: 0.067).

3.4. Spatial Prediction of the Performance of *C. javanica* Against *B. tabaci* at Different Regions of Brazil

The predicted performance of the bioinsecticide Lalguard C99 against *B. tabaci* biotype B in Brazil is shown in Figure 6. The model based on constant temperatures predicted that the application of *C. javanica* in Brazil would result in high performance (more than 60% mortality of *B. tabaci*) in the country's northern, northeastern, and central–western regions and low to moderate performance (10–25% mortality) for southern and southeastern regions (Figure 6A). However, when *C. javanica* was exposed to different temperatures for 6 h only or for 6 h daily followed by 7 days of fluctuating temperatures (mean of 28.4–30.2 °C), the model predicted a high performance of Lalguard C99 (over 60% mortality in *B. tabaci*) across all regions of Brazil (Figure 6B).

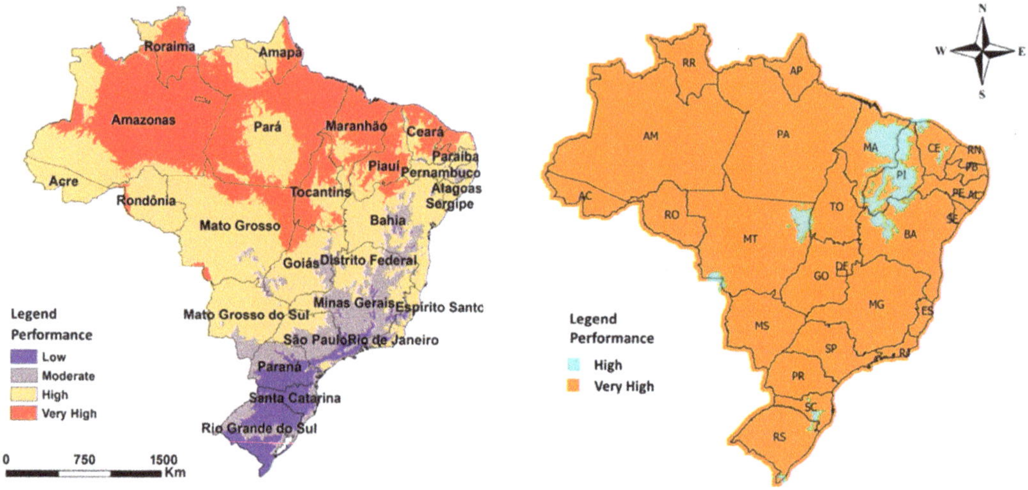

(A) (B)

Figure 6. A spatial model of the predicted performance of Lalguard C99 for the control of *B. tabaci* in Brazil according to the virulence experiments at constant temperature for 7 days (**A**), 6 h only, or for 6 h daily at 15, 20, 25, 30, and 35 °C followed by 7 days of fluctuating temperatures (mean of 28.4–30.2 °C) for 7 days (**B**). The figures were generated using QGIS 3.16 software. The names in the figure are the Brazilian states.

4. Discussion

The majority of bioassays designed to assess the development and virulence of entomopathogenic fungi (EPF), including *Cordyceps* sp., have been conducted in vitro at constant temperature regimes [33,36,49–53]. These studies have indicated that the optimum temperature for vegetative growth of *Cordyceps* sp. ranged from 25 to 30 °C, with growth being delayed at 15 °C and completely inhibited at 35 °C. These findings are consistent with our study under constant temperatures, where we observed that the ideal temperature range for the growth of *C. javanica* was between 25 and 30 °C, with optimal growth occurring between 28 and 29.1 °C. In addition, at constant temperatures, conidial production was higher at 25 and 30 °C and reduced at 15 and 20, and no production occurred at 35 °C. Xie et al. [54] reported that conidial production of *C. javanica* Pf04 on rice was higher at 25 °C than at 30 °C, with reduced production at 20 °C. However, in previous studies on rice, *C. javanica* BRM27666 also showed higher conidial production at 30 °C after 10 days [Quintela et al., unpublished data], suggesting that this strain has a higher optimal sporulation temperature. This highlights significant variation in temperature tolerance among strains [36].

Temperatures are not constant under field conditions. After field applications, conidia may be exposed to fluctuating temperatures, which could negatively impact their viability and virulence. In the field, temperatures fluctuate both throughout the year and between day and night. In this study, we created graphs that illustrate the number of hours that temperatures remain within specific ranges over the course of the year. For example, during the summer season, temperatures fluctuate significantly and can remain above 35 °C or below 15 °C for about 6 h, depending on the region. To simulate this effect, we first exposed the fungus in vitro to temperatures of 15, 20, 25, 30, and 35 °C for 6 h only, followed by 27.4 °C over the course of 10 days. Our study revealed that during the initial stages of germination, these temperatures did not affect mycelial growth on PDA. In addition, daily exposure to constant temperatures of 15, 20, 25, 30, and 35 °C for 6 h on a PDA medium, alternating with 18 h of fluctuating temperatures (mean of 28.5 °C) over a period of 8 days, reduced mycelial growth at the upper-temperature limit of 35 °C. This likely occurred because hyphae are less tolerant to high temperatures than conidia, which are naturally more heat-resistant structures [55–58]. Hyphal growth takes place inside the host, where the hyphae benefit from a degree of protection against hostile environmental conditions.

The infection process by EPF initiates with conidia germination and penetration in the insect cuticle and consequently develops into hyphae and mycelia in the hemocoel, effectively colonizing the insect's body [59–61]. After the fungus invades and kills the host, it produces new conidia under favorable temperature and high humidity conditions, which are essential for disease transmission to other hosts and maintaining its inoculum in the environment [62–64]. When our fungus was exposed to 15, 20, 25, 30, or 35 °C for 6 h only, followed by 10 days at 27.4 °C, conidial production was not affected. However, after exposure for 6 h daily at these temperatures followed by 10 days at room fluctuating temperatures (mean of 28.5 °C), the highest conidial production ($\approx 10^8$ conidia/colony) occurred at 20, 25, and 30 °C, while production at 15 and 35 °C was slightly lower ($\approx 10^7$ conidia/colony). These results show that probably six hours daily in regions with high-temperature regimes in Central, Northern, and Northeastern Brazil will reduce conidial production on the cadavers. In Southern and Southeastern Brazil, conidia production will be reduced only at temperatures below 15 °C.

In addition to our in vitro studies using artificial culture media, we also conducted in vivo studies with second instar whitefly nymphs, *B. tabaci*. At constant temperatures, the mortality rates of whitefly nymphs by *C. javanica* were low at 15, 20, and 35 °C (range of 5.3 to 14.9%). The highest level of nymphal mortality was observed at 25 and 30 °C (range of 60.9 to 68.5%), exhibiting a lower LT_{50} at 30 (5.4 days) compared to 25 °C (6.3 days). When second instar nymphs were exposed to the fungus for 6 h only or 6 h daily at different temperatures followed by seven days at fluctuating temperatures (mean of 28.4–30.2 °C), nymphal mortality was similar for all temperatures, and small differences were observed for LT_{50} values. This indicates that, after penetration, the development of the fungus within the insect is not influenced by temperature.

Cabanillas et al. [33] observed that the mycelial growth of *Cordyceps* sp. on PDA was higher at 30 °C than at 25 °C, indicating that this fungus has a higher optimal growth temperature than many other EPF. Fargues et al. [34] reported that the optimal temperature for growth is not necessarily the same as that for fungal infection of insects. On the other hand, Moorehouse et al. [65] showed that EPF tends to kill the host faster at temperatures that are optimum for fungal mycelial growth. In our study, fungal growth and infection of *B. tabaci* nymphs by *C. javanica* were similar at a temperature range of 25 to 30 °C, and the optimum temperature was between 28 and 29.1 °C. Furthermore, constant temperatures outside this range affected the development of the fungus both in the PDA medium and in the insect host. For this reason, we also tested the virulence of *C. javanica* to whitefly nymphs under fluctuating temperature conditions.

In the summer season, field temperatures often exceed 35 °C. However, these high temperatures for relatively short periods did not affect its virulence against whitefly nymphs, as observed in this study. On the insect's cuticle, fungal conidia typically invade the bodies of their hosts within 24 h [20]. After this period, most of the conidia in the field are lost, mainly due to exposure to solar radiation, high temperatures, and precipitation. Studies have shown that most *C. javanica* conidia lose viability in the field within 24 h after application due to the incidence of solar radiation [52] [Quintela et al., unpublished data]. Thus, the initial 24 h after the conidium contacts the insect's cuticle is the most critical period for the fungus to infect the host. Therefore, we strongly recommend applying *C. javanica* in the late afternoon in the field because there is an interaction between solar radiation and temperature in the persistence of conidia, where higher temperature aggravates the detrimental effect of solar radiation on conidial viability [66]. This approach is particularly important when managing mobile insects, such as whitefly adults or corn leafhoppers, which may come into contact with the conidia on the leaf after application.

Whitefly outbreaks typically occur at high temperatures during the summer crop season, demonstrating that the fungus can be used to control this pest when temperatures > 35 °C do not exceed 6 h per day. Under extreme temperatures during the growing season, the fungus, insects, and plants may be adversely affected [67]. Temperature has profound effects on the physiology and development of both the insect host and fungal pathogen (which simultaneously affect host susceptibility and pathogen virulence), and these effects are influenced, in turn, by factors related to the host plant. Evapotranspiration, for example, can lower leaf surface temperatures to levels substantially below ambient [68,69]. In the same way that high temperatures (35 °C) for prolonged periods affect the growth of the fungus, these conditions can influence the survival of *B. tabaci* and the development of soybean plants. Low temperatures also limit the growth and development of both whitefly and soybeans, similar to the fungus [70–72].

Mathematical models have become an important tool for understanding and predicting the virulence and suitable application areas for entomopathogenic fungi against insect pests often occurring in different environmental conditions [42,73]. Considering the virulence data under constant temperature, spatial prediction revealed that in some regions of Brazil, the fungus may effectively control whiteflies, while in other areas, the performance may be very low. For example, in the northern, northeastern, and central–western regions, the models predicted a higher virulence of *C. javanica*, while low to moderate virulence was predicted in the southern and southeastern regions of the country. However, the lowest temperatures observed in these regions of the country generally occur between May and July when whitefly populations are smaller [10]. Furthermore, soybeans are not usually cultivated during this period [74]. The temperature-dependent modeling was carried out considering the soybean growing season in Brazil. Based on these conditions, the spatial prediction revealed that *C. javanica* could be highly virulent to *B. tabaci* in all regions of Brazil. Thus, even in tropical soybean-producing regions, where higher average temperatures are recorded (>35 °C), such as a large part of MATOPIBA (states of Maranhão, Tocantins, Piauí, and Bahia) and the central–west region (states of Mato Grosso, Mato Grosso do Sul, and Goiás), the Lalguard C99 mycoinsecticide could be an excellent tool for managing *B. tabaci* populations. In the southern and southeastern regions, average temperatures remain close to the optimal temperature range (25 to 30 °C) for *C. javanica*, which suggests the product's high effectiveness.

Agbessenou et al. [75] used constant temperatures to establish temperature-dependent models to identify geographically suitable areas for deploying *M. anisopliae* for *Tuta absoluta* (Meyrick, 1917) control in East Africa. Their study revealed that the fungal performance was location-specific. Similarly, when we modeled our data to develop spatial predictions of potential areas for *C. javanica* using constant temperatures, Brazil was divided into regions with low, moderate, high, and very high performance. However, when we used fluctuating temperatures (more representative of field conditions), our results showed that the fungus can be used across all regions of Brazil with high to very high performance. Our spatial prediction results indicate that Lalguard C99 could become a key tool for reducing *B. tabaci* biotype B populations in the field, both at low densities (during the initial whitefly infestations from November to December) and at high densities with overlapping generations (from January to March) under field conditions. The environmental conditions, including temperature and high humidity due to the rainy season, support the efficacy of Lalguard C99 for whitefly control across Brazil.

Other factors can also be considered that will guarantee the effectiveness of Lalguard C99 in controlling whiteflies in soybeans. In previous studies, Boaventura et al. [6] observed that temperature was more detrimental to *C. javanica* virulence against *Bemisia* nymphs than relative humidity. This is probably because many pathogens find sufficient moisture for germination and host penetration within the leaf or insect microclimate boundary layer [29,30]. However, to promote horizontal transmission in whitefly populations that lead to epizootics, widespread infection of *C. javanica* must occur under high humidity conditions, ideal temperatures (25–30 °C), frequent rainfall, and high host densities [76]. *C. javanica* BRM27666 was isolated from a whitefly adult from epizootic on soybeans in Central–Western Brazil. Therefore, in addition to its greater tolerance to high temperatures, probably linked to its geographic origin [36], this strain is also highly adapted to soybean cultivation conditions. Another factor that may benefit the development of epizootics caused by *C. javanica* in whitefly populations in soybean is the microclimate conditions achieved by dew formation, creating a humid environment near the leaf surface. When plants are sown with 45 cm between rows, dew formation reaches between 1.000 and 1.500 cm^2 of leaf area [31], and this microclimate is excellent for promoting fungal development.

Soybean is the main crop that hosts whiteflies as well as one of the main commodities grown in Brazil [11,77]. Therefore, the spatial prediction of suitable areas for the use of Lalguard C99 in this study was based on climatic conditions, mainly temperature, observed during the soybean growing season in the main producing regions of the country. For other economically important crops that are also highly affected by *B. tabaci*, the climatic conditions during each growing season must be considered. For example, *B. tabaci* is one of the main pests affecting melon crops, particularly concentrated in the northeastern region in the states of Ceará and Rio Grande do Norte. In these regions, melon is cultivated throughout the year due to favorable climatic conditions, including high temperatures (ranging from 25 to 35 °C), low relative air humidity, and long periods of exposure to the sun, in addition to minimal rainfall, which is less of a limiting factor in production [78–80].

5. Conclusions

Our study provides a comprehensive understanding of the effects of both constant and fluctuating temperatures on the growth, conidial production, and virulence of *C. javanica* against *B. tabaci*. While constant temperatures clearly delineate the optimal range for fungal growth and infection (25–30 °C), fluctuating temperatures that replicate field conditions do not compromise fungal virulence, even when exposed to extreme temperatures (15 °C and 35 °C) for 6 h daily. However, it is important to recognize that the success of fungal infection is influenced by other factors, including solar radiation, humidity, rainfall, the insect's immune response, and its associated microbiota. Considering the soybean growing season in Brazil, temperature-dependent models predicted that the bioinsecticide Lalguard C99 BRM27666 will be effective against *B. tabaci* in all Brazilian regions based on its efficacy across a wide range of temperature regimes (for growth and conidiogenesis), speed of kill (LT_{50}), and virulence against this pest. The results of this study replicate field conditions, as Lalguard C99 has demonstrated high efficacy in controlling whiteflies in commercial areas in all Brazilian regions. To further support the findings of our research, monthly experiments are being conducted under varying solar radiation, temperature, and moisture conditions.

Supplementary Materials: The following supporting information can be downloaded at https://www.mdpi.com/article/10.3390/jof11020125/s1. Figure S1: Number of hours (represented by bars) that temperatures remain within the ranges of <15 °C, 16–20 °C, 21–25 °C, 26–30 °C, 31–35 °C, and >35 °C, and relative humidity (illustrated by a blue line), for the months of January through December 2017 in Brazil for the city and state of Vilhena, Rondônia (A); Rio Verde, Goiás (B); Sorriso,

Mato Grosso (C); Formoso do Araguaia, Tocantins (D); Balsas, Maranhão (E); Bom Jesus, Piauí (F); Luis Eduardo Magalhães, Bahia (G); Castro, Paraná (H); Itapeva, São Paulo (I); Tupanciretã, Rio Grande do Sul (J); Unaí, Minas Gerais (K); and Brasília, Distrito Federal (L); Figure S2: Temperature (°C) and relative humidity (%) recorded at one-hour intervals during the mycelial growth and conidial production experiments at constant temperatures of 10 (A), 15 (B), 20 (C), 25 (D), 30 (E) and 35 °C (F); Figure S3: Temperature (°C) and relative humidity (%) recorded at one-hour intervals during the mycelial growth and conidial production experiments held six hours at 15 (A), 20 (B), 25 (C), 30 (D) and 35 °C (E) and then transferred for BOD to 27.4 °C (F) for 10 days; Figure S4: Temperature (°C) and relative humidity (%) recorded at one-hour intervals during the mycelial growth and conidial production experiments held six hours at 15 (A), 20 (B), 25 (C), 30 (D) and 35 °C (E) alternating with 18 h at room temperature (F) for 10 days; Figure S5: Temperature (°C) and relative humidity (%) recorded at one-hour intervals during the virulence experiment held six hours at 15 (A), 20 (B), 30 (C) and 35 °C (D) and then transferred for room temperature (F) for 7 days; Figure S6: Temperature (°C) and relative humidity (%) recorded at one-hour intervals during the virulence experiment held six hours at 15 (A), 20 (B), 25 (C), 30 (D) and 35 °C (E) alternating with 18 h at room temperature (F) for 7 days; Figure S7: Temperature (°C) and relative humidity (%) recorded at one-hour intervals during the virulence experiment at constant temperatures of 15 (A), 20 (B), 25 (C), 30 (D), 35 °C (E) and room temperature (F); Table S1: p-values ($p \leq$ value) of the comparisons of confirmed mortality curves for *Bemisia tabaci* nymphs after treatment by *Cordyceps javanica* at different temperatures. Curves were considered significant different at $p \leq 0.05$.

Author Contributions: H.A.B., E.D.Q., T.A.G., K.P. and A.F.M. designed the studies; H.A.B. and L.A.Q. performed the bioassays; H.A.B. and J.F.A.S. analyzed the data; H.A.B. and E.D.Q. wrote the manuscript; T.A.G., K.P. and A.F.M. reviewed the manuscript. All authors have read and agreed to the published version of the manuscript.

Funding: This work was made possible through financial support from the FAPED/EMBRAPA/LALLEMAND grant n° 20200.20/0118-2-03, under the project "Development of a *Cordyceps (=Isaria) javanica*-based biopesticide for the control of whitefly, *Bemisia tabaci*, and other pests", and the National Council for Scientific and Technological Development (CNPq) grant n° 141708/2020-4, which supported the scholarship awarded to the first author.

Institutional Review Board Statement: Not applicable.

Informed Consent Statement: Not applicable.

Data Availability Statement: The original contributions presented in this study are included in the article/Supplementary Materials. Further inquiries can be directed to the corresponding authors.

Acknowledgments: The authors thank Ritter Yvan Guimapi for his invaluable technical support with modeling using the EPFA software and for preparing the temperature maps. Additionally, we extend our gratitude to Edson Djalma for his essential assistance in conducting the experiments.

Conflicts of Interest: The authors declare no conflicts of interest.

References

1. Wang, W.; Wang, Y.; Dong, G.; Chen, F. Development of *Cordyceps javanica* BE01 with enhanced virulence against *Hyphantria cunea* using polyethylene glycol-mediated protoplast transformation. *Front. Microbiol.* **2022**, *13*, 972425. [CrossRef] [PubMed]
2. Quintela, E.D.; Abreu, A.G.; Lima, J.F.D.S.; Mascarin, G.M.; Santos, J.B.D.; Brown, J.K. Reproduction of the whitefly *Bemisia tabaci* (Hemiptera: Aleyrodidae) B biotype in maize fields (*Zea mays* L.) in Brazil. *Pest Manag. Sci.* **2016**, *72*, 2181–2187. [CrossRef] [PubMed]
3. Lopes, R.B.; de Souza, D.A.; Inglis, P.W.; Faria, M. Diversity of anamorphic *Cordyceps* (formerly *Isaria*) isolated from Brazilian agricultural sites. *J. Invertebr. Pathol.* **2023**, *200*, 107956. [CrossRef] [PubMed]
4. Santos, T.T.M.D.; Quintela, E.D.; Mascarin, G.M.; Santana, M.V. Enhanced mortality of *Bemisia tabaci* nymphs by *Isaria javanica* combined with sublethal doses of chemical insecticides. *J. Appl. Entomol.* **2018**, *142*, 598–609. [CrossRef]

5. Mascarin, G.M.; Pereira-Junior, R.A.; Fernandes, É.K.K.; Quintela, E.D.; Dunlap, C.A.; Arthurs, S.P. Phenotype responses to abiotic stresses, asexual reproduction and virulence among isolates of the entomopathogenic fungus *Cordyceps javanica* (Hypocreales: Cordycipitaceae). *Microbiol. Res.* **2018**, *216*, 12–22. [CrossRef]
6. Boaventura, H.A.; Quintela, E.D.; Santos, E.N.; Silva, J.F.A.; Humber, R.A. Susceptibility of all nymphal stages of *Bemisia tabaci* Biotype B (Hemiptera: Aleyrodidae) to three Brazilian isolates of *Cordyceps* sp. (Hypocreales: Cordycipitaceae) in a screenhouse under variable temperature and moisture conditions. *Neotrop. Entomol.* **2021**, *50*, 100–113. [CrossRef]
7. AGROFIT. Sistemas de Agrotóxicos Fitossanitários. Available online: http://agrofit.agricultura.gov.br/agrofit_cons/principal_agrofit_cons (accessed on 6 April 2023).
8. Inoue-Nagata, A.K.; Lima, M.F.; Gilbertson, R.L. A review of geminivirus diseases in vegetables and other crops in Brazil: Current status and approaches for management. *Hortic. Bras.* **2016**, *34*, 8–18. [CrossRef]
9. Paredes-Montero, J.R.; Rizental, M.; Quintela, E.D.; Abreu, A.G.D.; Brown, J.K. Earlier than expected introductions of the *Bemisia tabaci* B mitotype in Brazil reveal an unprecedented, rapid invasion history. *Ecol. Evol.* **2022**, *12*, e8557. [CrossRef]
10. Paredes-Montero, J.R.; Haq, Q.I.; Mohamed, A.A.; Brown, J.K. Phylogeographic and SNPs analyses of *Bemisia tabaci* B mitotype populations reveal only two of eight haplotypes are invasive. *Biology* **2021**, *10*, 1048. [CrossRef]
11. Bergamin Filho, A.; Inoue-Nagata, A.K.; Bassanezi, R.B.; Belasque, J.; Amorim, L.; Macedo, M.A.; Barbosa, J.C.; Willocquet, L.; Savary, S. The importance of primary inoculum and area-wide disease management to crop health and food security. *Food Secur.* **2016**, *8*, 221–238. [CrossRef]
12. CONAB. Companhia Nacional de Abastecimento. Boletim da Safra de Grãos. Available online: https://www.conab.gov.br/info-agro/safras/graos (accessed on 24 May 2023).
13. Navas-Castillo, J.; Fiallo-Olivé, E.; Sánchez-Campos, S. Emerging virus diseases transmitted by whiteflies. *Annu. Rev. Phytopathol.* **2011**, *49*, 219–248. [CrossRef] [PubMed]
14. Almeida, Á.M.; Piuga, F.F.; Marin, S.R.; Kitajima, E.W.; Gaspar, J.O.; Oliveira, T.G.D.; Moraes, T.G.D. Detection and partial characterization of a Carlavirus causing stem necrosis of soybean in Brazil. *Fitopatol. Bras.* **2005**, *30*, 191–194. [CrossRef]
15. Zanardo, L.G.; Carvalho, C.M. Cowpea mild mottle virus (Carlavirus, Betaflexiviridae): A review. *Trop. Plant Pathol.* **2017**, *42*, 417–430. [CrossRef]
16. Bello, V.H.; Watanabe, L.F.M.; Fusco, L.M.; De Marchi, B.R.; Da Silva, F.B.; Gorayeb, E.S.; Moura, M.F.; Souza, I.M.; Muller, C.; Sanhueza, S.; et al. Outbreaks of *Bemisia tabaci* Mediterranean species in vegetable crops in São Paulo and Paraná States, Brazil. *Bull. Entomol. Res.* **2020**, *110*, 487–496. [CrossRef] [PubMed]
17. Vieira, S.S.; Bueno, A.F.; Boff, M.I.C.; Bueno, R.C.O.F.; Hoffman-Campo, C.B. Resistance of soybean genotypes to *Bemisia tabaci* (Genn.) biotype B (Hemiptera: Aleyrodidae). *Neotrop. Entomol.* **2011**, *40*, 117–122. [CrossRef]
18. Albergaria, N.M.; Cividanes, F.J. Thermal Requirements of *Bemisia tabaci* (Genn.) B-biotype (Hemiptera: Aleyrodidae). *Neotrop. Entomol.* **2002**, *31*, 359–363. [CrossRef]
19. McCoy, C.; Quintela, E.D.; Faria, M. Environmental persistence of entomopathogenic fungi. In *Factors Affecting the Survival of Entomopathogens*; Baur, M.E., Fuxa, J.R., Eds.; Southern Cooperative Series Bulletin: Oklahoma City, OK, USA, 2002; Volume 400, pp. 1–9.
20. Jaronski, S.T. Ecological factors in the inundative use of fungal entomopathogens. *BioControl* **2010**, *55*, 159–185. [CrossRef]
21. Fernandes, E.K.K.; Rangel, D.E.N.; Braga, G.U.L.; Roberts, D.W. Tolerance of entomopathogenic fungi to ultraviolet radiation: A review on screening of strains and their formulation. *Curr. Genet.* **2015**, *61*, 427–440. [CrossRef]
22. Abdul Qayyum, M.; Bilal, H.; Ali, H.; Raza, H.; Wajid, M. Factors affecting the epizootics of entomopathogenic fungi-A review. *J. Bioresour. Manag.* **2021**, *8*, 5. [CrossRef]
23. Quesada-Moraga, E.; González-Mas, N.; Yousef-Yousef, M.; Garrido-Jurado, I.; Fernández-Bravo, M. Key role of environmental competence in successful use of entomopathogenic fungi in microbial pest control. *J. Pest Sci.* **2024**, *97*, 1–15. [CrossRef]
24. Ferron, P. Influence of relative humidity on the development of fungal infection caused by *Beauveria bassiana* (Fungi Imperfecti, Moniliales) in imagines of *Acanthoscelides obtectus* (Col.: Bruchidae). *Entomophaga* **1977**, *22*, 393–396. [CrossRef]
25. Riba, G.; Marcandier, S. Influence de l'humidité relative sur l'agressivité et la viabilité des souches de *Beauveria bassiana* (Bals.) Vuillemin et de *Metarhizium anisopliae* (Metsch.) Sorokin, hyphomycétes pathogènes de la pyrale du maïs, *Ostrinia nubilalis* Hubn. *Agronomie* **1984**, *4*, 189–194. [CrossRef]
26. Ramoska, W.A. The influence of relative humidity on *Beauveria bassiana* infectivity and replication in the chinch bug, *Blissus leucopterus*. *J. Invertebr. Pathol.* **1984**, *43*, 389–394. [CrossRef]
27. Marcandier, S.; Khachatourians, G.G. Susceptibility of the migratory grasshopper, *Melanoplus sanguinipes* (Fab.) (Orthoptera: Acrididae), to *Beauveria bassiana* (Bals.) Vuillemin (Hyphomycete): Influence of relative humidity. *Can. Entomol.* **1987**, *119*, 901–907. [CrossRef]
28. Fargues, J.; Goettel, M.S.; Smits, N.; Ouedraogo, A.; Rougier, M. Effect of temperature on vegetative growth of *Beauveria bassiana* isolates from different origins. *Mycologia* **1997**, *89*, 383–392. [CrossRef]

29. Boulard, T.; Mermier, M.; Fargues, J.; Smits, N.; Rougier, M.; Roy, J.C. Tomato leaf boundary layer climate: Implications for microbiological whitefly control in greenhouses. *Agr. Forest Meteorol.* **2002**, *110*, 159–176. [CrossRef]
30. Wraight, S.P.; Carruthers, R.I.; Jaronski, S.T.; Bradley, C.A.; Garza, C.J.; Galaini-Wraight, S. Evaluation of the entomopathogenic fungi *Beauveria bassiana* and *Paecilomyces fumosoroseus* for microbial control of the silverleaf whitefly, *Bemisia argentifolii*. *Biol. Control* **2000**, *17*, 203–217. [CrossRef]
31. Sosa-Gómez, D.R. Intraspecific variation and population structure of the velvetbean caterpillar, *Anticarsia gemmatalis* Hubner 1818 (Insecta: Lepidoptera Noctuidae). *Genet. Mol. Biol.* **2004**, *27*, 378–384. [CrossRef]
32. Zimmermann, G. Effect of high temperatures and artificial sunlight on the viability of conidia of *Metarhizium anisopliae*. *J. Invertebr. Pathol.* **1982**, *40*, 36–40. [CrossRef]
33. Cabanillas, H.E.; Jones, W.A. Effects of temperature and culture media on vegetative growth of an entomopathogenic fungus *Isaria* sp. (Hypocreales: Clavicipitaceae) naturally affecting the whitefly, *Bemisia tabaci* in Texas. *Mycopathologia* **2009**, *167*, 263–271. [CrossRef]
34. Fargues, J.; Maniania, N.K.; Delmas, J.C.; Smits, N. Influence de la température sur la croissance in vitro d'hyphomycétes entomopathogénes. *Agronomie* **1992**, *12*, 557–564. [CrossRef]
35. Mietkiewski, R.; Tkaczuk, C.; Zurek, M.; Van der Geest, L.P.S. Temperature requirement of four entomopathogenic fungi. *Acta Mycol.* **1994**, *29*, 109–120. [CrossRef]
36. Vidal, C.; Fargues, J.; Lacey, L.A. Intraspecific variability of *Paecilomyces fumosoroseus*: Effect of temperature on vegetative growth. *J. Invertebr. Pathol.* **1997**, *70*, 18–26. [CrossRef]
37. Fargues, J.; Bon, M.C. Influence of temperature preferences of two *Paecilomyces fumosoroseus* lineages on their co-infection pattern. *J. Invertebr. Pathol.* **2004**, *87*, 94–104. [CrossRef]
38. INMET. Instituto Nacional de Meteorologia. Boletim Meteorológico. Available online: https://www.inmet.gov.br/portal/ (accessed on 25 September 2024).
39. INMET. Instituto Nacional de Meteorologia. Estação Meteorológica de Observação de Superfície Automática. Available online: https://portal.inmet.gov.br/servicos/esta%C3%A7%C3%B5es-autom%C3%A1ticas (accessed on 22 August 2017).
40. IBGE. Instituto Brasileiro de Geografia e Estatística. Available online: https://sidra.ibge.gov.br/home/lspa/brasil (accessed on 30 August 2017).
41. Quintela, E.D. Manejo integrado dos insetos e outros invertebrados pragas do feijoeiro. *Inf. Agropec.* **2004**, *25*, 113–136.
42. Guimapi, R.A.; Mohamed, S.A.; Biber-Freudenberger, L.; Mwangi, W.; Ekesi, S.; Borgemeister, C.; Tonnang, H.E.Z. Decision support system for fitting and mapping nonlinear functions with application to insect pest management in the biological control context. *Algorithms* **2020**, *13*, 104. [CrossRef]
43. Ritz, C.; Baty, F.; Streibig, J.C.; Gerhard, D. Dose-response analysis using R. *PLoS ONE* **2015**, *10*, e0146021. [CrossRef]
44. Archontoulis, S.V.; Miguez, F.E. Nonlinear regression models and applications in agricultural research. *Agron. J.* **2015**, *107*, 786–798. [CrossRef]
45. Ratkowsky, D.A.; Olley, J.; McMeekin, T.A.; Ball, A. Relationship between temperature and growth rate of bacterial cultures. *J. Bacteriol.* **1982**, *149*, 1–5. [CrossRef]
46. Ratkowsky, D.A.; Lowry, R.K.; McMeekin, T.A.; Stokes, A.N.; Chandler, R.E. Model for bacterial culture growth rate throughout the entire biokinetic temperature range. *J. Bacteriol.* **1983**, *154*, 1222–1226. [CrossRef]
47. Shi, P.; Ge, F.; Sun, Y.; Chen, C. A simple model for describing the effect of temperature on insect developmental rate. *J. Asia-Pacific Entomol.* **2011**, *14*, 15–20. [CrossRef]
48. Shi, P.J.; Reddy, G.V.; Chen, L.; Ge, F. Comparison of thermal performance equations in describing temperature-dependent developmental rates of insects: (I) empirical models. *Ann. Entomol. Soc. Am.* **2015**, *109*, 211–215. [CrossRef]
49. Ekesi, S.; Maniania, N.K.; Ampong-Nyarko, K. Effect of temperature on germination, radial growth and virulence of *Metarhizium anisopliae* and *Beauveria bassiana* on *Megalurothrips sjostedti*. *Biocontrol Sci. Technol.* **1999**, *9*, 177–185. [CrossRef]
50. Shimazu, M. Effects of temperature on growth of *Beauveria bassiana* F-263, a strain highly virulent to the Japanese pine sawyer, *Monochamus alternatus*, especially tolerance to high temperatures. *Appl. Entomol. Zool.* **2004**, *39*, 469–475. [CrossRef]
51. Borisade, O.A.; Magan, N. Growth and sporulation of entomopathogenic *Beauveria bassiana*, *Metarhizium anisopliae*, *Isaria farinosa* and *Isaria fumosorosea* strains in relation to water activity and temperature interactions. *Biocontrol Sci. Technol.* **2014**, *24*, 999–1011. [CrossRef]
52. Wu, S.; Toews, M.D.; Oliveira-Hofman, C.; Behle, R.W.; Simmons, A.M.; Shapiro-Ilan, D.I. Environmental tolerance of entomopathogenic fungi: A new strain of *Cordyceps javanica* isolated from a whitefly epizootic versus commercial fungal strains. *Insects* **2020**, *11*, 711. [CrossRef]
53. Lee, J.H.; Lee, Y.S.; Kim, Y.C. Effects of temperature and culture media composition on sporulation, mycelial growth, and antifungal activity of *Isaria javanica* pf185. *Res. Plant Dis.* **2021**, *27*, 99–106. [CrossRef]
54. Xie, L.; Han, J.H.; Kim, J.J.; Lee, S.Y. Effects of culture conditions on conidial production of the sweet potato whitefly pathogenic fungus *Isaria javanica*. *Mycoscience* **2015**, *57*, 64–70. [CrossRef]

55. Crisan, E.V. Current concepts of thermophilism and the thermophilic fungi. *Mycologia* **1973**, *65*, 1171–1198. [CrossRef]
56. Rangel, D.E.N.; Braga, G.Ú.L.; Anderson, A.J.; Roberts, D.W. Variability in conidial thermotolerance of Metarhizium anisopliae isolates from different geographic origins. *J. Invertebr. Pathol.* **2005**, *88*, 116–125. [CrossRef]
57. Rangel, D.E.N.; Fernandes, E.K.K.; Dettenmaier, S.J.; Roberts, D.W. Thermotolerance of germlings and mycelium of the insect-pathogenic fungus Metarhizium spp. and mycelial recovery after heat stress. *J. Basic Microbiol.* **2010**, *50*, 344–350. [CrossRef] [PubMed]
58. Souza, R.K.F.; Azevedo, R.F.F.; Lobo, A.O.; Rangel, D.E.N. Conidial water affinity is an important characteristic for thermotolerance in entomopathogenic fungi. *Biocontrol Sci. Technol.* **2014**, *24*, 448–461. [CrossRef]
59. Mora, M.A.E.; Castilho, A.M.C.; Fraga, M.E. Classification and infection mechanism of entomopathogenic fungi. *Arq. Inst. Biol.* **2017**, *84*, e0552015. [CrossRef]
60. Ma, M.; Luo, J.; Li, C.; Eleftherianos, I.; Zhang, W.; Xu, L. A life-and-death struggle: Interaction of insects with entomopathogenic fungi across various infection stages. *Front. Immunol.* **2024**, *14*, 1329843. [CrossRef]
61. Hong, S.; Shang, J.; Sun, Y.; Tang, G.; Wang, C. Fungal infection of insects: Molecular insights and prospects. *Trends Microbiol.* **2024**, *32*, 302–316. [CrossRef]
62. Hajek, A.E.; Eilenberg, J. Ecological Basis for Use of Predators, Parasitoids, and Pathogens to Control Pests. In *Natural Enemies*; Cambridge University Press: Cambridge, UK, 2018; Volume 2, pp. 109–136.
63. Bustillo, A.E.; Cárdenas, R.; Posada, F.J. Natural enemies and competitors of *Hypothenemus hampei* (Ferrari) (Coleoptera: Scolytidae) in Colombia. *Neotrop. Entomol.* **2002**, *31*, 35–639. [CrossRef]
64. Estrada, M.E.; Romero, M.; Rivero, M.J.; Barroso, F. Presencia natural de *Beauveria bassiana* (Bals.) Vuill. en el cultivo de la caña de azúcar (*Saccharum* sp. híbrido) en Cuba. *Rev. Iberoam. Micol.* **2004**, *21*, 42–43.
65. Moorhouse, E.R.; Gillespie, A.T.; Charnley, A.K. The influence of temperature on the susceptibility of vine weevil, *Otiorhynchus sulcatus* (Fabricius)(Coleoptera: Curculionidae), larvae to *Metarhizium anisopliae* (Deuteromycotina: Hyphomycetes). *Ann. Appl. Biol.* **1994**, *124*, 185–193. [CrossRef]
66. Smits, N.; Fargues, J.; Rougier, M.; Goujet, R.; Itier, B. Effects of temperature and solar radiation interactions on the survival of quiescent conidia of the entomopathogenic hyphomycete *Paecilomyces fumosoroseus* (Wize) Brown and Smith. *Mycopathologia* **1996**, *135*, 163–170. [CrossRef]
67. Rosenzweig, C.; Iglesius, A.; Yang, X.B.; Epstein, P.R.; Chivian, E. *Climate Change and Extreme Weather Events-Implications for Food Production, Plant Diseases, and Pests*; NASA: Greenbelt, MA, USA, 2001.
68. Willmer, P.G.; Hughes, J.P.; Woodford, J.A.T.; Gordon, S.C. The effects of crop microclimate and associated physiological constraints on the seasonal and diurnal distribution patterns of raspberry beetle (*Byturus tomentosus*) on the host plant *Rubus idaeus*. *Ecol. Entomol.* **1996**, *21*, 87–97. [CrossRef]
69. Kersebaum, K.C.; Stöckle, C.O. Simulation of climate change effects on evapotranspiration, photosynthesis, crop growth, and yield processes in crop models. In *Modeling Processes and Their Interactions in Cropping Systems: Challenges for the 21st Century*; Wiley: Hoboken, NJ, USA, 2022; pp. 291–331.
70. Qiu, B.; Ren, S.; Mandour, N.S.; Lin, L. Effect of temperature on the development and reproduction of *Bemisia tabaci* B biotype (Homoptera: Aleyrodidae). *Insect Sci.* **2003**, *10*, 43–49.
71. Reis, L.; Santos e Silva, C.M.; Bezerra, B.; Mutti, P.; Spyrides, M.H.; Silva, P.; Magalhães, T.; Ferreira, R.; Rodrigues, D.; Andrade, L. Influence of climate variability on soybean yield in MATOPIBA, Brazil. *Atmosphere* **2020**, *11*, 1130. [CrossRef]
72. Guan, J.; Gai, Y.; Guan, Y.; Adnan, R.; Zhao, Q.; Xie, Z.; Athar, M.; Zhang, S.; Zhang, Z.; Zhao, Z.; et al. Improvement of heat stress tolerance in soybean (*Glycine max* L), by using conventional and molecular tools. *Front. Plant Sci.* **2022**, *13*, 993189.
73. Klass, J.I.; Blanford, S.; Thomas, M.B. Development of a model for evaluating the effects of environmental temperature and thermal behaviour on biological control of locusts and grasshoppers using pathogens. *Agric. For. Entomol.* **2007**, *9*, 189–199. [CrossRef]
74. Guimarães, F.D.S.; Rezende, P.M.D.; Castro, E.M.D.; Carvalho, E.D.A.; Andrade, M.J.B.D.; Carvalho, E.R. Soybean cultivars (*Glycine max* (L.) Merrill) for summer culture in Lavras MG region. *Ciênc. e Agrotec.* **2008**, *32*, 1099–1106. [CrossRef]
75. Agbessenou, A.; Akutse, K.S.; Yusuf, A.A.; Wekesa, S.W.; Khamis, F.M. Temperature-dependent modelling and spatial prediction reveal suitable geographical areas for deployment of two *Metarhizium anisopliae* isolates for *Tuta absoluta* management. *Sci. Rep.* **2021**, *11*, 23346. [CrossRef]
76. Hajek, A.E.; Meyling, N.V. Fungi. In *Ecology of Invertebrate Diseases*, 1st ed.; Hajek, A.E., Shapiro-Ilan, D.I., Eds.; John Wiley & Sons: Hoboken, NJ, USA, 2018; pp. 327–377.
77. Schutze, I.X.; Yamamoto, P.T.; Malaquias, J.B.; Herritt, M.; Thompson, A.; Merten, P.; Naranjo, S.E. Correlation-based network analysis of the influence of *Bemisia tabaci* feeding on photosynthesis and foliar sugar and starch composition in soybean. *Insects* **2022**, *13*, 56. [CrossRef]
78. Nunes, G.H.D.S.; Andrade Neto, R.D.C.; Costa Filho, J.H.D.; Melo, S.B.D. Influência de variáveis ambientais sobre a interação genótipos x ambientes em meloeiro. *Rev. Bras. Frutic.* **2011**, *33*, 1194–1199. [CrossRef]

79. Vanomark, G.M.M.D.S.; Espínola Sobrinho, J.; Bezerra, J.R.C.; Santos, C.A.C.D.; Azevedo, P.V.D.; Silva, S.T.A.D.; Bezerra, B.G. Energy balance partitioning and evapotranspiration from irrigated muskmelon under semi-arid conditions. *Bragantia* **2018**, *77*, 168–180. [CrossRef]
80. Melo, T.K.D.; Espínola Sobrinho, J.O.S.E.; Medeiros, J.F.D.; Figueiredo, V.B.; Silva, J.S.D.; Sá, F.V.D.S. Impacts of climate change scenarios in the Brazilian semiarid region on watermelon cultivars. *Rev. Caatinga* **2020**, *33*, 794–802. [CrossRef]

Journal of *Fungi*

MDPI

Article

MaNrtB, a Putative Nitrate Transporter, Contributes to Stress Tolerance and Virulence in the Entomopathogenic Fungus *Metarhizium acridum*

Jia Wang [1,2,3,4], Yuneng Zou [1,2,3,4], Yuxian Xia [1,2,3,4,*] and Kai Jin [1,2,3,4,*]

1 Genetic Engineering Research Center, School of Life Sciences, Chongqing University, Chongqing 401331, China; 202326021033@cqu.edu.cn (J.W.)
2 Chongqing Engineering Research Center for Fungal Insecticide, Chongqing 401331, China
3 Key Laboratory of Gene Function and Regulation Technologies Under Chongqing Municipal Education Commission, Chongqing 401331, China
4 National Engineering Research Center of Microbial Pesticides, Chongqing 401331, China
* Correspondence: yuxianxia@cqu.edu.cn (Y.X.); jinkai@cqu.edu.cn (K.J.)

Abstract: Nitrogen is an essential nutrient that frequently determines the growth rate of fungi. Nitrate transporter proteins (Nrts) play a crucial role in the cellular absorption of nitrate from the environment. Entomopathogenic fungi (EPF) have shown their potential in the biological control of pests. Thus, comprehending the mechanisms that govern the pathogenicity and stress tolerance of EPF is helpful in improving the effectiveness and practical application of these fungal biocontrol agents. In this study, we utilized homologous recombination to create *MaNrtB* deletion mutants and complementation strains. We systematically investigated the biological functions of the nitrate transporter protein gene *MaNrtB* in *M. acridum*. Our findings revealed that the disruption of *MaNrtB* resulted in delayed conidial germination without affecting conidial production. Stress tolerance assays demonstrated that the *MaNrtB* disruption strain was more vulnerable to UV-B irradiation, hyperosmotic stress, and cell wall disturbing agents, yet it exhibited increased heat resistance compared to the wild-type strain. Bioassays on the locust *Locusta migratoria manilensis* showed that the disruption of *MaNrtB* impaired the fungal virulence owing to the reduced appressorium formation on the insect cuticle and the attenuated growth in the locust hemolymph. These findings provide new perspectives for understanding the pathogenesis of EPF.

Keywords: entomopathogenic fungi; *Metarhizium acridum*; *MaNrtB*; stress tolerance; virulence

1. Introduction

Numerous insect pests are prevalent in the natural environment, presenting significant challenges to agriculture, ecological balance, and human well-being [1]. Entomopathogenic fungi (EPF), a group of fungi that can cause infections and death in insects and other arthropods, have emerged as a promising and eco-friendly alternative to chemical insecticides and are extensively utilized as biocontrol agents against insect pests [2]. Well-known EPFs, such as *Metarhizium acridum* and *Beauveria bassiana*, have shown their potential and advantages in the biological control of pests within agricultural ecosystems [3]. Therefore, understanding the mechanisms that govern the pathogenicity and stress tolerance of EPF is helpful in improving the effectiveness and practical application of these fungal biocontrol agents.

In fungi, detecting and effectively utilizing nutrients are crucial for the successful establishment and rapid expansion of fungal colonies, and nitrogen, a critical nutrient, often dictates the growth rate of fungi [4]. The metabolism of nitrogen sources encompasses a range of biochemical pathways and mechanisms, such as the assimilation of ammonia, the reduction of nitrates, the fixation of nitrogen, the synthesis of amino acids, and so on [5]. Within this metabolic network, nitrate metabolism is an important part of nitrogen source metabolism in fungi.

Nitrate transporter proteins (Nrts) are pivotal in the cellular uptake of nitrate from the surroundings, playing a critical role in the assimilation, metabolism, and utilization of nitrate. The genetic exploration of these transporters in eukaryotes began in 1991 with the discovery of the *CrnA* gene, which encodes a transmembrane protein, in *Aspergillus nidulans* [6]. The *CrnA*-deficient strains exhibited nitrate uptake defects in both the conidial and the mycelial growth phases [7]. These high-affinity transport systems are part of the nitrate/nitrite porter family, a unique subset within the major facilitator superfamily (MFS), the largest group of secondary transporters [8]. Most eukaryotic organisms possess multiple Nrt protein variants, each encoded by distinct genes [9]. For instance, in the yeast *Hansenula polymorpha* (syn *Ogataea angusta*), the *YNT1* mutants exhibits a deficiency in the uptake of nitrate [10]. In the pathogenic fungus *Fusarium oxysporum* f. sp. *Lycopersici*, the targeted deletion of the nitrate-transporting gene *Ntr1* significantly impairs fungal growth when nitrate serves as the nitrogen source [11]. Similarly, in *Ustilago maydis* (syn *Mycosarcoma maydis*), the growth of nitrate transporter protein mutants is hindered in media where nitrate is the sole nitrogen source and their pathogenicity is diminished [12]. *Trichoderma atroviride*, a prevalent plant growth-promoting fungus, has been shown to rely on the nitrate transporter CHL1 for nitrogen uptake in MS medium supplemented with NH_4NO_3 [13]. Collectively, these findings underscore the influence of nitrate transporter proteins on fungal growth and pathogenicity. Despite their importance, the characterization of nitrate transporter genes in entomopathogenic fungi remains unexplored. Here, we investigated some biological processes that may be affected by the activity of a putative nitrate transporter, MaNrtB, in *M. acridum*.

2. Materials and Methods

2.1. Strains and Culture Conditions

The wild-type strain of *M. acridum* (WT) has been archived at the China General Microbiological Culture Collection Center (CGMCC) with the accession number 0877. In conducting the genetic engineering and fungal transformation for this study, we employed *Escherichia coli* DH5α (Solarbio, Beijing, China) and *Agrobacterium tumefaciens* AGL1 (Solarbio, Beijing, China), respectively. The fungal strains employed in this investigation were cultivated on a modified Sabouraud dextrose agar medium, specifically a one-quarter-strength formulation (1/4 SDAY). This medium is composed of 1% dextrose, 0.25% mycological peptone, 0.5% yeast extract, and 2% agar, measured by the weight to volume ratio. The cultures were maintained at a temperature of 28 °C for a period of 15 days to ensure the conidia reached maturity [14].

2.2. Bioinformatic Analyses

All the protein sequences were sourced from the NCBI genome database, accessible at https://www.ncbi.nlm.nih.gov/ (accessed on 25 December 2024). The specific nucleic acid sequence (MAC_03189) and protein sequence (XP_007809529.2) were obtained in FASTA format. Utilizing these protein sequences, a selection of diverse filamentous and non-filamentous fungi were made for BLASTp sequence alignment to identify similarities. Protein 3D structure maps were retrieved from the AlphaFold Protein Structure Database at https://alphafold.ebi.ac.uk/ (accessed on 25 December 2024). For constructing the phylogenetic tree, MEGA v7.0 was employed with the neighbor-joining approach. The sequences were aligned using the ClustalW algorithm integrated within MEGA and employing a bootstrap test with 1000 replicates. Another alignment parameter was set to default settings. The aligned sequences were then exported to GeneDoc 2.7 for homology analysis. Domain prediction was conducted using the SMART tool, available at http://smart.embl.de/ (accessed on 25 December 2024), and TMH prediction was performed using TMHMM at https://services.healthtech.dtu.dk/ (accessed on 25 December 2024). Additionally, BioXM v2.7.1 software was used to calculate the molecular weights and isoelectric points of the proteins.

2.3. Fungal Mutant Generation

The *MaNrtB* nucleic acid sequence (2535 bp) and an additional 2000 bp flanking the *MaNrtB* gene on both sides were obtained from the NCBI database. Among them, a 1.3 kb (left border) and a 1.4 kb (right border) flanking sequence of the *MaNrtB* gene locus were amplified using the primer pairs of *MaNrtB*-LF/*MaNrtB*-LR and *MaNrtB*-RF/*MaNrtB*-RR, respectively (Supplementary Table S1). The fragments were then successively ligated into the *Hind*III/*Xba*I and *Eco*RV/*Eco*RI-restricted pK2-PB vectors and verified by PCR using the primers LF/Pt-R and Bar-F/RR [15], respectively. Fungal transformation and validation of the transformants followed the methods described previously [16]. Potential Δ*MaNrtB* mutants were identified on Czapek–Dox (CZA) medium supplemented with 500 μg/mL glufosinate ammonium (Sigma, St. Louis, MO, USA), and the presence of the desired recombinant sequences was confirmed by PCR using the primer pairs of NrtB-VF/Pt-R and NrtB-VR/Bar-F. The pK2-sur vector was employed to generate the *MaNrtB* complemented strains (CPs). Initially, the full-length *eGFP* and *TtrpC* genes were cloned and ligated into the pK2-sur vector to create the pK2-eGFP-sur vector. Then, the promoter and ORF fragments were amplified from the WT genome using the primers NrtB-CP-F and NrtB-CP-R, and ligated to the *Hind*III/*Bam*HI-restricted pK2-eGFP-sur vector by one-step cloning to create the pK2-sur-*MaNrtB* construct [17], which was introduced into the Δ*MaNrtB* strain. Candidate CP strains were selected on CZA medium with 20 μg/mL chlorimuron ethyl (Sigma, Bellefonte, PA, USA), and recombinant events were verified by PCR using the primer pair of NrtB-CP-VF/GFP-VR. Reverse transcription quantitative PCR (RT-qPCR) analysis was performed to confirm the Δ*MaNrtB* and CP strains. The schematic diagrams of the pK2-PB and pK2-sur vectors are shown in Figure S1. A complete list of the primers utilized in this study is provided in Table S1.

2.4. Conidial Germination and Conidial Yield Assays

Conidial suspensions at a concentration of 1×10^7 conidia/mL were prepared with mature and freshly harvested conidia and a 0.05% Tween-80 solution. An aliquot of 50 μL was uniformly spread onto 1/4 SDAY plates. Over a period of 10 h, at 2-h intervals, three 1 cm^2 sections were excised from each strain's plate. The number of germinated conidia was counted for every 100 conidia observed. Germination was defined as the emergence of a bud whose length was more than one-third of the width of the conidium. The germination rate was determined by the following formula: (number of germinated conidia/total number of conidia) $\times$ 100%. For assessing the conidial yield, 2 μL of the conidial suspension (1×10^6 conidia/mL) was deposited into 24-well plates with 1 mL of 1/4 SDAY medium in each plate well. Every three days, the yield of conidia was measured. Briefly, the fungal cultures were transferred to a 2 mL centrifuge tube. After adding 1 mL of 0.05% Tween-80 solution for grinding, the mixture was vortexed. The suspension was then diluted, and the conidial concentration was determined using a hemocytometer. Each experiment was independently performed a minimum of three times to ensure reliability.

2.5. Stress Tolerance Assays

The 1/4 SDAY medium was amended with 0.5 M NaCl, 1 M sorbitol (SOR), 500 μg/mL Congo red (CR), 0.01% *w/v* sodium dodecyl sulfate (SDS), 50 μg/mL Calcofluor white (CFW), and 60 mmol/L H_2O_2 to simulate adverse conditions. A 2 μL aliquot of conidial suspension (1×10^6 conidia/mL) from each strain was spread onto the modified culture plates and incubated at 28 °C for 6 days. The colony growth was then photographed and measured. To evaluate the heat-shock sensitivity of the fungal strains, the conidial suspensions (1×10^7 conidia/mL) were subjected to a water bath at 45 °C for durations of 2, 4, 6, and 8 h. Post-treatment, the suspensions were spread onto 1/4 SDAY medium and incubated at 28 °C for 20 h to determine the germination rates. For the UV-B tolerance of the fungal strains, the conidial suspensions (1×10^7 conidia/mL) were spread on 1/4 SDAY medium and exposed to UV-B radiation at a dose of 1350 mW/m^2 for periods of 0.5, 1.0, 1.5, and 2.0 h. Following exposure, the cultures were incubated at 28 °C for 20 h to assess the conidial germination. Each experiment was conducted in triplicate.

2.6. Bioassays

Two distinct bioassay methods were employed as previously described [15]. Briefly, in the topical inoculation approach, suspensions of the WT, Δ*MaNrtB*, and CP strains, prepared in paraffin oil at a concentration of 1×10^7 conidia/mL, were applied to the fifth-instar nymphs of locusts, *Locusta migratoria manilensis* (5 μL per nymph). For the intrahemocoel injection method, the conidia from the fungal strains were suspended in double-distilled water (ddH_2O) at a concentration of 1×10^6 conidia/mL and injected directly into the locust hemocoel (5 μL per nymph). For both inoculation techniques, each experimental group contained 20 fifth-instar nymphs, and the negative control groups were treated with pure paraffin oil and ddH_2O, respectively. The survival rates of the locusts were monitored and documented at intervals of 12 h throughout the experiment.

To investigate the nutrient utilization efficiency of the various fungal strains, we employed liquid CZA medium along with two modified versions of CZA, SPM and TPM, to assess the fungal development under distinct nutritional conditions. The CZA medium was supplemented with 3% sucrose and 0.3% $NaNO_3$, while the SPM contained 3% sucrose and 0.5% peptone, and the TPM was formulated with 3% trehalose and 0.5% peptone. Among the three media, CZA has nitrate as the main nitrogen source donor and thus can be used to assess the nitrate transport activity of NrtBp in *M. acridum*. TPM is used to mimic insect hemolymph, as described previously [18]. These media were designed to replicate and evaluate the growth of fungi in a variety of nutrient-rich environments. Briefly, a 30 μL sample of a conidial suspension at a concentration of 10^7 conidia/mL was introduced into each 30 mL medium, which was then incubated for 60 h at 28 °C with a shaker speed of 200 rpm to determine the biomass. This procedure was conducted in triplicate for each condition. To evaluate the host immune response, the total RNA was extracted from the fat bodies in the nymphs of locusts after injection to assess the expression levels of two key antimicrobial peptide (AMP) genes, *Attacin* and *Defensin*, using the protocol outlined in a previous study [19]. Each bioassay experiment was repeated three times.

2.7. Appressorium Formation Assays

Conidial suspensions for each strain were prepared in sterilized ddH_2O to a density of 1×10^7 conidia/mL. The sterilized locust hindwings were first coated with the conidial suspensions and then carefully spread onto pristine microscope slides for incubation at a controlled temperature of 28 °C. The conidia germination rate was monitored and recorded every two hours. Additionally, after a 12 h incubation period, the appressorium formation rate was also counted. This experimental procedure was conducted in triplicate to ensure reliability.

To assess the fungal penetration capability, we placed sterilized locust hindwings on 1/4 SDAY plates. A 2 μL aliquot of conidial suspension from each fungal strain, diluted to a concentration of 1×10^6 conidia/mL, was dropped on the wings carefully. These were then incubated at 28 °C for 44 h. Following photography and documentation, the wings were removed to allow continued incubation for an additional four days, with photographs taken to document the colony morphology. As a control, conidial suspensions were also directly inoculated onto 1/4 SDAY medium and incubated at 28 °C for a duration of 6 days. Each treatment contained 10 wings and these assays were repeated three times.

2.8. Data Analyses

A one-way ANOVA test was conducted to evaluate the phenotypic estimates, followed by Tukey's honestly significant difference (HSD) test to ascertain the significance of any observed differences among the groups.

3. Results

3.1. Features of MaNrtB and Mutant Generation

Bioinformatics analysis revealed that there is only one putative Nrt gene, designated as *MaNrtB*, in *M. acridum*. The *MaNrtB* gene spans 1612 base pairs and encodes a protein of 502 amino acids, distributed across two exons. The calculated molecular weight for the *MaNrtB* protein is 53.75 kDa, with an isoelectric point of 7.8. Domain analysis, conducted via the SMART online tool, indicated the presence of an MFS domain. As with other reported Nrt homologues [6,11,12,20–22], the tertiary structure of NrtB in *M. acridum* putatively contains 12 transmembrane-spanning alpha helices (TMHs)that are colored in blue with a less conserved intracellular central loop (Figures 1A and S2). The sequence alignment of the orthologous NrtB proteins revealed that these TMH domains are highly conserved, whereas the central loop region exhibits a lack of order (Figure S3). Additionally, phylogenetic analysis grouped *MaNrtB* together with several of its homologs that have been evidenced to possess nitrate transport capacity, such as *A. nidulans* [6] and *F. oxysporum* [11] (Figure 1B), and these domains are notably conserved in filamentous fungi and exhibit lower conservation in yeast (Figure 1B). Overall, the significant similarity in the tertiary structures of these proteins suggests that they likely have similar functions. *MaNrtB*-knockout transformants were generated through homologous recombination. The CP strain was derived using the ectopic strategy (Figure 2B). The Δ*MaNrtB* and CP strains were validated by PCR (Figure S4). Furthermore, the expression levels of the *MaNrtB* gene in the WT, Δ*MaNrtB*, and CP strains were quantified using RT-qPCR. The expression levels of *MaNrtB* in the WT and CP strains were markedly higher than in the Δ*MaNrtB* strain, confirming the successful generation of both the Δ*MaNrtB* and CP strains (Figure 2C).

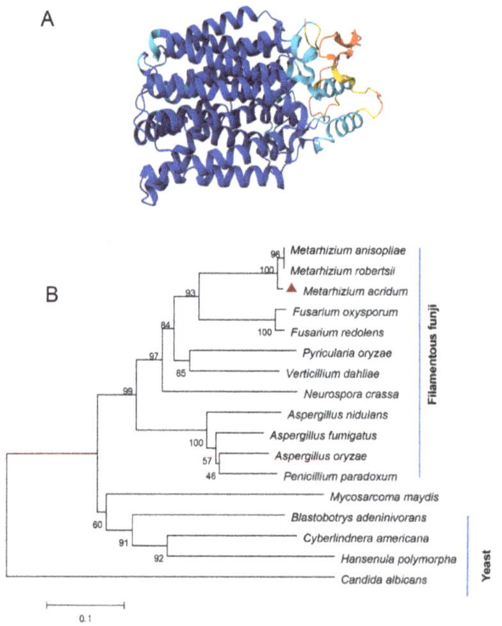

Figure 1. Bioinformatics of *MaNrtB*. (**A**) Domain analysis of *MaNrtB*, which contains 12 transmembrane (TM) spanning alpha helices marked in a blue color. (**B**) A phylogenetic tree was constructed with the NrtB protein sequences of *Metarhizium acridum* (EFY90826.1), *Metarhizium anisopliae* (KAF5125400.1), *Metarhizium robertsii* (XP_007820472.2), *Fusarium oxysporum* (XP_018232784.1), *Fusarium redolens* (XP_046056618.1), *Pyricularia oryzae* (XP_003710887.1), *Verticillium dahlia* (KAF3360447.1), *Neurospora crassa* (XP_957430.2), *Aspergillus nidulans* (XP_658612.1), *Aspergillus fumigatus* (KAH1494629.1), *Aspergillus oryzae* (EIT78908.1), *Penicillium paradoxum* (XP_057035394.1), *Mycosarcoma maydis* (XP_011390345.1), *Blastobotrys adeninivorans* (CAQ77149.1), *Cyberlindnera americana* (QFR37177.1), *Hansenula polymorpha* (XP_018213600.1), and *Candida albicans* (XP_717110.1). The red triangle represents the NrtB homologous protein in *M. acridum*.

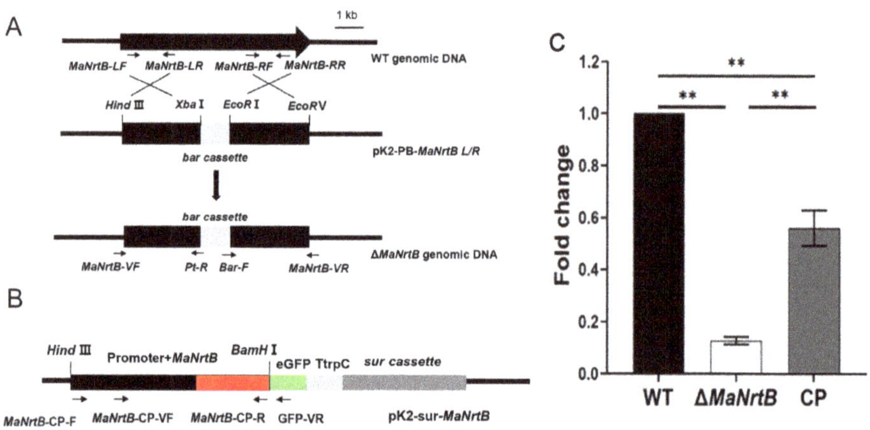

Figure 2. Disruption and complementation of *MaNrtB*. (**A**,**B**) Schematic diagrams of the knockout and complementation vector constructions. Black arrows indicate the positions of the primers. (**C**) Verification of transformants by RT-qPCR. Error bars = mean ± SEM. Asterisks indicate a significant difference at (**) $p < 0.01$.

3.2. Disruption of MaNrtB Impaired Conidial Germination but Not Conidial Yield

Our initial assessment focused on the germination rates and conidial yields of the different fungal strains. It was observed that the Δ*MaNrtB* strain exhibited a considerably reduced germination rate compared to both the WT and CP strains after 4 h of incubation (Figure 3A). Furthermore, the Δ*MaNrtB* strain required a significantly longer time to reach 50% germination (GT_{50}) (Figure 3B). In contrast, the conidial yield among the three strains was found to be statistically indistinguishable (Figure 3C). Collectively, these findings suggest that the disruption of the *MaNrtB* gene in *M. acridum* led to the retardation of conidial germination without impacting the conidial production.

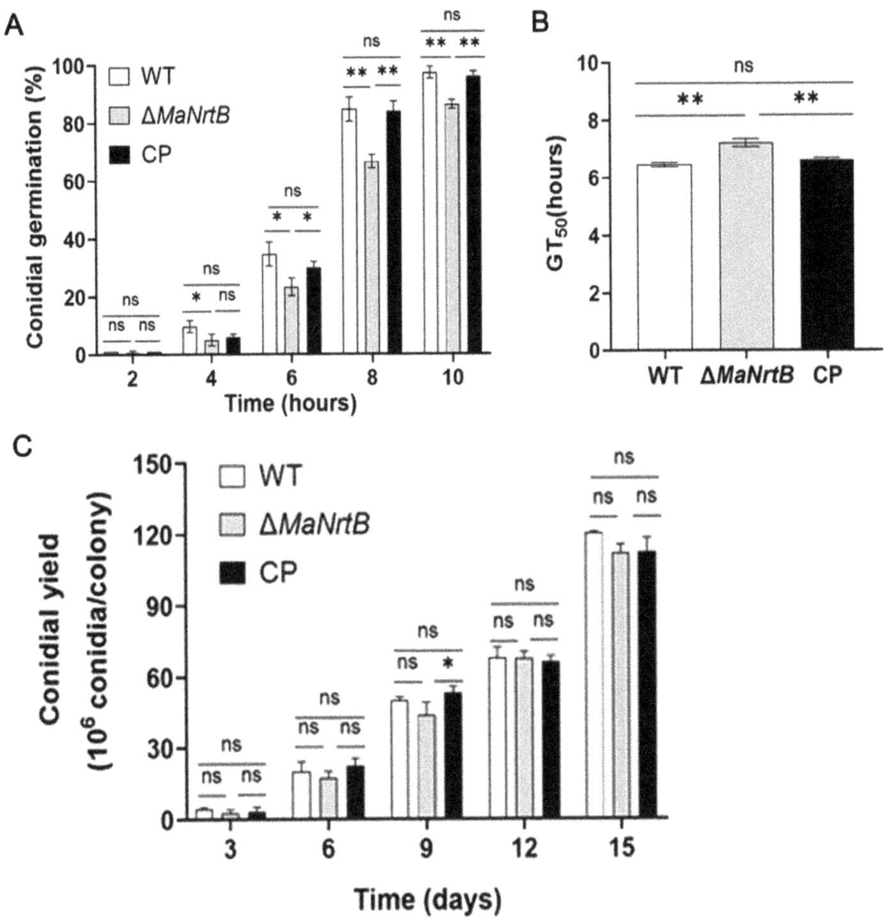

Figure 3. Deletion of *MaNrtB* affected the conidial germination but not the conidial yield. (**A**) The germination rate of each strain of *M. acridum* at different times. (**B**) The GT_{50} of each strain during germination. (**C**) The conidia yield of each strain on 1/4 SDAY medium for 3 d, 6 d, 9 d, 12 d, and 15 d. Error bars = mean ± SEM. Asterisks indicate a significant difference at (*) $p < 0.05$, (**) $p < 0.01$, or (ns) $p > 0.05$.

3.3. Disruption of MaNrtB Affected Fungal Stress Tolerances

The ability to withstand environmental stress is crucial for EPF. Thus, we examined the UV-B and heat-shock stress tolerances of the WT, Δ*MaNrtB*, and CP strains. Post-UV-B irradiation, the Δ*MaNrtB* strain exhibited a markedly reduced conidial germination rate after one hour and a significantly lower GT_{50} compared to the WT and CP strains (Figure 4A,B). Conversely, under heat-shock conditions, the Δ*MaNrtB* strain displayed a higher germination rate after four hours of treatment, with a notably elevated GT_{50} (Figure 4C,D). These findings suggest that the disruption of *MaNrtB* in *M. acridum* compromised its UV-B resistance but enhanced its heat tolerance. Furthermore, spot assays revealed that the Δ*MaNrtB* strain formed smaller colonies relative to the WT and CP strains on 1/4 SDAY medium supplemented with SOR, NaCl, H_2O_2, CR, and CFW (Figure 5), indicating that the Δ*MaNrtB* strain was more susceptible to hyperosmotic, oxidative, and cell wall stresses.

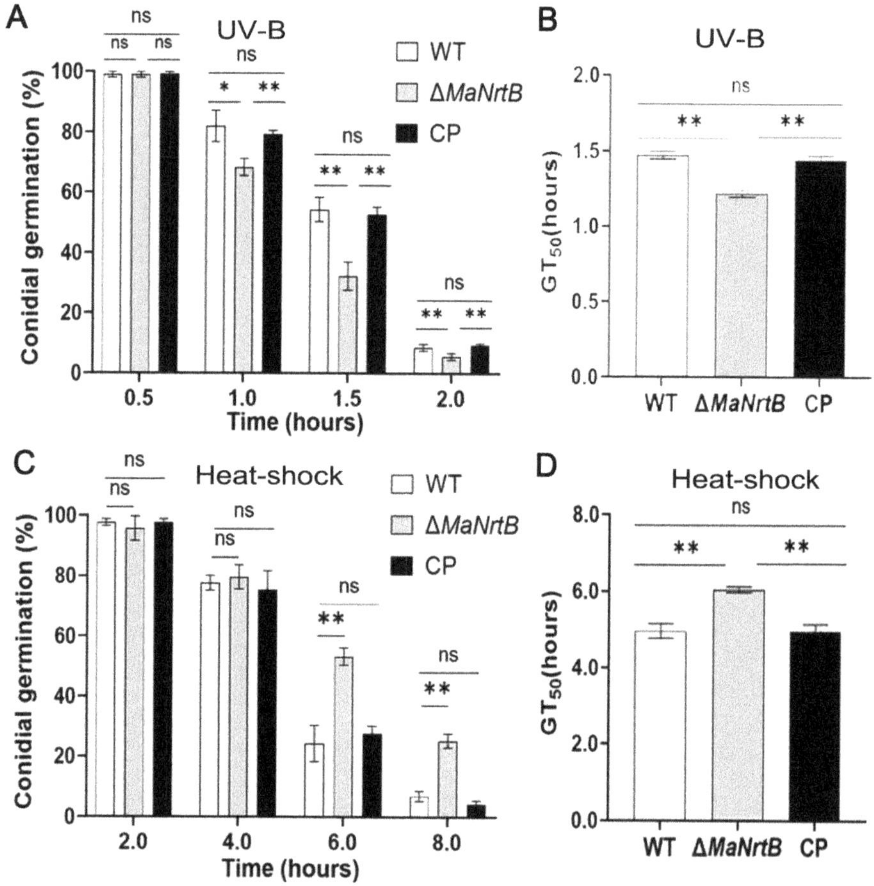

Figure 4. Deletion of *MaNrtB* reduced the fungal tolerance to UV-B irradiation but increased the fungal tolerance to heat shock. (**A**) Conidial germination treated with UV-B irradiation for 0.5 h, 1.0 h, 1.5 h, and 2.0 h. (**B**) The GT_{50} under UV-B irradiation. (**C**) Germination rates treated with heat shock for 2.0 h, 4.0 h, 6.0 h, and 8.0 h. (**D**) GT_{50} under heat shock. Error bars = mean ± SEM. Asterisks indicate a significant difference at (*) $p < 0.05$, (**) $p < 0.01$, or (ns) $p > 0.05$.

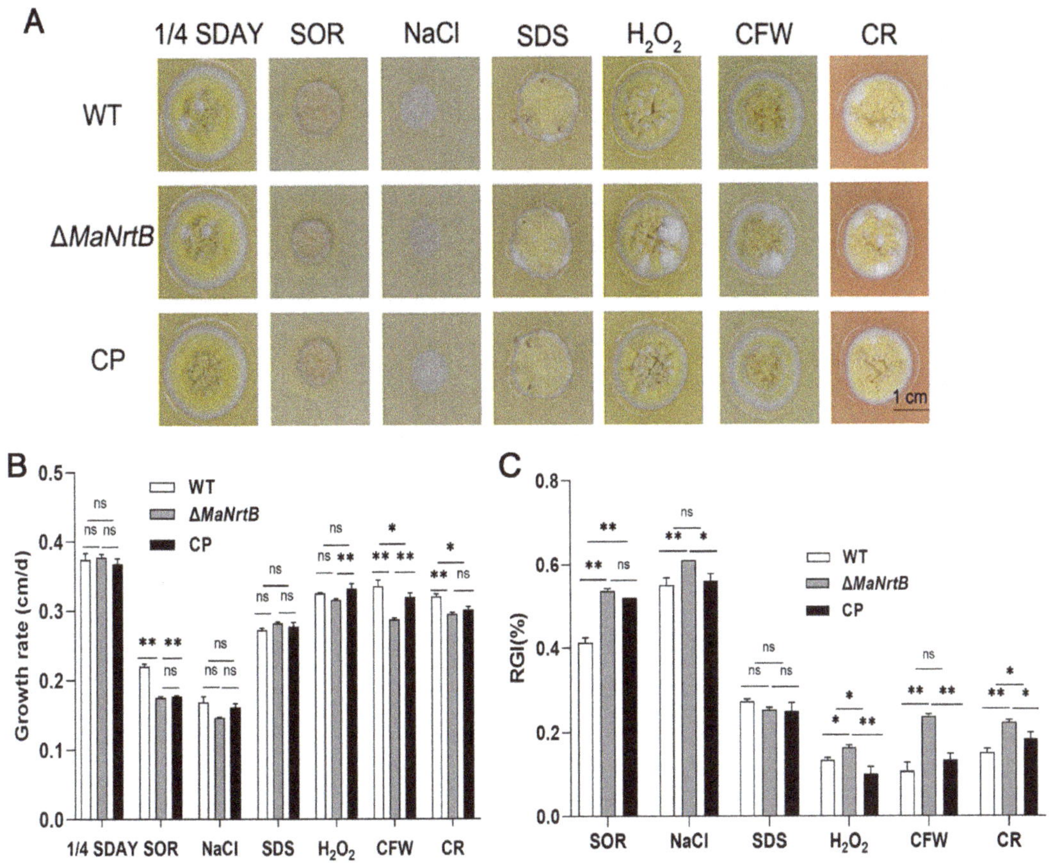

Figure 5. Disruption of *MaNrtB* reduced tolerances to multiple chemical reagents. (**A**) Colony morphologies of *M. acridum* strains grown for 6 days on 1/4 SDAY with different chemical reagents. (**B**) The growth rates of *M. acridum* strains under different chemical reagents. (**C**) Relative growth inhibition rates of *M. acridum* strains under different chemical reagents. Error bars = mean ± SEM. Asterisks indicate a significant difference at (*) $p < 0.05$, (**) $p < 0.01$, or (ns) $p > 0.05$.

3.4. Disruption of MaNrtB Attenuated Fungal Virulence

To assess the role of *MaNrtB* in virulence, we conducted bioassays using two infection methods, topical application and direct intrahemocoel injection. With topical application of the WT or CP strains, all the locusts succumbed by 8.5 days post-inoculation (dpi), whereas those inoculated with the Δ*MaNrtB* strain died after 10.0 dpi (Figure 6A). The median lethal time (LT_{50}) for the Δ*MaNrtB* strain was 6.26 ± 0.80 days, which was significantly longer than the 4.90 ± 0.69 days for the WT and 5.32 ± 0.73 days for the CP strain ($p < 0.05$; Figure 6B). Similarly, when using the intrahemocoel injection method, the survival rates of the locusts infected with the Δ*MaNrtB* strain were also significantly higher than those infected with the WT or CP strains (Figure 6C). Likewise, the LT_{50} for the Δ*MaNrtB* strain, at 4.64 ± 0.67 days, was significantly longer than the 3.55 ± 0.56 days for the WT and 3.97 ± 0.60 days for the CP strains ($p < 0.05$; Figure 6D). Collectively, these findings indicate that *MaNrtB* plays a significant role in the virulence of *M. acridum*.

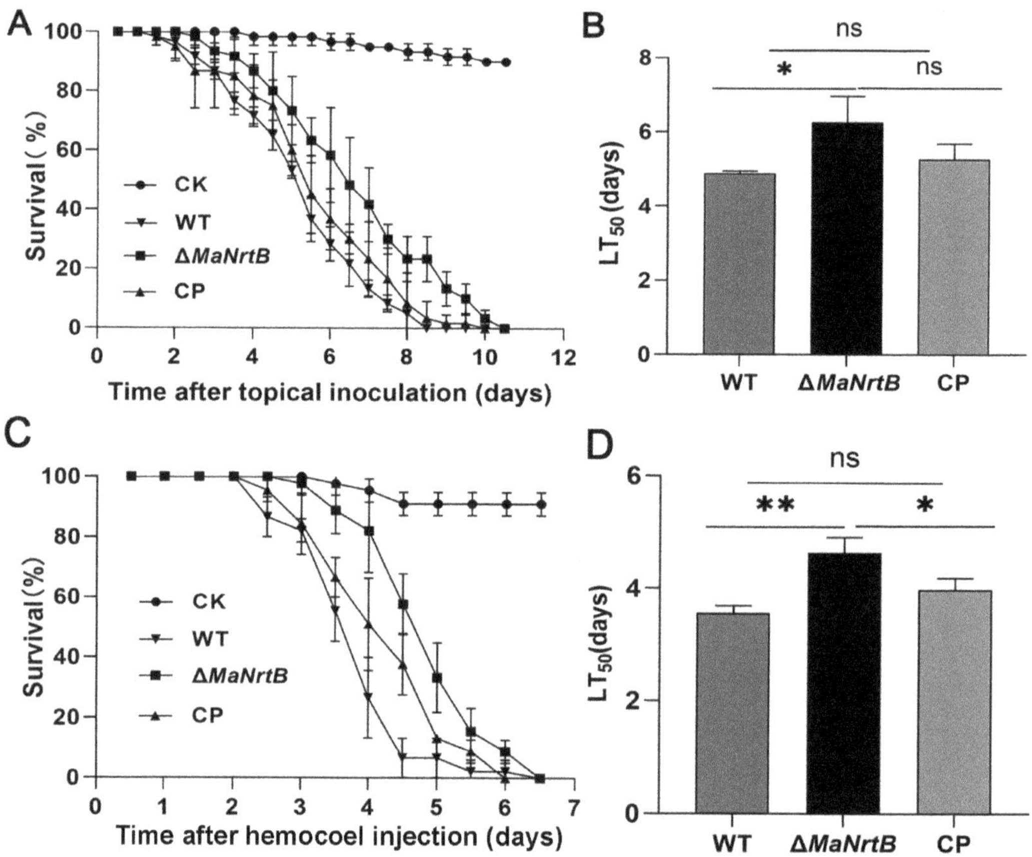

Figure 6. Disruption of *MaNrtB* impaired fungal virulence. (**A**) Survival rates of locusts after topical application of fungal conidia. Liquid paraffin oil as a control. (**B**) LT_{50} of fungal strains against locusts by topical application. (**C**) Survival rates of locusts after hemocoel injection of conidia. Sterile water as a control. (**D**) LT_{50} of fungal strains against locusts by hemocoel injection. Error bars = mean ± SEM. Asterisks indicate a significant difference at (*) $p < 0.05$, (**) $p < 0.01$, or (ns) $p > 0.05$.

To explore whether the *MaNrtB* deletion impacts fungal penetration, we inoculated *M. acridum* on the hind wings of locusts and analyzed the penetration ability of each strain. The results showed that compared to the WT and CP, the Δ*MaNrtB* strain formed the smallest colony with the lowest penetration ability (Figure 7A). To elucidate the functions of *MaNrtB* in the infectious process, we analyzed the germination of conidia and the formation of appressoria on the locust hind wings for the WT, Δ*MaNrtB*, and CP strains. The absence of *MaNrtB* led to slower conidial germination on the locust hind wings (Figure 7B,C). Microscopic observation showed that all three strains formed normal appressorium, but the *MaNrtB* mutation also resulted in significantly decreased appressorium formation ($p < 0.01$; Figures S5 and 7D). After a 24 h induction on locust wings, the Δ*MaNrtB* strain formed about 30% normal appressoria, in contrast to the over 50% observed in both the WT and CP strains (Figure 7D).

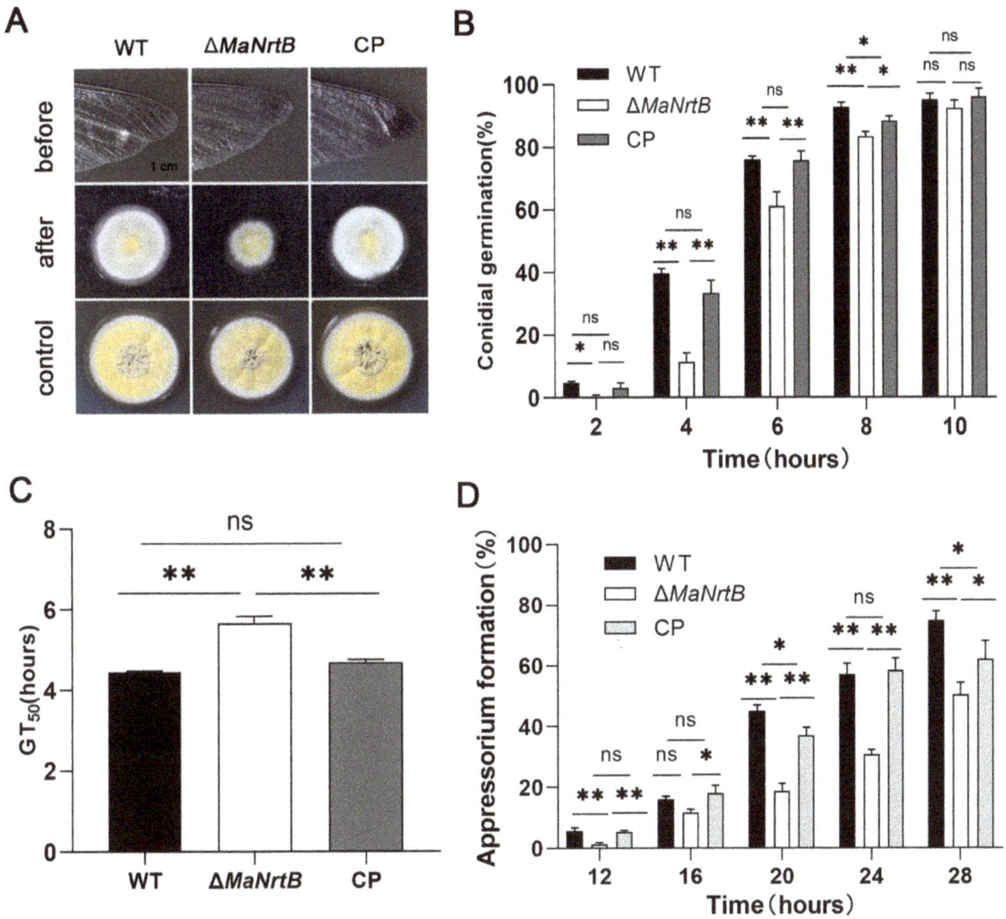

Figure 7. Disruption of *MaNrtB* affected locust cuticle penetration of *M. acridum*. (**A**) Penetration assays. (**B**) The germination rate of the WT, Δ*MaNrtB* and CP strains growing on the hind wings of locust for 2 h, 4 h, 6 h, 8 h and 10 h. (**C**) The GT_{50} of each strain on the hind wings of locust. (**D**) The appressorium formation rates of the WT, Δ*MaNrtB* and CP strains growing on the hind wings of locust for 12 h, 16 h, 20 h, 24 h, and 28 h. Error bars = mean ± SEM. Asterisks indicate a significant difference at (*) $p < 0.05$, (**) $p < 0.01$, or (ns) $p > 0.05$.

To verify whether the *MaNrtB* deletion impacts fungal colonization in hemolymph, we first simulated fungal growth in different media, then quantified the biomass, and analyzed the uptake of carbon and nitrogen sources. It was found that the fungi accumulated the most biomass in SPM medium containing sucrose and peptone, which were the preferable choice for both nitrogen and carbon sources, and the Δ*MaNrtB* mutant accumulated about 22% biomass less than the WT (Figure 8A). Replacing sucrose with trehalose widened the gap to about 43% in TPM (Figure 8A). In the nutrient-poor CZA standard medium, the gap in accumulated biomass reached 56% (Figure 8A). Furthermore, to probe whether the *MaNrtB* gene contributed to the ability of the fungus to evade host immune defenses, the expression levels of *Defensin* and *Attacin*, Toll- and/or Imd-activating antimicrobial peptide genes involved in fungal recognition and immune response in the insect fat bodies were analyzed after injection with conidial suspension from different fungal strains. The results displayed that lower levels of expression of *Attacin* (32%) and *Defensin* (43%) were found in the locusts infected by the *MaNrtB* strain compared to those infected by the WT after 24 dpi (Figure 8B).

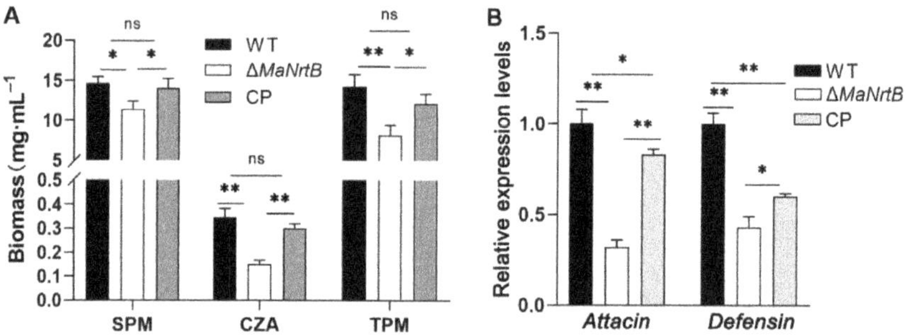

Figure 8. Disruption of *MaNrtB* impaired fungal growth in locust hemolymph. (**A**) The biomass levels were quantified from the 3-day-old submerged cultures in CZA and three amended media, sucrose–peptone medium (SPM) and trehalose–peptone medium (TPM). (**B**) The relative expression of *Attacin* and *Defensin* in locust fat bodies was determined at 24 h after injection by RT-qPCR. Error bars = mean ± SEM. Asterisks indicate a significant difference at (*) $p < 0.05$, (**) $p < 0.01$, or (ns) $p > 0.05$.

4. Discussion

NrtB, a conserved nitrate transporter in filamentous fungi, has been the subject of limited research regarding its molecular mechanisms and their relation to fungal growth. In our study, we employed homologous recombination to generate *MaNrtB* deletion and complementation transformants, conducting a thorough analysis of the *MaNrtB* functions. Our results indicated that *MaNrtB*, the nitrate transporter gene, plays important roles in the conidial germination, stress tolerances, and virulence of the entomopathogenic fungus *M. acridum*. These findings provide new perspectives for understanding the biological properties and pathogenesis of EPF.

Nitrogen sources play a pivotal role in the morphological development, production of secondary metabolites, and virulence of EPF [23]. Typically, fungi prioritize certain nitrogen sources, such as ammonium, and repress the assimilation of others, such as nitrate, a phenomenon termed nitrogen catabolite repression (NCR) [24]. In Ascomycetes, including *A. nidulans*, the key regulators AreA and AreB govern nitrogen metabolism, with nitrate metabolism gene activation primarily under the control of the transcription factors AreA and NirA [25–27]. AreA acts as a positive regulator that can alleviate NCR, while NirA is a transcription factor specific to the nitrate assimilation pathway, induced by nitrate or nitrite [28]. NmrA is another central regulator within the NCR pathway, modulating fungal nitrogen metabolism by interacting with and suppressing the activity of AreA and AreB [29–32]. In our previous research, we established a library of differentially expressed genes responsive to nitrate [33]. Among the genes, *MaNCP1*, a C2H2-type transcription factor gene, was identified to modulate nitrate metabolism by interacting with the *MaNmrA* in *M. acridum* [34,35]. *MaNCP1* influences nitrate metabolism and impacts nitric oxide (NO) synthesis [34], which can engage in various signal transduction pathways, leading to phenotypic alterations in the fungus [36]. Nitrate can trigger the upregulation of *MaNCP1* and *MaNmrA*, and the expression of the nitrate transporter gene *MaNrtB* is downregulated following the deletion of *MaNCP1* or *MaNmrA* [34]. This suggests reduced transport of nitrate from the extracellular environment into the cell, yet the nitrate content in the Δ*MaNCP1* and Δ*MaNmrA* strains was significantly higher than in the wild-type strain, indicating a severe disruption in nitrate catabolism [34]. As a key component of the conserved nitrate assimilation pathway, Nrts regulate cellular nitrogen utilization and further influence fungal growth and development [37]. In our research, the Δ*MaNrtB* strain displayed slower growth, reduced stress resistance, and decreased virulence against locusts, just like the mutant of Δ*MaNCP1* [38] and Δ*MaNmrA* strains [39], underscoring the important role of nitrate metabolism in sustaining fungal growth.

The capacity of conidia to withstand environmental stress is critical for the efficacy of EPF in the field, underscoring the importance of identifying robust strains for practical applications [40]. Our research revealed that the Δ*MaNrtB* strain showed reduced tolerances to UV-B and cell wall stress but enhanced resistance to heat shock. The fungal cell wall is the primary defense against external stressors [41]. Compounds such as CFW and CR are commonly utilized to challenge fungal cell walls in vitro [42]. CFW interferes with chitin assembly, and CR inhibits β-glucan synthesis, both leading to cell wall damage and triggering cell wall stress responses [43]. In *Arabidopsis thaliana*, a co-expression analysis of cell wall remodeling genes with nitrate and ammonium transporters revealed that there is a notably strong correlation between the regulation of cell wall remodeling and nitrate assimilation processes, while ammonium transporters appeared to have a more limited role in these co-expression networks [44]. Thus, the inhibited growth of Δ*MaNrtB* on 1/4 SDAY medium containing CR and CFW suggests that *MaNrtB* also plays a role in maintaining cell wall integrity.

The conidial germination rate of the Δ*MaNrtB* strain was diminished, echoing observations in *U. maydis* [12] and *F. oxysporum* [11]. However, the production of conidia remained unaffected in the Δ*MaNrtB* strain, suggesting that the *MaNrtB* gene is pivotal for conidial germination but exerts minimal influence on the process of conidiation. In *F. fujikuroi*, a reduction in biomass was noted in mutants of the nitrate transporter protein compared to the wild type [45]. Consistent with prior fungal studies, the growth rate of *F. graminearum* is dependent on the availability and type of nitrogen sources [46]. In *B. bassiana*, deprivation of nitrogen nutrients leads to reduced conidium production [47]. Disrupting the gene encoding the nitrate-responsive transcription factor NirA in *B. bassiana* causes reduced growth rates and diminished efficiency in utilizing nitrate and urea [48]. This supports the notion that the highly conserved nitrate assimilation pathway is crucial for promoting growth across various filamentous fungi.

Elucidating the impact of nitrogen sources on fungal virulence is helpful in developing novel pest control strategies. A successful infection hinges on the ability of a pathogen to overcome host defenses and secure vital nutrients necessary to complete its life cycle [49]. The nitrate assimilation pathway plays a significant role in fungal virulence. In *Colletotrichum acutatum*, a strain with a deletion of the *NirA1* gene could not grow on media using nitrate or nitrite as the sole nitrogen source and displayed reduced virulence on strawberry leaves [50]. In *N. crassa*, growth tests with nitrate as the sole nitrogen source showed that the homologous protein *nit-10* mutant was unable to grow with a low nitrate concentration [51]. In the plant pathogen *Magnaporthe oryzae*, the trehalose-6-phosphate synthase enzyme (Tps1) links glucose-6-phosphate metabolism to nitrogen source utilization, modulating nitrate reductase activity [52]. Moreover, a study showed that there is a genetic switch that consists of Tps1 and the nitrogen metabolite repressor gene *NMR1* [53]. This genetic switch controls the expression of certain nitrogen-utilization GATA-factor transcriptional factors downstream, including ASD4 (the homologous protein to AreB in *M. oryzae*), and it is necessary for regulation of a set of genes that are expressed during appressorium-mediated infection in *M. oryzae* [53]. Further research has reported that targeted deletion of *Asd4* led to an increase in glutamine levels in *M. oryzae*, which can further activate the TOR pathway and promote autophagy, thereby promoting hyphal growth while impeding appressorium formation [54,55]. The phenotypic observations of the Δ*MaAreB* strain [55] were consistent with the *M. oryzae ASD4* mutant [56]. In our study, the disruption of the *MaNrtB* gene led to delayed conidial germination, reduced appressorium formation, and attenuated growth within the insect host; thus, we speculate that the disorder of the nitrate assimilation pathways may have led to changes in the glutamine levels in vivo and influenced the TOR pathway, ultimately resulting in impaired growth and virulence in *M. acridum*.

For entomopathogenic fungi, the capacity to penetrate the insect cuticle and proliferate in the hemolymph is crucial for fungal virulence [57]. Fungi not only need to overcome the host immune response but also compete for nutrients with other microorganisms within the host; for instance, *M. rileyi* generates compounds that inhibit the proliferation of competing microorganisms, thereby facilitating its own swift expansion and contributing to the demise of the host [58]. In our study, media containing different carbon and nitrogen sources were used to simulate the growth pattern of fungi facing different nutrient environments. Correspondingly, the Δ*MaNrtB* strain showed significantly decreased total biomass in each media compared to the WT and CP strains, suggesting that the deletion of *MaNrtB* contributes to trehalose utilization, which would further affect fungal growth and dimorphic transformation within the host haemocoele, and hence fungal virulence [56]. In the nutrient-poor CZA standard medium, the gap in accumulated biomass reached 56%, while the absence of Nrts greatly limited fungal growth in nitrate-containing environments, which is consistent with previous studies [7,10–12]. Furthermore, fungal cells must overcome the host immune activity to propagate by yeast-like budding in the host haemocoele [59]. We assessed the immune response of locusts by quantifying the expression levels of two AMP genes, *Attacin* and *Defensin*. In this study, the locusts infected by Δ*MaNrtB* displayed decreased AMP gene expression, implicating an impaired ability of the Δ*MaNrtB* mutant in the fungal evasion from locust immunity defense. Given that the mutant strain exhibits a reduced growth rate in media with varying nutrient compositions, this could potentially impair its development under diverse conditions. Such developmental constraints might, in turn, account for the insect's diminished defensive response.

In summary, our study sheds light on the multifaceted functions of the *NrtB* gene in the entomopathogenic fungus *M. acridum*. We have demonstrated that NrtB plays a pivotal role in multi-stress tolerance, nutrient assimilation, and fungal pathogenicity. These insights contribute to a deeper understanding of the molecular mechanisms underlying fungal virulence.

Supplementary Materials: The following supporting information can be downloaded at https://www.mdpi.com/article/10.3390/jof11020111/s1, Figure S1: The schematic diagrams of the pK2-PB and pK2-sur vectors. (A) *Hin*dIII/*Xba*I and *Eco*RV/*Eco*RI-restricted pK2-PB vectors used to insert the left border and right border of *MaNrtB*, respectively. Pt-R and Bar-F are universal primers located in the *bar* cassette. (B) The eGFP and TtrpC sequences were enzymatically ligated to the 5′ end of the sur cassette, resulting in the formation of the PK2-eGFP-SUR vector. This construct can be used for the construction of any gene C-terminal fusion EGFP vector. Figure S2: Structure prediction of the *MaNrtB* protein. (A) Comparison of the three-dimensional structure of the nitrate transporter protein Nrt in different organisms. The three-dimensional structures of these Nrt homologous proteins are similar. Accession ID: *Aspergillus nidulans* CrnA (XP_658612.1), *A. nidulans* NrtB (AAL50818.1), *N. crassa* Nit-10 (XP_957430.2), *M. maydis* Nrt (XP_011390345.1), *H. polymorpha* (XP_018213600.1), *A. thaliana* NRT 2.1 (NP_172288.1), and *Escherichia coli* (CAD6015891.1). (B) Transmembrane structure analysis of the *MaNrtB* protein. The 12 transmembrane domains are highlighted with a red color. Figure S3: Alignments of the amino acid sequences of the NrtB protein with those in other fungi. The MFS structural domain is marked by a red line. The transmembrane domain is marked by a black line. Accession numbers are as follows: *M. acridum* (XP_007809529.2), *M. anisopliae* (KAF5125400.1), *M. robertsii* (XP_007820472.2), *A. fumigatus* (KAH1494629.1), *F. redolens* (XP_046056618.1), *N. crassa* (XP_957430.2), *P. oryzae* (XP_003710887.1), *P. paradoxum* (XP_057035394.1), and *V. dahliae* (KAF3360447.1). Figure S4: Validations of the Δ*MaNrtB* and CP strains by PCR using the primer pairs of NrtB-VF/Pt-R (A), Bar-F/NrtB-VR (B), NrtB-CP-VF/GFP-VR (C), and TEF-F/TEF-R (D). Figure S5: Photomicrographs of the appressoria formed by WT, Δ*MaNrtB*, and CP. All three strains were able to form normal appressoria, but there were no obvious differences in their morphology. Table S1: Primers used in this study. Table S2: Data of the fungal biomass measurements in different media.

Author Contributions: Conceptualization, Y.X. and K.J.; methodology, Y.X. and K.J.; software, J.W.; validation, K.J.; formal analysis, J.W. and K.J.; investigation, J.W. and Y.Z.; resources, Y.X. and K.J.; data curation, J.W. and Y.Z.; writing—original draft preparation, J.W. and K.J.; writing—review and editing, K.J.; visualization, J.W.; supervision, Y.X. and K.J.; project administration, Y.X. and K.J.; funding acquisition, Y.X. and K.J. All authors have read and agreed to the published version of the manuscript.

Funding: This research was funded by the National Natural Science Foundation of China (32172479), the Natural Science Foundation Project of Chongqing (CSTB2022NSCQ-MSX1185), the technology innovation and application development project of Chongqing (CSTB2023TIAD-KPX0045), and the fundamental research funds for the central universities (2024CDJXY016).

Institutional Review Board Statement: Not applicable.

Informed Consent Statement: Not applicable.

Data Availability Statement: Data are contained within the article and Supplementary Materials.

Conflicts of Interest: The authors declare no conflicts of interest.

References

1. Savary, S.; Willocquet, L.; Pethybridge, S.J.; Esker, P.; McRoberts, N.; Nelson, A. The global burden of pathogens and pests on major food crops. *Nat. Ecol. Evol.* **2019**, *3*, 430–439. [CrossRef] [PubMed]
2. Butt, T.M.; Coates, C.J.; Dubovskiy, I.M.; Ratcliffe, N.A. Entomopathogenic fungi: New insights into host-pathogen interactions. *Adv. Genet.* **2016**, *94*, 307–364. [PubMed]
3. Thomas, M.B.; Read, A.F. Can fungal biopesticides control malaria? *Nat. Rev. Microbiol.* **2007**, *5*, 377–383. [CrossRef] [PubMed]
4. Morschhäuser, J. Nitrogen regulation of morphogenesis and protease secretion in *Candida albicans*. *Int. J. Med. Microbiol.* **2011**, *301*, 390–394. [CrossRef]
5. Magasanik, B.; Kaiser, C.A. Nitrogen regulation in *Saccharomyces cerevisiae*. *Gene* **2002**, *290*, 1–18. [CrossRef]
6. Unkles, S.E.; Hawker, K.L.; Grieve, C.; Campbell, E.I.; Montague, P.; Kinghorn, J.R. crnA encodes a nitrate transporter in *Aspergillus nidulans*. *Proc. Natl. Acad. Sci. USA* **1991**, *88*, 204–208. [CrossRef]
7. Zhou, J.J.; Trueman, L.J.; Boorer, K.J.; Theodoulou, F.L.; Forde, B.G.; Miller, A.J. A high affinity fungal nitrate carrier with two transport mechanisms. *J. Biol. Chem.* **2000**, *275*, 39894–39899. [CrossRef] [PubMed]
8. Pellizzaro, A.; Alibert, B.; Planchet, E.; Limami, A.M.; Morère-Le Paven, M.C. Nitrate transporters: An overview in legumes. *Planta* **2017**, *246*, 585–595. [CrossRef]
9. Akhtar, N.; Karabika, E.; Kinghorn, J.R.; Glass, A.D.; Unkles, S.E.; Rouch, D.A. High-affinity nitrate/nitrite transporters NrtA and NrtB of *Aspergillus nidulans* exhibit high specificity and different inhibitor sensitivity. *Microbiology* **2015**, *161*, 1435–1446. [CrossRef]
10. Machín, F.; Medina, B.; Navarro, F.J.; Pérez, M.D.; Veenhuis, M.; Tejera, P.; Lorenzo, H.; Lancha, A.; Siverio, J.M. The role of Ynt1 in nitrate and nitrite transport in the yeast *Hansenula polymorpha*. *Yeast* **2004**, *21*, 265–276. [CrossRef]
11. Gomez-Gil, L.; Camara Almiron, J.; Rodriguez Carrillo, P.L.; Olivares Medina, C.N.; Bravo Ruiz, G.; Romo Rodriguez, P.; Corrales Escobosa, A.R.; Gutierrez Corona, F.; Roncero, M.I. Nitrate assimilation pathway (NAP): Role of structural (nit) and transporter (ntr1) genes in *Fusarium oxysporum* f.sp. lycopersici growth and pathogenicity. *Curr. Genet.* **2018**, *64*, 493–507.
12. Khanal, S.; Schroeder, L.; Nava-Mercado, O.A.; Mendoza, H.; Perlin, M.H. Role for nitrate assimilatory genes in virulence of *Ustilago maydis*. *Fungal Biol.* **2021**, *125*, 764–775. [CrossRef]
13. López-Bucio, J.; Esparza-Reynoso, S.; Pelagio-Flores, R. Nitrogen availability determines plant growth promotion and the induction of root branching by the probiotic fungus *Trichoderma atroviride* in *Arabidopsis* seedlings. *Arch. Microbiol.* **2022**, *204*, 380. [CrossRef] [PubMed]
14. Cao, Y.; Zhu, X.; Jiao, R.; Xia, Y. The *Magas1* gene is involved in pathogenesis by affecting penetration in *Metarhizium acridum*. *J. Microbiol. Biotechnol.* **2012**, *22*, 889–893. [CrossRef]

15. Jin, K.; Ming, Y.; Xia, Y.X. *MaHog1*, a Hog1-type mitogen-activated protein kinase gene, contributes to stress tolerance and virulence of the entomopathogenic fungus *Metarhizium acridum*. *Microbiology* **2012**, *158*, 2987–2996. [CrossRef]
16. Wen, Z.; Tian, H.; Xia, Y.; Jin, K. MaPmt1, a protein *O*-mannosyltransferase, contributes to virulence through governing the appressorium turgor pressure in *Metarhizium acridum*. *Fungal Genet. Biol.* **2020**, *145*, 103480. [CrossRef]
17. Du, Y.; Jin, K.; Xia, Y. Involvement of MaSom1, a downstream transcriptional factor of cAMP/PKA pathway, in conidial yield, stress tolerances, and virulence in *Metarhizium acridum*. *Appl. Microbiol. Biotechnol.* **2018**, *102*, 5611–5623. [CrossRef] [PubMed]
18. Cai, Q.; Wang, J.J.; Xie, J.T.; Jiang, D.H. Functional characterization of BbEaf6 in *Beauveria bassiana*: Implications for fungal virulence and stress response. *Virulence* **2024**, *15*, 2387172. [CrossRef]
19. Zhang, J.J.; Jiang, H.; Du, Y.; Keyhani, N.O.; Xia, Y.; Jin, K. Members of chitin synthase family in *Metarhizium acridum* differentially affect fungal growth, stress tolerances, cell wall integrity and virulence. *PLoS Pathog.* **2019**, *15*, e1007964. [CrossRef]
20. Sano, M. *Aspergillus oryzae* nrtA affects kojic acid production. *Biosci. Biotechnol. Biochem.* **2016**, *80*, 1776–1780. [CrossRef] [PubMed]
21. Pérez, M.D.; González, C.; Avila, J.; Brito, N.; Siverio, J.M. The YNT1 gene encoding the nitrate transporter in the yeast *Hansenula polymorpha* is clustered with genes YNI1 and YNR1 encoding nitrite reductase and nitrate reductase, and its disruption causes inability to grow in nitrate. *Biochem. J.* **1997**, *321 Pt 2*, 397–403. [CrossRef] [PubMed]
22. Gao-Rubinelli, F.; Marzluf, G.A. Identification and characterization of a nitrate transporter gene in *Neurospora crassa*. *Biochem. Genet.* **2004**, *42*, 21–34. [CrossRef] [PubMed]
23. Snoeijers, S.S.; Pérez-García, A.; Joosten, M.H.; De Wit, P.J.G.M. The effect of nitrogen on disease development and gene expression in bacterial and fungal plant pathogens. *Eur. J. Plant Pathol.* **2000**, *106*, 493–506. [CrossRef]
24. Tudzynski, B. Nitrogen regulation of fungal secondary metabolism in fungi. *Front. Microbiol.* **2014**, *5*, 656. [CrossRef] [PubMed]
25. Marzluf, G.A. Genetic regulation of nitrogen metabolism in the fungi. *Microbiol. Mol. Biol. Rev.* **1997**, *61*, 17–32. [PubMed]
26. Wong, K.H.; Hynes, M.J.; Todd, R.B.; Davis, M.A. Transcriptional control of nmrA by the bZIP transcription factor MeaB reveals a new level of nitrogen regulation in *Aspergillus nidulans*. *Mol. Microbiol.* **2007**, *66*, 534–551. [CrossRef]
27. Ravagnani, A.; Gorfinkiel, L.; Langdon, T.; Diallinas, G.; Adjadj, E.; Demais, S.; Gorton, D.; Arst, H.N.; Scazzocchio, C. Subtle hydrophobic interactions between the seventh residue of the zinc finger loop and the first base of an HGATAR sequence determine promoter-specific recognition by the *Aspergillus nidulans* GATA factor AreA. *EMBO J.* **1997**, *16*, 3974–3986. [CrossRef] [PubMed]
28. Burger, G.; Tilburn, J.; Scazzocchio, C. Molecular cloning and functional characterization of the pathway-specific regulatory gene *nirA*, which controls nitrate assimilation in *Aspergillus nidulans*. *Mol. Cell. Biol.* **1991**, *11*, 795–802.
29. Pan, H.; Feng, B.; Marzluf, G.A. Two distinct protein-protein interactions between the NIT2 and NMR regulatory proteins are required to establish nitrogen metabolite repression in *Neurospora crassa*. *Mol. Microbiol.* **1997**, *26*, 721–729. [CrossRef] [PubMed]
30. Hubermana, L.B.; Wu, V.W.; Kowbela, D.J.; Lee, J.; Daum, C.; Grigorieva, I.V.; O'Malley, R.C.; Glassa, N.L. DNA affinity purification sequencing and transcriptional profiling reveal new aspects of nitrogen regulation in a filamentous fungus. *Proc. Natl. Acad. Sci. USA* **2021**, *118*, e2009501118. [CrossRef]
31. Berger, H.; Basheer, A.; Bock, S.; Reyes-Dominguez, Y.; Dalik, T.; Altmann, F.; Strauss, J. Dissecting individual steps of nitrogen transcription factor cooperation in the *Aspergillus nidulans* nitrate cluster. *Mol. Microbiol.* **2008**, *69*, 1385–1398. [CrossRef] [PubMed]
32. Kotaka, M.; Johnson, C.; Lamb, H.K.; Hawkins, A.R.; Ren, J.; Stammers, D.K. Structural analysis of the recognition of the negative regulator NmrA and DNA by the zinc finger from the GATA-type transcription factor AreA. *J. Mol. Biol.* **2008**, *381*, 373–382. [CrossRef]
33. Wang, Z.L.; Jin, K.; Xia, Y.X. Transcriptional analysis of the conidiation pattern shift of the entomopathogenic fungus *Metarhizium acridum* in response to different nutrients. *BMC Genom.* **2016**, *17*, 586. [CrossRef] [PubMed]
34. Li, C.C.; Xu, D.X.; Hu, M.W.; Zhang, Q.P.; Xia, Y.X.; Jin, K. MaNCP1, a C2H2 zinc finger protein, governs the conidiation pattern shift through regulating the reductive pathway for nitric oxide synthesis in the filamentous fungus *Metarhizium acridum*. *Microbiol. Spectr.* **2022**, *10*, e00538-22. [CrossRef] [PubMed]
35. Li, C.C.; Xia, Y.X.; Jin, K. The C2H2 zinc finger protein MaNCP1 contributes to conidiation through governing the nitrate assimilation pathway in the entomopathogenic fungus *Metarhizium acridum*. *J. Fungi* **2022**, *8*, 942. [CrossRef] [PubMed]
36. Zhao, Y.X.; Lim, J.Y.; Xu, J.Y.; Yu, J.H.; Zheng, W. Nitric oxide as a developmental and metabolic signal in filamentous fungi. *Mol. Microbiol.* **2020**, *113*, 872–882. [CrossRef]
37. Unkles, S.E.; Zhuo, D.; Siddiqi, M.Y.; Kinghorn, J.R.; Glass, A.D.M. Apparent genetic redundancy facilitates ecological plasticity for nitrate transport. *EMBO J.* **2001**, *22*, 6246–6255. [CrossRef]
38. Li, C.C.; Xia, Y.X.; Jin, K. N-terminal zinc fingers of MaNCP1 contribute to growth, stress tolerance, and virulence in *Metarhizium acridum*. *Int. J. Biol. Macromol.* **2022**, *216*, 426–436. [CrossRef]
39. Li, C.C.; Zhang, Q.P.; Xia, Y.X.; Jin, K. MaNmrA, a negative transcription regulator in nitrogen catabolite repression pathway, contributes to nutrient utilization, stress resistance and virulence in entomopathogenic fungus *Metarhizium acridum*. *Biology* **2021**, *10*, 1167. [CrossRef] [PubMed]
40. Xie, X.Q.; Li, F.; Ying, S.H.; Feng, M.G. Additive contributions of two manganese-cored superoxide dismutases (MnSODs) to antioxidation, UV tolerance and virulence of *Beauveria bassiana*. *PLoS ONE* **2012**, *7*, e30298. [CrossRef]

41. Ibe, C.; Munro, C.A. Fungal cell wall proteins and signaling pathways form a cytoprotective network to combat stresses. *J. Fungi* **2021**, *7*, 739. [CrossRef]
42. Perez-Nadales, E.; Di Pietro, A. The transmembrane protein Sho1 cooperates with the mucin Msb2 to regulate invasive growth and plant infection in *Fusarium oxysporum*. *Mol. Plant Pathol.* **2015**, *16*, 593–603. [CrossRef] [PubMed]
43. Ram, A.F.; Klis, F.M. Identification of fungal cell wall mutants using susceptibility assays based on Calcofluor white and Congo red. *Nat. Protoc.* **2006**, *1*, 2253–2256. [CrossRef]
44. Landi, S.; Esposito, S. Nitrate Uptake Affects Cell wall synthesis and modeling. *Front. Plant Sci.* **2017**, *8*, 1376. [CrossRef] [PubMed]
45. Pfannmüller, A.; Boysen, J.M.; Tudzynski, B. Nitrate assimilation in *Fusarium fujikuroi* is controlled by multiple levels of regulation. *Front. Microbiol.* **2017**, *8*, 381. [CrossRef]
46. Gardiner, D.M.; Kazan, K.; Manners, J.M. Nutrient profiling revealspotent inducers of trichothecene biosynthesis in *Fusarium graminearum*. *Fungal Genet. Biol.* **2009**, *46*, 604–613. [CrossRef] [PubMed]
47. Qi, Y.X.; Chen, F.X.; Li, Z.Z. Effects of carbon sources and nitrogen sources on the biological characteristics and antifungal activity of *Beauveria bassiana*. *Acta Laser Biol. Sin.* **2011**, *20*, 38–44.
48. Liu, Y.; Li, Y.X.; Tong, S.; Wang, J.Y.; Zhu, S.A.; Zhang, L.Y.; Fan, Y.H. Effects of *NirA1* gene on growth, stress resistance and virulence of *Beauveria bassiana*. *Acta Microbiol. Sin.* **2021**, *61*, 2469–2480.
49. Brown, N.A.; Urban, M.; Van de Meene, A.M.L.; Hammond-Kosack, K.E. The infection biology of *Fusarium graminearum*: Defining the pathways of spikelet to spikelet colonisation in wheat ears. *Fungal Biol.* **2010**, *114*, 555–571. [CrossRef]
50. Horowitz, S.; Freeman, S.; Zveibil, A.; Yarden, O. A defect in nir1, a nirA-like transcription factor, confers morphological abnormalities and loss of pathogenicity in *Colletotrichum acutatum*. *Mol. Plant Pathol.* **2006**, *7*, 341–354. [CrossRef]
51. Wilson, R.A.; Jenkinson, J.M.; Gibson, R.P.; Littlechild, J.A.; Wang, Z.Y.; Talbot, N.J. Tps1 regulates the pentose phosphate pathway, nitrogen metabolism and fungal virulence. *EMBO J.* **2007**, *26*, 3673–3685. [CrossRef] [PubMed]
52. Wilson, R.A.; Gibson, R.P.; Quispe, C.F.; Littlechild, J.A.; Talbot, N.J. An NADPH-dependent genetic switch regulates plant infection by the rice blast fungus. *Proc. Natl. Acad. Sci. USA* **2010**, *107*, 21902–21907. [CrossRef] [PubMed]
53. Marroquin-Guzman, M.; Wilson, R.A. GATA-dependent glutaminolysis drives appressorium formation in *Magnaporthe oryzae* by suppressing TOR inhibition of cAMP/PKA signaling. *PLoS Pathog.* **2015**, *11*, e1004851. [CrossRef] [PubMed]
54. Li, G.; Gong, Z.; Dulal, N.; Marroquin-Guzman, M.; Rocha, R.O.; Richter, M.; Wilson, R.A. A protein kinase coordinates cycles of autophagy and glutaminolysis in invasive hyphae of the fungus *Magnaporthe oryzae* within rice cells. *Nat. Commun.* **2023**, *14*, 4146. [CrossRef]
55. Li, C.C.; Zhang, Q.P.; Xia, Y.X.; Jin, K. MaAreB, a GATA transcription factor, is involved in nitrogen source utilization, stress tolerances and virulence in *Metarhizium acridum*. *J. Fungi* **2021**, *7*, 512. [CrossRef] [PubMed]
56. van de Veerdonk, F.L.; Gresnigt, M.S.; Romani, L.; Romani, L.; Netea, M.G.; Latgé, J.P. *Aspergillus fumigatus* morphology and dynamic host interactions. *Nat. Rev. Microbiol.* **2017**, *15*, 661–674. [CrossRef] [PubMed]
57. Wang, G.; Xu, S.; Chen, L.; Zhan, T.; Zhang, X.; Liang, H.; Chen, B.; Peng, Y. Gut microbial diversity reveals differences in pathogenicity between *Metarhizium rileyi* and *Beauveria bassiana* during the early stage of infection in *Spodoptera litura* larvae. *Microorganisms* **2024**, *12*, 1129. [CrossRef]
58. Jin, K.; Peng, G.X.; Liu, Y.C.; Xia, Y.X. The acid trehalase, ATM1, contributes to the in vivo growth and virulence of the entomopathogenic fungus, *Metarhizium acridum*. *Fungal Genet. Biol.* **2015**, *77*, 61–67. [CrossRef] [PubMed]
59. Wang, C.S.; St Leger, R.J. A collagenous protective coat enables *Metarhizium anisopliae* to evade insect immune responses. *Proc. Natl. Acad. Sci. USA* **2006**, *103*, 6647–6652. [CrossRef]

Journal of *Fungi*

MDPI

Article

Visualizing Oral Infection Dynamics of *Beauveria bassiana* in the Gut of *Tribolium castaneum*

Lautaro Preisegger [1], Juan Cruz Flecha [1], Fiorella Ghilini [2], Daysi Espin-Sánchez [1], Eduardo Prieto [2,3], Héctor Oberti [4,5], Eduardo Abreo [4], Carla Huarte-Bonnet [1], Nicolás Pedrini [1,*] and Maria Constanza Mannino [1,*]

1 Instituto de Investigaciones Bioquímicas de La Plata (INIBIOLP), CCT La Plata Consejo Nacional de Investigaciones Científicas y Técnicas (CONICET), Universidad Nacional de La Plata (UNLP), Calles 60 y 120, La Plata 1900, Argentina; lpreisegger@med.unlp.edu.ar (L.P.); juancruzflecha16@gmail.com (J.C.F.); despinsanchez@med.unlp.edu.ar (D.E.-S.); carlahb@biol.unlp.edu.ar (C.H.-B.)

2 Instituto de Investigaciones Fisicoquìmicas Teóricas y Aplicadas (INIFTA), CCT La Plata Consejo Nacional de Investigaciones Científicas y Técnicas (CONICET), Universidad Nacional de La Plata (UNLP), Diagonal 113 y 64 S/N, La Plata 1900, Argentina; fiorelag@inifta.unlp.edu.ar (F.G.); edprieto@quimica.unlp.edu.ar (E.P.)

3 Instituto de Ciencias de la Salud, Universidad Nacional Arturo Jauretche (ICS-UNAJ), Avenida Calchaqui, Florencio Varela 6200, Argentina

4 Laboratorio de Bioproducción, Plataforma de Bioinsumos, Instituto Nacional de Investigación Agropecuaria, Estación Experimental Wilson Ferreira Aldunate, Ruta 48 Km 10, Canelones 90000, Uruguay; hoberti@inia.org.uy (H.O.); eabreo@inia.org.uy (E.A.)

5 Departamento de Biotecnología, Universidad ORT Uruguay, Montevideo 10129, Uruguay

* Correspondence: npedrini@med.unlp.edu.ar (N.P.); mmannino@med.unlp.edu.ar (M.C.M.)

Abstract: The ability of entomopathogenic fungi, such as *Beauveria bassiana*, to infect insects by penetrating their cuticle is well documented. However, some insects have evolved mechanisms to combat fungal infections. The red flour beetle (*Tribolium castaneum*), a major pest causing significant economic losses in stored product environments globally, embeds antifungal compounds within its cuticle as a protective barrier. Previous reports have addressed the contributions of non-cuticular infection routes, noting an increase in mortality in beetles fed with conidia. In this study, we further explore the progression and dynamics of oral exposure in the gut of *T. castaneum* after feeding with an encapsulated *B. bassiana* conidia formulation. First, we characterized the formulation surface using atomic force microscopy, observing no significant topological differences between capsules containing and not containing conidia. Confocal microscopy confirmed uniform conidia distribution within the hydrogel matrix. Then, larvae and adult insects fed with the conidia-encapsulated formulation exhibited *B. bassiana* distributed throughout the alimentary canal, with a higher presence of conidia before the pyloric chamber. More conidia were found in the larval midgut and hindgut compared to adults, but no germinated conidia were observed in the epithelium. These results suggest that the presence of conidia obstructs the gut, impairing the insect's ability to ingest, process, and absorb nutrients. This disruption may weaken the host, increasing its susceptibility to infections and, ultimately, leading to death. By providing the first direct observation of fungal conidia within the alimentary canal of *T. castaneum*, this study highlights a novel aspect of fungal–host interaction and opens new avenues for advancing fungal-based pest control strategies by exploiting stage-specific vulnerabilities.

Keywords: entomopathogenic fungi; insect gut; oral infection; coleoptera; EGFP; alginate; AFM

1. Introduction

Entomopathogenic fungi such as *Beauveria bassiana* and *Metarhizium* spp. infect insects by penetrating the cuticle, a route not used by other microbial pathogens like protozoa, bacteria, or viruses [1]. During infection, these fungi release enzymes and toxins that compromise the host's immune system [2,3]. Several studies have identified specific fungal genes involved in these processes, although the molecular mechanisms involved remain under investigation [4–9]. Also, entomopathogenic fungi can interact with the insect gut microbiome, affecting the insect's immunity and gut permeability [10,11]. These interactions are pivotal in overcoming host defenses and facilitating infection. Insects exhibit various defensive behaviors and physiological adaptations against fungal infections. Some insects alter their feeding or grooming behavior to avoid spores, while others strengthen their immune defenses [9]. One example of evolved immunity strategies is the production of antifungal compounds, which are incorporated into their cuticles, making them less susceptible to fungal infections [12–19]. Using this strategy, adults of the red flour beetle *Tribolium castaneum* exhibit tolerance to *B. bassiana* infection via cuticular penetration [18–21].

Among non-canonical infection routes, ingestion of fungal propagules was first proposed more than a century ago [22], and there is more recent evidence demonstrating that entomopathogenic fungi can achieve infection through routes other than cuticular penetration [23]. In various insect species, the ultimate destination of ingested conidia varies, as does the potential for internal adhesion and germination [22]. Research on oral infection in Coleoptera suggests that fungal invasion could start through the mouthparts of different beetle species, often with hyphal growth, but there are no reported cases of germination inside the digestive tracts [22]. Thus, the fate of conidia following ingestion in *T. castaneum* remains largely unknown. Recently, we investigated dual routes of infection—combining cuticular penetration with the ingestion of conidia incorporated into the insect's diet. This approach highlighted the significant contribution of alternative infection pathways, leading to increased mortality and activation of immunity-related gene clusters [3,24]. *T. castaneum* feeding habits do not differ drastically between larvae and adults; however, there are key anatomical differences in the alimentary canal related to cell renewal and food storage [25]. Aspects of the beetle's feeding behavior and alimentary canal anatomy may facilitate the establishment of fungal conidia ingested with the diet, either by enabling infection or by sufficiently disrupting nutrient absorption to adversely affect the insect's life cycle, ultimately resulting in reduced fitness [26].

Fungal formulations incorporating gelling agents, such as sodium alginate, are widely used to protect the active agent from environmental stressors, thereby extending the release and efficacy of fungal conidia. This characteristic presents a valuable opportunity to ensure that conidia come into contact only through ingestion after these formulations are offered in the beetle diet.

In this study, we propose the use of a conidial hydrogel formulation to characterize the distribution and behavior of conidia within larvae and adult *T. castaneum*. Feeding insects with conidia embedded in the diet enabled tracking of conidial movement through the alimentary canal using GFP-tagged *B. bassiana*. Fungal spores were detected throughout the alimentary canal in both larvae and adult insects, with a uniform distribution but a higher concentration in the hindgut (ileum) region of both larvae and adults. These findings indicate that ingested conidia are unable to germinate within the alimentary canal but exert a measurable impact on the insects' fitness, increasing susceptibility to fungal infection and eventual mortality.

2. Materials and Methods

2.1. Fungi

The entomopathogenic fungus *B. bassiana* strain ARSEF 2860 (ARSEF is WDCM collection number 112) was selected for this study. The strain was provided by R. Humber's Lab (Plant Soil and Nutrition Laboratory, USDA-ARS, Ithaca, NY, USA). To obtain conidia, the fungus was maintained and routinely grown at 26 °C for 15 days in Potato Dextrose Agar (PDA; Britania, Buenos Aires, Argentina) medium with 1% ampicillin plates. Sporulating plates were harvested, and fungal propagules were processed for downstream assays.

2.1.1. Enhanced Green Fluorescent Protein (EGFP) *Beauveria bassiana*

The binary vector pCGEN-GFP [27] containing the enhanced green fluorescent protein (EGFP) [28] and the geneticin resistance gene (neo) was used for *Agrobacterium tumefaciens*-mediated transformation of EGFP in *B. bassiana* ARSEF 2860, as described in Oberti et al. [29]. The best transformant strain was selected by fluorescence intensity quantitation in the 488 channel under a Carl Zeiss Observer LSM 800 Confocal Microscope (Oberkochen, Germany) with a 40X NA 0.75 Objective. The stability of the selected transformants was evaluated through serial passaging in Czapek-Dox medium (Merk, Darmstadt, Germany) with G-418 (neoR) (700 μg/mL) (Sigma-Aldrich, St. Louis, MO, USA) for ten consecutive generations to ensure their ability to maintain the genetic modification. Following this, the transformants were transferred to PDA medium without G418 to assess phenotypic stability in comparison to the wild-type strains. The best transformant was selected based on the results of virulence bioassays, which compared the GFP strain to the wild-type strain using model insects, as well as the quantification of fluorescence intensity in the 488 channel under a Carl Zeiss Observer LSM 800 Confocal Microscope with a 40× NA 0.75 Objective.

2.1.2. Conidia Powder Preparation

Harvested conidia were dried for 7 days at 26 °C in a 35 mm Petri dish with sterile filter paper, which was placed inside a 90 mm Petri dish filled with silica gel, and allowed to dry in an incubator. Once dried, the propagule was ground with a mortar and pestle into a uniform powder and stored at 4 °C. The number of conidia per gram of conidia powder was determined in a Neubauer chamber after resuspending 2 mg of conidia in a sterile solution of 0.05% Tween 80. Viability was verified using the quantified conidia suspension.

2.2. Insects

Larvae and adults of *T. castaneum* from colonies maintained at the Instituto de Investigaciones Bioquímicas de La Plata (INIBIOLP) were used. Insects were reared on white wheat flour, with 5% non-fat dried milk, 5% brewer's yeast, and 5% wheat germ, under a 12L:12D photoperiod at 27 ± 2 °C and 70 ± 5% relative humidity. All larvae were selected as last instar larvae (4.5 ± 0.5 mm length), unsexed insects. All adults used in the assays were 2-week-old, unsexed insects.

2.3. Conidia-Encapsulated Formulation

Hydrogel capsules were prepared using a mixture of 1 mL of 1% sodium alginate, 0.1 g of wheat flour, and conidia powder, with the concentration adjusted to 1×10^9 conidia/mL of mixture. To form the capsules, a 10 mL medical-grade syringe with a 21 G × 1″ hypodermic needle was used. The mixture was extruded dropwise over a 0.2 M calcium chloride solution in a beaker, which was stirred continuously to ensure uninterrupted capsule formation. The mixture droplets instantaneously formed capsules upon contact with the calcium chloride solution and were left to consolidate under agitation for 40 min.

The capsules were then rinsed three times in sterile distilled water and placed on sterile filter paper to absorb excess water. Immediately afterward, the capsules were dried for 3 days as described before for conidia. Controls were prepared similarly, but with sterile distilled water replacing the conidia powder. Yield was calculated as the number of capsules obtained per ml of sodium alginate used. The size of the encapsulated formulation was evaluated using a Zeiss Primo Star optical microscope (Jena, Germany) (40×, NA 0.75). A total of 100 capsules were used, and three independent experiments were conducted.

2.4. Capsules Characterization

2.4.1. Surface Topography

For formulation topography characterization, atomic force microscopy (AFM) images of freshly prepared capsules were obtained in air using a MultiMode Scanning Probe Microscope (Veeco, Plainview, NY, USA) equipped with a Nanoscope V controller (Veeco). All measurements were performed in Tapping® mode. Images were captured at 1 Hz with silicon tips (RTESP, 215–254 kHz and 20–80 N/m). Roughness data (Ra) were obtained from four separate images taken from different regions of each substrate using Nanoscope V Software 9.7. Ra was calculated as the arithmetic average of the absolute values of the surface height deviations from the mean plane.

2.4.2. Conidial Distribution

Confocal microscopy images of capsules were used to assess the distribution of conidia inside the hydrogel matrix. The capsules were made with *B. bassiana* ARSEF 2860 EGFP conidia, hydrated, sectioned using a stereoscopic microscope (Zeiss, Jena, Germany, Stemi 305, 0.8x–5x) with a scalpel blade at 0.5 mm, and mounted in 100 μL of sterile physiological solution on a microscope slide. A coverslip was placed on top and sealed with nail polish. Controls with no conidia and non-EGFP conidia were treated similarly. Confocal microscopy images were acquired using a Carl Zeiss Observer LSM 800 Confocal Microscope (Objectives: 10X NA 0.3, 40X NA 0.75, 63X oil immersion NA 1.4). A total of five images per condition were taken for representative and quantitative purposes.

2.4.3. Capsules Stress Tolerance

Thermal tolerance: Individual capsules were placed in 100 μL of 0.05% Tween 80 and incubated for 1, 3, and 5 h at 45 °C in a heat block. Controls were prepared using capsules without conidia. Afterward, capsules were homogenized using a pellet pestle, and 100 μL of the homogenate was plated onto a Sabouraud Dextrose Yeast Extract Agar (SDYA) plate. Samples were incubated for 10 days at 26 °C. Conidia viability was assessed by collecting fungal sporulation from each plate in 5 mL of 0.01% Tween 80, diluting, and quantifying the conidia in a Neubauer chamber. The experiment was repeated three times.

Oxidative stress tolerance: Individual capsules were placed in 1 mL of 12 mM H_2O_2 solution and kept at 4 °C for 72 h. As controls, the same amount of conidia powder and capsules without conidia in 1 mL of 12 mM H_2O_2 were used. Each sample was centrifuged at 13,000 rpm for 1 min, and 900 μL of the supernatant was discarded. The remaining solution was homogenized. Next, 100 μL of each sample was plated onto SDYA with ampicillin (50 μg/mL) and chloramphenicol (34 μg/mL). The samples were incubated for 10 days at 26 °C. Conidia viability was assessed by collecting fungal sporulation from each plate in 5 mL of 0.05% Tween 80, diluting, and quantifying the conidia in a Neubauer chamber. The experiment was repeated three times.

2.4.4. Capsules Pathogenicity Assay

Three groups of 10 *T. castaneum* individuals (either adults or larvae) were placed in sterile plates containing capsules, accounting for a total of 1×10^9 conidia. Control plates were prepared similarly, using capsules without conidia. The individuals were maintained at 26 °C for 15 days. Mortality was checked daily, and all dead adults and larvae were removed after each count. Subsequently, the dead insects were washed in 10% bleach for 30 s, then in 70% ethanol for 30 s, rinsed in sterile distilled water, allowed to dry on sterile filter paper, and placed in individual humid chambers at 26 °C to confirm fungal infection. The experiment was repeated three times, and mortality data were corrected for control mortality using Abbott's formula [30].

2.4.5. Conidia Ingestion Assay and Dissection

Capsules used for feeding were made with *B. bassiana* ARSEF 2860 EGFP, as described in Section 2.4. *T. castaneum* adults and larvae were placed in groups of 10 insects in sterile plates containing capsules, accounting for 1×10^9 conidia/g. Control plates were prepared similarly, using capsules without conidia. The individuals were maintained at 26 °C for 72 h. Afterward, the insects were dissected in sterile physiological solution on a clean microscope slide. The alimentary canals were extracted and placed in 50 µL of physiological solution.

2.4.6. Cuticular Infection Assay and Dissection

Wheat flour and fungal powder (1×10^9 conidia), without encapsulation, were used to feed the insects under the conditions described in Section 2.4.5. Controls were prepared by replacing the conidia with wheat flour. The same experimental procedures were followed for dissecting the alimentary canal for observation.

2.5. Colony-Forming Unit Counts

The alimentary canals were placed in physiological solution, vigorously vortexed for 1 min, diluted to a final volume of 1 mL, and plated onto PDA leaving the alimentary canal out. All samples were incubated at 26 °C for 7 days, and colony-forming unit (CFU) counts were recorded. Controls were treated in the same manner. The experiments were repeated three times. All CFUs were examined for fluorescence using confocal microscopy.

2.6. Confocal Microscopy

Dissected alimentary canals or insect heads were placed in 20 µL of physiological solution on a clean microscope slide, sealed with a coverslip and nail polish for observation. Imaging was performed using a Carl Zeiss Observer LSM 800 Confocal Microscope, equipped with 10X, 40X plan-neofluar air objectives and a 63X apochromat oil objective (NA 0.3, 0.75, and 1.4, respectively), controlled by ZEN software 3.11. Three images were captured for each region and condition. Alimentary canal sections were imaged with the different objectives and digital zoom at 488 nm and 561 nm channels. Laser power settings were 2.51% for the 561 nm channel and 2% for the 488 nm channel (pinhole 459 nm).

2.7. Data Analysis

All statistical analyses and graphs were performed using GraphPad Prism version 10.0.0 for Windows (GraphPad Software Inc., San Diego, CA, USA; www.graphpad.com). Image analysis was conducted using Fiji (ImageJ 2.14.0; http://imagej.net/ij (accessed on 15 December 2024)) [31]. Intensity quantification was performed by setting five lines (Regions of Interest, ROI) across the image in the 488 nm and 561 nm channels, and the intensity distribution was graphed against distance. Conidia counts were performed manually using the rectangle tool to define four areas per image from three independent images. Counts were made using the Multi-point tool and were normalized by area.

3. Results

3.1. Morphological Characterization of Capsules and Conidia Distribution Within the Matrix

The morphological characterization of alginate capsule formulations was performed using optical microscopy. While initially spherical, the capsules flattened into a disc shape upon drying (Figure S1A,B). The hydrated and dried capsules, both control and conidia-containing, showed no significant differences in size (Figure S1A–D) or yield (97 ± 13 capsules/mL).

Surface topology analysis using AFM revealed no significant differences in any of the analyzed parameters between control and conidia-containing capsules (Figures 1A,B and S2), indicating that conidia do not alter the surface properties relevant for beetle feeding, at least within the resolution limits of the imaging method. It also confirms that conidia are trapped in the hydrogel matrix, reducing the risk of cuticular infection.

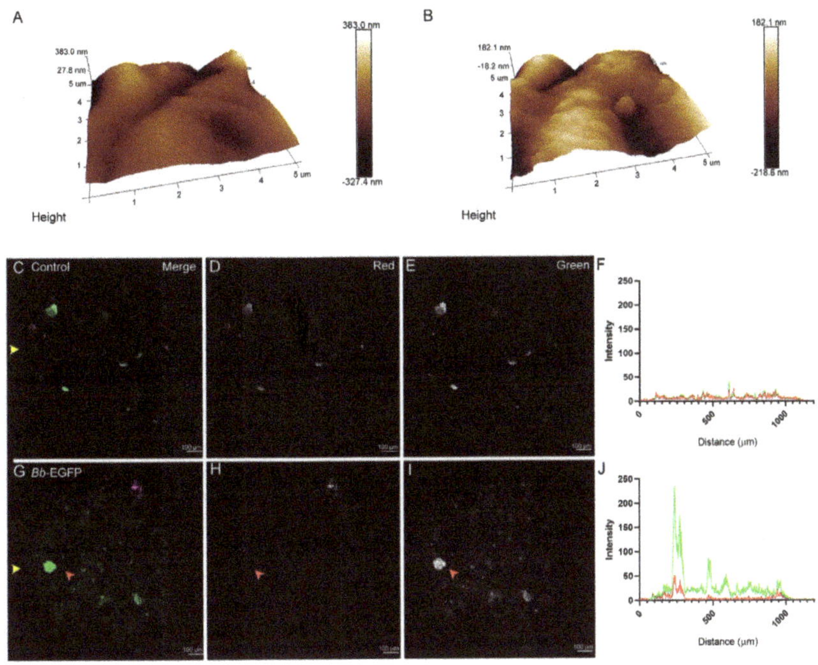

Figure 1. Microscopic characterization of *B. bassiana* alginate capsules. Control (**A**) and conidia-containing (**B**) capsule topologies were characterized using atomic force microscopy (AFM) over a 5 μm area. No significant differences were found. Confocal images of control (**C–E**) and conidia-containing (**G–I**) capsules are shown. Intensity quantification confirms the absence of green signal peaks (**F**), while (**J**) shows significantly higher green peaks in conidia-containing capsules. In (**F–J**), the green and red lines represent the green and red channel signals, respectively. Channels are indicated in the top-right corner of the top panels. Red arrows indicate conidia, and yellow arrows indicate the horizontal line where signal quantification was measured. Scale bar for confocal microscopy images represents 100 μm. Sample treatment is indicated at the top left corner.

Longitudinal sections of the hydrogel capsules confirmed uniform conidia distribution throughout the matrix. Both control (Figure 1C–F) and *B. bassiana*-containing capsules (Figure S1G–J) exhibited minimal autofluorescence. Fluorescence was much more pronounced in EGFP-tagged *B. bassiana* (Figure 1G–J). The EGFP-tagged conidia displayed peaks in the 488 nm (green) channel, with additional autofluorescence detected at 561 nm (red), consistent with fungal autofluorescence (Figure S3A–H).

Thermal and oxidative stress assays further demonstrated that the alginate formulation significantly outperformed the dried conidia powder under all tested conditions, highlighting its enhanced protective capabilities (Figure S2).

Potential cuticular infection during feeding was also tested by imaging the heads of *T. castaneum* after they were fed hydrogel-flour conidia capsules and flour-(naked) conidia powder. There was no evidence of conidia adhesion in the head cuticle in controls (Figure 2A–D) nor on the cuticle of beetles fed with hydrogel conidia capsules (Figure 2E–H), in contrast to those fed with naked-conidia powder (Figure 2I–L), where a significant green signal was shown (Figure 2I–L, red arrow), allowing the green signal to be positively identified as conidia.

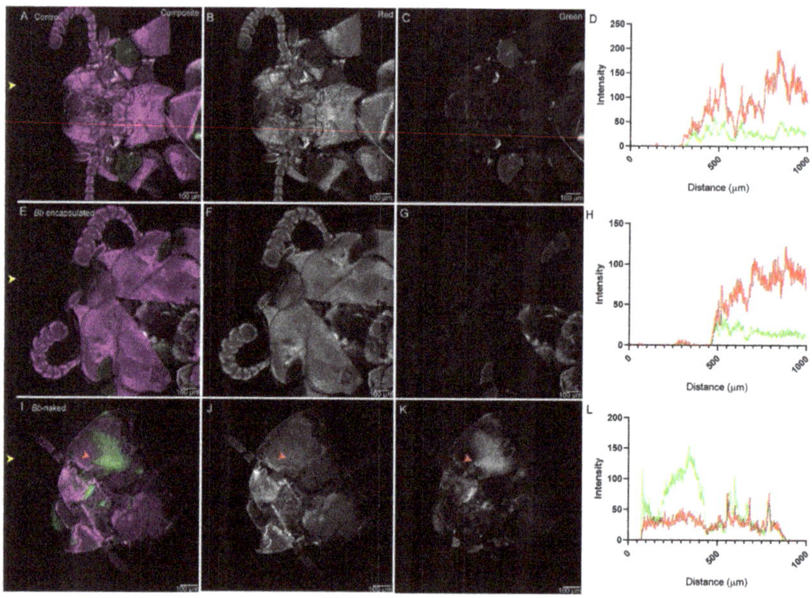

Figure 2. Adult *T. castaneum* heads after feeding. Autofluorescence in both the red and the green channels in control samples that were not in contact with fungi (**A–D**). Adults fed encapsulated conidia showed no green signal on the head cuticle (**E–H**), indicating that no conidia were detected. For beetles fed naked-conidia powder (**I–L**), an increase in the green signal was detected on the cuticle, indicating conidia adhesion (red arrow). Fluorescent signal quantification is represented in panels (**D**,**H**,**L**), green lines represent signals in the green channel, while red lines represent signals in the red channel. The channels are indicated in the top-right corner of the top panels. Yellow arrows indicate the horizontal line where signal quantification was measured. Scale bar for confocal microscopy images represents 100 μm. Sample treatment is indicated at the top left corner.

3.2. Capsules Virulence and Conidia Presence in the Alimentary Canal of T. castaneum

To evaluate the availability and virulence of the alginate hydrogel capsules, cumulative mortality assays were conducted. Larval and adult *T. castaneum* individuals were fed either conidia-containing capsules or control capsules. Larvae began to exhibit mortality as early as day 2 of feeding, whereas adult mortality was observed from day 5 onwards (Figure 3A). Cumulative mortality was also measured for the dried conidia powder used to prepare the capsules (Figure 3B). For larvae, mortality rates were comparable between the encapsulated and powder formulations, demonstrating that the hydrogel capsules effectively delivered *B. bassiana* conidia to this life stage. However, in adults, dried conidial powder induced twice the cumulative mortality compared to the capsules, suggesting that conidia powder provides greater efficacy in adult *T. castaneum*. Consistent with previous findings, adult beetles showed greater tolerance to *B. bassiana* infection, with cumulative mortality rates lower than those of larvae across both delivery methods. This disparity likely reflects differences in susceptibility between life stages, such as variations in immune defenses or cuticle composition. Despite the relatively lower mortality in adults, humid chamber assays performed on cadavers revealed fungal growth in all dead insects, indicating that *B. bassiana* confined to the hydrogel capsule successfully infected and killed both larvae and adults.

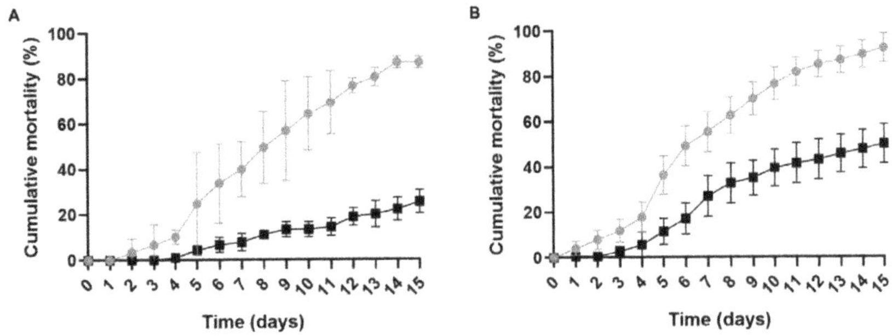

Figure 3. Cumulative mortality of larvae and adults of *T. castaneum* fed with *B. bassiana*. (**A**) Cumulative mortality for capsules. (**B**) Cumulative mortality for conidia powder. Larvae are represented in grey, and adults in black. Mortality rates were corrected using Abbott's formula.

To confirm the ingestion and localization of conidia within the alimentary canal, dissections were performed, and the alimentary canal contents were plated to evaluate colony-forming units (CFUs). EGFP-tagged colonies were recovered in both larvae and adults, demonstrating effective ingestion (Figure S4A). The recovery of EGFP-tagged *B. bassiana* from the alimentary canal confirms that the alginate hydrogel capsules served as an effective vehicle for conidial delivery.

3.3. Conidia Visualization in *T. castaneum* Alimentary Canal

The insect alimentary canal is divided into three main regions: the foregut, midgut, and hindgut, each lined by a single layer of epithelial cells (Figure 4A,B). To better understand the fate of conidia following ingestion, the alimentary canal of *T. castaneum* was imaged using confocal microscopy.

Autofluorescence in the gut tissue was initially characterized, revealing strong signals, predominantly in the 561 nm (red) channel (Figure S3I–L). Among the gut structures, the pyloric chamber cells (PyC) were distinctly identifiable in both larval and adult *T. castaneum* and were selected as reference points for imaging.

To examine the distribution of EGFP-tagged conidia within the alimentary canal of *T. castaneum* larvae, confocal images were acquired after feeding with conidia-containing hydrogel capsules. After identification of the PyC, imaging was performed on the midgut (MG) and ileum (IL) regions (blue arrows, Figure 4B). EGFP-tagged conidia were observed in large quantities surrounding the PyC, appearing as individual spherical cells or cell clusters (red arrows, Figure 4C–E). Fluorescence intensity quantification revealed distinct peaks in the green channel, corresponding to the presence of conidia, while the red channel peaks indicated autofluorescence of the PyC (Figure 4F). Conidia were clearly visible on the green channel (Figure 4E) but absent in the red channel (Figure 4D) in regions highlighted by red arrows. In control larvae, minimal to no signal was detected in the green channel, although the PyC autofluorescence was evident in the red channel (Figure 4G–J).

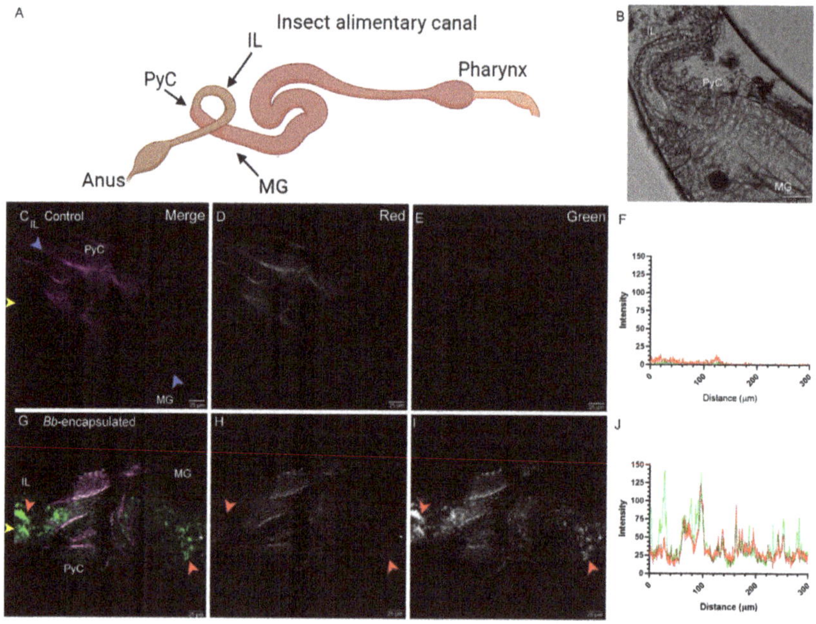

Figure 4. Presence of conidia in the alimentary canal of larval *T. castaneum* fed alginate capsules containing *B. bassiana* conidia. (**A**) Diagram of the insect alimentary canal structure with references to imaging locations. (**B**) Bright-field image of *T. castaneum* larvae showing pyloric chamber cells (PyC), midgut (MG), and ileum (IL) sections. (**C–F**) Control larvae show signal only in the red channel, while in (**G–J**), larvae fed conidia exhibit a clear green signal throughout the different regions of the alimentary canal. In (**F**,**J**), fluorescent signal quantification graphs are shown, and green and red lines indicate signals in the green and red channels, respectively. Channel information is provided in the top-right of the top panels. Red arrows indicate conidia, blue arrows indicate anatomical regions of the gut, and yellow arrows indicate the horizontal line where signal quantification was performed. Scale bar for confocal microscopy images represents 25 μm and 100 μm for brightfield, respectively. Sample treatment is indicated at the top left corner.

To further assess the localization and morphology of conidia, higher magnification images of the MG and IL regions were obtained from both control and conidia-fed larvae (Figure 5). As expected, MG control samples showed no detectable conidia in the MG (Figure 5A–C), as is reflected in the absence of signal in the green channel (Figure 5D). In contrast, larvae fed with EGFP-tagged conidia exhibited abundant fungal presence in the MG, with individual conidia clearly visible (Figure 5E–G, red arrow). Quantitative fluorescence analysis showed peaks in the green channel corresponding to the observed conidia (Figure 5H). For the IL region control samples, no conidia can be detected (Figure 5I–K) and a minimal fluorescent signal was detected in the green channel (Figure 5L). Similarly to the observations for the MG region, in the IL region of larvae fed with EGFP-tagged conidia, individual spherical conidia were detected (Figure 5M–P), with an increase in the green signal as well (Figure 5P). Conidia counts in the analyzed alimentary canal regions revealed significant differences between the MG and hindgut (HG) for both larvae and adults (Figure S4B). In larvae, conidia tended to accumulate more in the HG compared to the MG, while in adults, a higher number of conidia were found in the MG compared to the HG (Figure S4B). A closer inspection of the alimentary canal lumen revealed no evidence of germinating conidia in either the MG or IL regions, suggesting that the conidia remained viable but ungerminated within the gut environment.

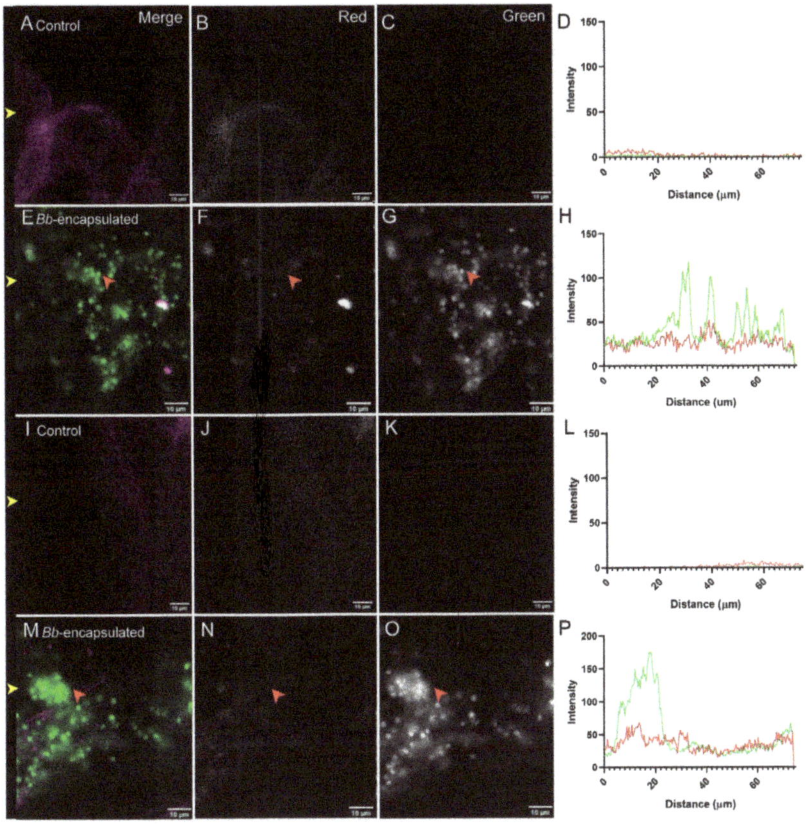

Figure 5. Conidia distribution in *T. castaneum* larvae in the midgut (MG) and ileum (IL) sections of the larval alimentary canal. In control MG (**A–D**) and IL (**I–L**), no significant green signal is observed or quantified (**D**,**L**).The MG from larvae fed conidia-containing capsules (**E–H**) displayed multiple clusters of fungal spores, where individual conidia could be clearly seen (red arrows). Similarly, in the IL from larvae fed conidia-containing capsules (**M–P**), clusters of fungal spores (red arrows) and individual conidia are visible. In (**D**,**H**,**L**,**P**), green and red lines indicate signals in the green and red channels, respectively. Channels are indicated in the top-right corner of the top panels. Yellow arrows indicate the horizontal line where signal quantification was measured. Scale bar for confocal microscopy images represents 10 μm. Sample treatment is indicated at the top left corner.

The alimentary canal of adult *T. castaneum* exhibits greater anatomical and secretory complexity compared to larvae, as previously described [24]. Notable structural features include a more melanized PyC (Figure 6A–C), the MG lined with finger-like projections known as regenerative crypts—clusters of regenerative cells (Figure 6B)—and the distinctly organized ileum (IL) and colon (CL) regions (Figure 6C). These areas were used as reference points for fluorescence imaging in adults.

In control adults, the PyC was the only distinct structure observed (Figure 6D–F), emitting low autofluorescence signals primarily detected in the red channel (Figure 6G). In contrast, adults fed EGFP-tagged conidia capsules displayed a significant accumulation of conidia within the MG. Groups of EGFP-tagged conidia were clearly visible, often mixed with the formulation matrix (Figure 6H–J). A specific conidia cluster, marked by a red arrow, is shown in the merged image (Figure 6H), absent in the red channel (Figure 6I), but prominent in the green channel (Figure 6J). Quantitative fluorescence analysis confirmed intensity peaks exclusively in the green channel (Figure 6K).

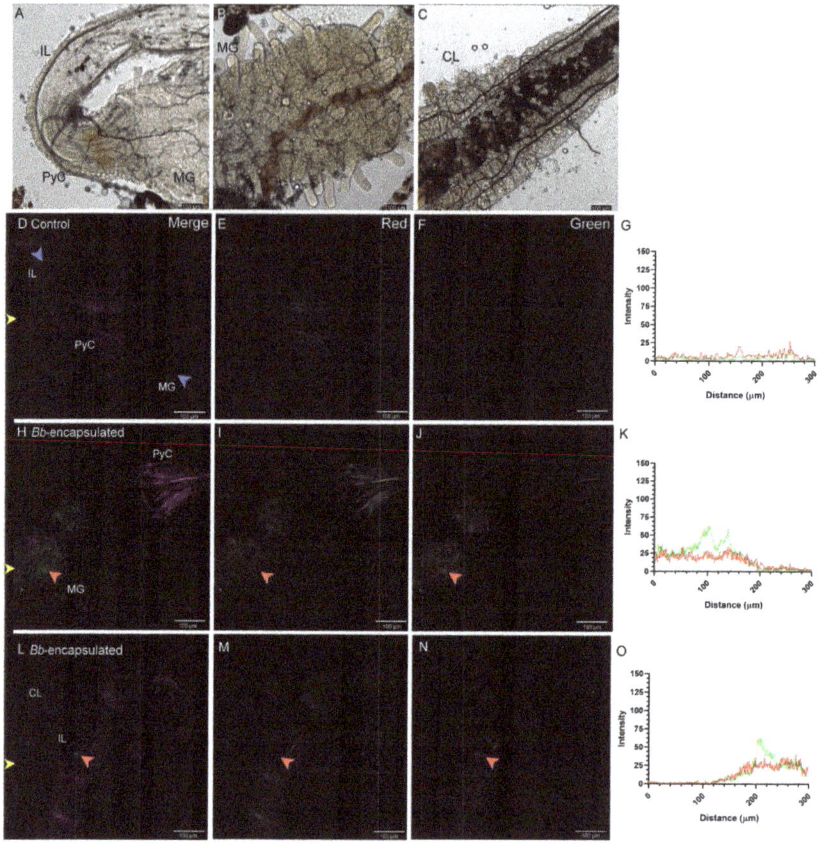

Figure 6. Presence of conidia in the alimentary canal of adult *T. castaneum* fed alginate capsules containing *B. bassiana* conidia. The adult *T. castaneum* alimentary canal has a more complex structure than that of the larval stage: (**A**) ileum (IL) and pyloric chamber (PyC); (**B**) midgut (MG) and (**C**) colon (CL). Control insects did not show significant signal in the green channel (**D–G**), whereas insects fed conidia-containing capsules displayed green signal and visible conidia (red arrows) in MG (**H–K**), and IL and CL (**L–O**). In (**G**,**K**,**O**), green and red lines indicate signals in the green and red channels, respectively. Channels are indicated in the top-right corner of the top panels. Blue arrows indicate anatomical regions of the gut, and yellow arrows indicate the horizontal line where signal quantification was performed. Scale bar for confocal microscopy images represents 100 μm. Sample treatment is indicated at the top left corner.

Similar observations were made in the ileum (IL) region (Figure 6L–N). Conidia clusters, marked by red arrows in panels L-N of Figure 6, were detected only in the green channel, with corresponding intensity quantification shown in Figure 6O. In Figure 7, higher magnification images of the MG and HG regions were analyzed. Controls for the MG (Figure 7A–D) and HG (Figure 7I–L) showed only the characteristic low autofluorescence previously observed in the insect. In the adult MG, several conidia clusters were visible (red arrows, Figure 7E–H). Conidia presence is also reflected in intensity quantification, with peaks in the green channel (Figure 7H). Similarly, in the HG, clusters and individual conidia were clearly identifiable (red arrows, Figure 7M–P), with corresponding green channel peaks observed (Figure 7P). Consistent with findings in larvae, the conidia count in adults did not show significant differences between the MG and IL regions (Figure

S4B). These results suggest that the distribution and accessibility of conidia are comparable across these alimentary canal regions in both developmental stages.

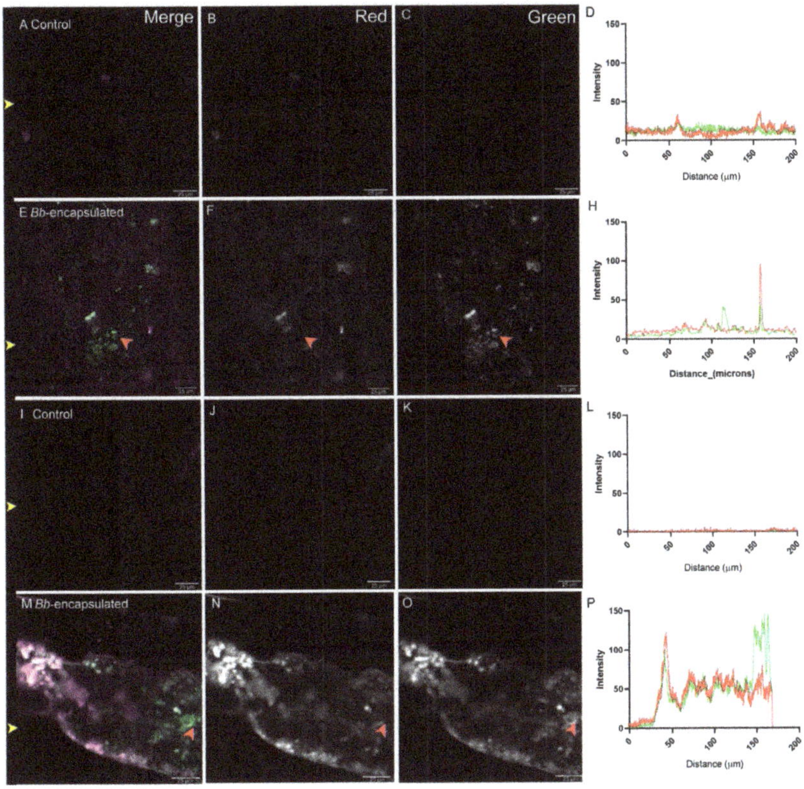

Figure 7. Conidia distribution in the adult *T. castaneum* midgut (MG) and ileum-colon (IL) sections of the alimentary canal. In control insects, no significant green signal is observed in either MG (**A–D**) or IL (**I–L**), nor is any signal quantified in panels (**D**,**L**). Adults fed conidia-containing capsules (**E–H**) displayed multiple clusters of fungal spores in the MG, with individual conidia clearly visible (red arrows). Similarly, in the IL, clusters of fungal spores (red arrows) and individual conidia were observed (**M–P**). This is also reflected in the intensity quantification, where peaks in the green channel were detected (**P**). In (**D**,**H**,**L**,**P**), green and red lines indicate signals in the green and red channels, respectively. Channel information is indicated in the top-right corner of the top panels. Red arrows point to conidia, and yellow arrows indicate the horizontal lines where signal quantification was performed. Scale bar for confocal microscopy images represents 25 μm. Sample treatment is indicated at the top left corner.

When beetles were exposed to a mixture of wheat flour and conidia powder without the gelling agent, a very different scenario emerged. As shown in Figure 8, there were no conidia in controls (Figure 8A–D). In contrast, conidia were only evident in the larval midgut (Figure 8E–H). The distribution of conidia appeared to differ from that observed when conidia were encapsulated (Figure 4), with no germination observed. For adults, no conidia were detected in the alimentary canal. Neither fungal presence nor significant differences between control (Figure 8I–L) and treated samples (Figure 8M–P) were observed. We can conclude that this delivery method is ineffective for oral infection in adults but may be relevant for larvae.

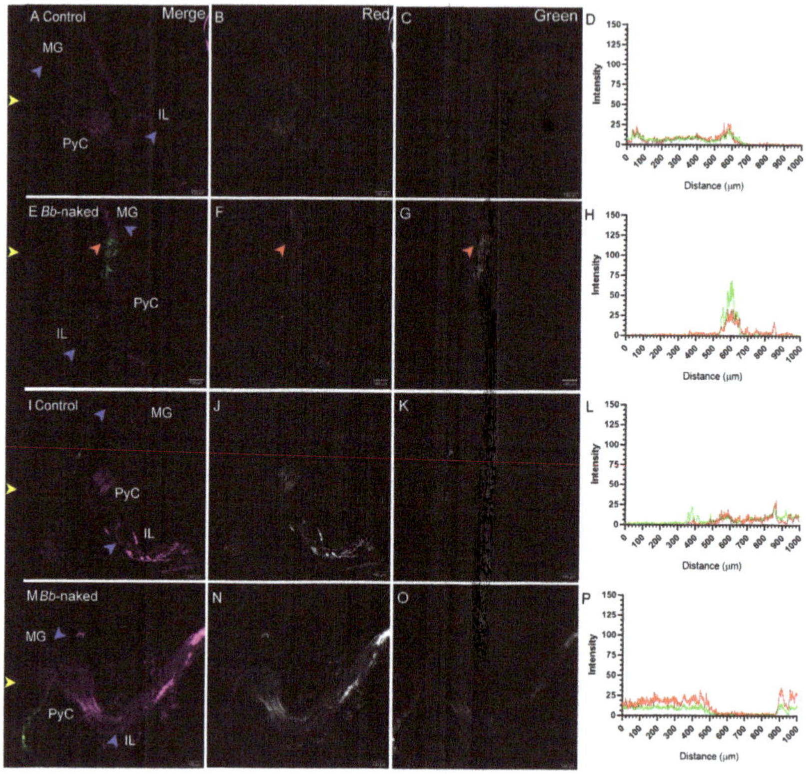

Figure 8. Larvae and adult *T. castaneum* fed with *B. bassiana* naked-conidia mixed into their diet. Control larvae did not show any signal (**A–C**) in the green channel (**D**). Larvae fed a conidia-containing diet clearly exhibited conidia in the midgut (MG) (red arrow) (**E–G**). This is also reflected in the quantification, where a green signal peak was detected (**H**). For adults, control (**I–L**) and fungus-treated samples (**M–P**) showed no significant differences. In (**D**,**H**,**L**,**P**), green and red lines indicate signals in the green and red channels, respectively. Channels are indicated in the top-right corner of the top panels. Blue arrows indicate anatomical regions of the gut, and yellow arrows indicate the horizontal line where signal quantification was performed. Scale bar for confocal microscopy images represents 100 μm. Sample treatment is indicated at the top left corner.

4. Discussion

Entomopathogenic fungi are well-known for infecting insects through cuticular adhesion and penetration, while oral infection via spore ingestion remains comparatively unexplored. This route becomes particularly relevant when fungal conidia are ingested along with food sources, such as grains and their derived products, by insect pests like the red flour beetle *T. castaneum*, which poses a significant threat to agricultural productivity, especially in developing countries. Previous studies have documented *B. bassiana's* potential to infect through oral ingestion, as its spores can bypass the alimentary canal defenses to reach internal tissues, as shown in other insect species [3,32–35]. This study aimed to investigate the potential of *B. bassiana* infection via the oral route in *T. castaneum*, focusing on its interaction within the insect's alimentary canal.

While oral fungal infections are often associated with spore germination and tissue invasion, recent studies suggest alternative mechanisms of mortality. For instance, in *Metarhizium anisopliae*, mortality in mosquito larvae has been attributed to protease-induced apoptosis rather than direct cuticular or gut invasion, highlighting the role of fungal enzymes in disrupting host defenses [36]. Similarly, *B. bassiana* may induce mortality through proteolytic enzymes, which degrade structural components of the insect gut and impair cellular immunity. This aligns with our observation that while conidia germination was not detected in the alimentary canal, mortality still occurred, potentially due to enzymatic activity or secondary metabolites disrupting nutrient absorption.

Hydrogel encapsulation, which is used to protect and extend the viability of fungal conidia, provided significant advantages for targeted pest control [37]. This formulation controlled conidial release, minimized premature germination, and provided a monitored environment for studying conidial distribution. The encapsulation process successfully ensured a uniform distribution of conidia, as confirmed by atomic force and confocal microscopies, and did not hinder their viability or infection potential. Encapsulated conidia were specifically accessible through ingestion, minimizing unintended exposure and preventing conidial adhesion to the beetle's cuticle. However, the reduced efficacy in adults suggests the need for further optimization of the capsule formulation to enhance its virulence in more resistant life stages. Notably, no fungal deposition was observed on the heads of beetles exposed to hydrogel capsules, contrasting with the conidia-flour mixture treatment, further highlighting the effectiveness of encapsulation in preventing cuticular adhesion. Although conidia ingestion via hydrogel capsules led to mortality in *T. castaneum*, the underlying mechanisms driving fungal infection following ingestion remain unclear. The current study traced the path of fluorescent *B. bassiana* conidia after ingestion and hypothesized whether these fungi might germinate within the gut. While previous studies on other insect species like *Bombyx mori* [38] documented hyphal growth from the gut, similar observations have not been reported in Coleoptera species such as *T. castaneum* [23,39].

In this study, conidia were more concentrated in the hindgut of larvae, with fewer conidia recovered from the midgut. The hindgut is a narrower section of the alimentary canal, which alone can account for a higher count of conidia per area. For adult beetles, the analysis was somewhat more complex. While the number of conidia counted in the midgut was similar to that in larvae, a reduction in the hindgut compared to both the larval hindgut and the adult midgut was observed. The observed reduction in hindgut conidia compared to larvae may reflect faster food passage and more efficient gut defenses, as seen in other insect–pathogen interactions [36]. This suggests that gut anatomy and feeding behavior may influence fungal uptake and distribution. The more developed musculature and faster food passage in adults likely contributed to a reduced fungal retention time in the gut, thus limiting fungal adhesion opportunities. Additionally, the adult midgut's regenerative crypts, which house intestinal stem cells (ISCs) that rapidly renew the gut epithelium, may complicate conidia adhesion and colonization. ISCs are known to play a role in immunity, yet their involvement in the response to entomopathogenic fungi in Coleoptera remains poorly understood [25]. Recent RNA-seq analysis of *B. bassiana*-infected beetles revealed Toll receptors and members of the JNK pathway but no evidence of antimicrobial peptides (AMPs), suggesting that the rapid passage of conidia might limit immune responses [3].

Another key point that has to be considered is the presence of secondary metabolites secreted by *B. bassiana*. After reaching the nutrient-rich inside of the host, the fungal mycelium transitions into a specialized yeast-like cell phenotype known as hyphal bodies, which secrete a diverse array of toxic secondary metabolites [2,40]. These compounds act as immunosuppressants and infection facilitators [41], with distinct roles depending on tissue (live or dead) and infection state [42–44]. For example, destruxins produced by *Metarhizium anisopliae* induce paralysis, disrupt antioxidant defenses, suppress the humoral immune system, and exhibit phagodepressant activity, contributing to host starvation [7,45]. Similarly, *B. bassiana* secondary metabolites such as beauvericins and beauverolides impair phagocytosis and immune cell movement, further weakening the host immune response [46–49]. The high abundace of conidia observed in the alimentary canal of both larvae and adults in this study, along with the associated mortality rates, may also be linked to the secretion of these toxic metabolites by fungal spores. Further research is required to identify the specific secondary metabolites involved in oral infection and to clarify their roles in the pathogenesis of *B. bassiana*.

The insect gut is protected by the peritrophic membrane, a chitin-based structure reinforced with proteins and glycoproteins [25], which may physically hinder fungal adhesion and limit contact with gut tissues. Beyond serving as an "exclusion sieve", the glycoproteins in the peritrophic membrane provide binding sites for various pathogens, restricting their access to the midgut epithelium [50]. Fungal entomopathogens such as *B. bassiana* have evolved mechanisms to breach this barrier, producing chitinases and enterotoxins, which bind to specific sites on the peritrophic membrane in Diptera and Lepidoptera [50,51]. However, this binding is influenced by the developmental stage and insect species, as the peritrophic membrane can sequester protein toxin receptors and limit the penetration of dissolved endotoxin aggregates [52,53]. Additionally, membrane-associated proteases can cleave protein toxins, further contributing to resistance [53,54].

Furthermore, digestive enzymes like proteases, lipases, and chitinases counteract fungal infection by degrading fungal cell walls, particularly the chitin matrix [55]. This could potentially explain the reduction in conidia number observed in the adult hindgut. In adult *T. castaneum*, genes involved in chitin metabolism were significantly affected when *B. bassiana* conidia was added to the insect diet [3]. This complex dynamic interaction between the fungal pathogen trying to breach the peritrophic membrane and the insect's defensive response likely contributes to the developmental stage differences observed in this study.

The gut environment also influences fungal pathogenesis. The alkaline environment of the insect gut may also exacerbate fungal degradation, as optimal fungal germination conditions typically fall within a pH range of 6–7, while the insect gut's pH may hinder fungal germination and pathogenesis [56]. Moreover, the microbiota present in the insect gut can play a critical role in immune response and resistance to entomopathogenic fungi [57]. For example, *B. bassiana* infection accelerates mortality in individuals with specific gut microbiota, while bacteria such as *Wolbachia* and antifungal-producing *Burkholderia* confer resistance to this fungus [58,59]. The role of microbiota in oral fungal infection also remains unclear in *T. castaneum*, as prior studies did not detect AMPs after oral infection, although activation of the Toll pathway was evident [3]. Despite the presence of specific insect defenses, *B. bassiana* has evolved strategies to overcome gut defenses, such as producing inhibitors of peptidases [3] and serine proteases [56] to neutralize digestive enzymes in the midgut. This resistance might explain the difference in conidial recovery between larvae and adults. As discussed below, the gut environment and the insect's immune system may influence fungal adhesion and colonization.

The differences observed when encapsulated conidia were fed to beetles, in contrast to naked-conidia powder, were more striking in adults. Although conidia adhered to the head of adult beetles were visualized, no conidia were detected in the alimentary canal. This could be due to adult *T. castaneum* repelling most of the conidia by preventing adhesion to the cuticle [18], or at least to an extent where it cannot be detected in the gut. The gut conditions may also influence the naked conidia, making them more susceptible to the intestinal environment, such as harsh pH conditions or the presence of digestive enzymes. This is an interesting question that requires further research to be answered.

The observed differences in conidial distribution and infection outcomes between larvae and adults of *T. castaneum* likely contribute to varying mortality rates in each developmental stage following fungal exposure. While both larvae and adults were susceptible to *B. bassiana* infection via oral ingestion, mortality rates were higher in larvae compared to adults. This could be attributed to several factors, including differences in immune response, gut anatomy, and digestive processes. The immune response at different insect stages may play a key role, as mortality rates from naked conidia via cuticular infection are also higher in larvae compared to adults [17–19]. The larvae immune defenses, including cellular responses, may be less effective at neutralizing fungal invaders, enabling a higher rate of infection and mortality. On the other hand, the alimentary canal of larvae is less developed, and their slower feeding behavior may allow for greater retention of conidia within the gut, increasing the likelihood of fungal colonization. Conversely, adult beetles exhibit more complex and efficient immune responses. As mentioned earlier, the presence of regenerative crypts and faster food passage could reduce the retention time of conidia in the gut, limiting fungal adhesion and colonization. The more developed immune system of adults may also play a crucial role in limiting fungal spread. For example, the activation of the Toll pathway, as observed in previous studies [3,22], could help adults mount a more rapid immune response to prevent fungal proliferation. Furthermore, the peritrophic membrane, which is more developed in adults, may serve as an additional physical barrier to fungal attachment and penetration, further protecting the beetles from infection. However, despite these defenses, adults still experience some level of mortality when exposed to *B. bassiana*. This suggests that the physiological mechanisms at play—such as immune responses, gut structure, and digestion—are not entirely sufficient to prevent fungal infection, particularly when the conidia are ingested via formulations like hydrogel capsules.

Overall, these findings highlight that fungal virulence and the effectiveness of fungal-based pest control strategies may be highly stage-dependent. The increased tolerance of adults may lead to lower mortality rates compared to larvae; however, this does not diminish the potential for significant pest control effects, particularly in larval populations. Further research into the immune dynamics, gut physiology, and microbial interactions across different developmental stages of *T. castaneum* will provide valuable insights for optimizing fungal formulations and enhancing pest management strategies.

These findings suggest that although conidia ingested through hydrogel capsules do not germinate within the alimentary canal of either larvae or adults, they still disrupt the insect's ability to process and absorb nutrients. This blockage or competition may contribute to mortality by compromising the insect's overall health. Future research should focus on elucidating the physiological mechanisms behind stage-dependent differences in fungal infections and optimizing fungal formulations for more effective pest control.

In conclusion, this study provides the first direct observation of conidia within the alimentary canal of *T. castaneum*. While conidia germination was not detected in the gut, their presence seems to contribute to mortality by impairing nutrient absorption. These results highlight the potential of encapsulated formulations in advancing fungal-based pest control strategies, opening a promising avenue for infection via fungal ingestion.

Supplementary Materials: The following supporting information can be downloaded at: https://www.mdpi.com/article/10.3390/jof11020101/s1, Figure S1: Characterization of alginate capsules. Figure S2: atomic force microscopy (AFM) measurements. Figure S3: Confocal microscopy characterization of *B. bassiana* and *T. castaneum*. Figure S4: Conidia quantification in the alimentary canal.

Author Contributions: Conceptualization, N.P. and M.C.M.; methodology, M.C.M., C.H.-B., L.P., F.G., J.C.F., E.P., D.E.-S., H.O. and E.A.; formal analysis, M.C.M., L.P., E.P. and J.C.F.; investigation, M.C.M., L.P., F.G., J.C.F., E.P. and D.E.-S.; resources, N.P., F.G., E.P., M.C.M. and C.H.-B.; writing—original draft preparation, M.C.M.; writing—review and editing, N.P., C.H.-B., E.P. and F.G.; visualization, M.C.M., L.P. and E.P.; supervision, N.P. and M.C.M.; project administration, M.C.M.; funding acquisition, N.P. All authors have read and agreed to the published version of the manuscript.

Funding: This research and the APC were funded by Agencia Nacional de Promoción Científica y Tecnológica, grant number PICT 2020 serie A 02932.

Institutional Review Board Statement: Not applicable.

Informed Consent Statement: Not applicable.

Data Availability Statement: The original contributions presented in this study are included in the article/Supplementary Materials. Further inquiries can be directed to the corresponding authors.

Conflicts of Interest: The authors declare no conflicts of interest. The funders had no role in the design of this study; in the collection, analyses, or interpretation of data; in the writing of the manuscript; or in the decision to publish the results.

References

1. Boucias, D.; Pendland, J. Attachment of mycopathogens to cuticle. The initial events of mycoses in arthropod hosts. In *The Fungal Spore and Disease Initiation in Plant and Animals*; Cole, G., Hoch, H., Eds.; Springer: New York, NY, USA, 1991; pp. 101–127.
2. Pedrini, N. The Entomopathogenic Fungus *Beauveria bassiana* Shows Its Toxic Side within Insects: Expression of Genes Encoding Secondary Metabolites during Pathogenesis. *J. Fungi* **2022**, *8*, 488. [CrossRef]
3. Mannino, M.C.; Davyt-Colo, B.; Huarte-Bonnet, C.; Diambra, L.; Pedrini, N. Transcriptomic Landscape of the Interaction between the Entomopathogenic Fungus *Beauveria bassiana* and Its Tolerant Host *Tribolium castaneum* Revealed by Dual RNA-seq. *Sci. Rep.* **2023**, *13*, 16506. [CrossRef] [PubMed]
4. Wang, Y.; Dong, Y.; Liu, K.; Li, G.; Cheng, J.; Cao, Y.; Yang, Y.; Qin, L.; Huang, B. Conserved Fungal Effector NLS1 Suppresses Lepidoptera Insect Immunity by Targeting the Host Defense Protein Hdd11. *Insect Sci.* 2024; *Early View*. [CrossRef]
5. Zhang, Q.; Wei, X.; Fang, W.; Huang, X.; Zhang, X. The Secretory Protein COA1 Enables *Metarhizium robertsii* to Evade Insect Immune Recognition during Cuticle Penetration. *Commun. Biol.* **2024**, *7*, 1220. [CrossRef]
6. Feng, P.; Shang, Y.; Cen, K.; Wang, C. Fungal Biosynthesis of the Bibenzoquinone Oosporein to Evade Insect Immunity. *Proc. Natl. Acad. Sci. USA* **2015**, *112*, 11365–11370. [CrossRef] [PubMed]
7. Pal, S.; St. Leger, R.J.; Wu, L.P. Fungal Peptide Destruxin A Plays a Specific Role in Suppressing the Innate Immune Response in *Drosophila melanogaster*. *J. Biol. Chem.* **2007**, *282*, 8969–8977. [CrossRef]
8. Li, Z.; Wang, C.; Xia, Y. Isolation of Two Locust Protein Targets of a Protein Tyrosine Phosphatase from *Metarhizium anisopliae* Strain CQMa102. *J. Invertebr. Pathol.* **2008**, *99*, 151–155. [CrossRef] [PubMed]
9. Korša, A.; Lo, L.K.; Gandhi, S.; Bang, C.; Kurtz, J. Oral Immune Priming Treatment Alters Microbiome Composition in the Red Flour Beetle *Tribolium castaneum*. *Front Microbiol.* **2022**, *13*, 793143. [CrossRef]
10. McFarlane, D.J.; Aitken, E.A.B.; Ridley, A.W.; Walter, G.H. Dietary Relationships of *Tribolium castaneum* (Herbst) with Microfungi. *J. Invertebr. Pathol.* **2020**, *173*, 107384. [CrossRef]
11. Pedrini, N. Molecular interactions between entomopathogenic fungi (Hypocreales) and their insect host: Perspectives from stressful cuticle and hemolymph battlefields and the potential of dual RNA sequencing for future studies. *Fungal Biol.* **2018**, *122*, 538–545. [CrossRef]
12. Aldrich, J. Chemical ecology of the Heteroptera. *Annu. Rev. Entomol.* **1998**, *33*, 211–238. [CrossRef]
13. Eisner, T. Beetle's spray discourages predators. *Nat. Hist.* **1966**, *75*, 42–47.
14. Ulrich, K.R.; Feldlaufer, M.F.; Kramer, M.; St. Leger, R.J. Inhibition of the entomopathogenic fungus *Metarhizium anisopliae* sensu lato in vitro by the bed bug defensive secretions (E)-2-hexenal and (E)-2-octenal. *BioControl* **2015**, *60*, 517–526. [CrossRef]
15. Sosa-Gómez, D.; Boucias, D.; Nation, J. Attachment of *Metarhizium anisopliae* to the southern green stink bug *Nezara viridula* cuticle and fungistatic effect of cuticular lipids and aldehydes. *J. Invertebr. Pathol.* **1997**, *69*, 31–39. [CrossRef] [PubMed]
16. da Silva, R.A.; Quintela, E.D.; Mascarin, G.M.; Pedrini, N.; Lião, L.M.; Ferri, P.H. Unveiling chemical defense in the rice stalk stink bug against the entomopathogenic fungus *Metarhizium anisopliae*. *J. Invertebr. Pathol.* **2015**, *127*, 93–100. [CrossRef] [PubMed]
17. Pedrini, N.; Villaverde, M.L.; Fuse, C.; Dal Bello, G.; Juárez, M.P. *Beauveria bassiana* infection alters colony development and defensive secretions of the beetles *Tribolium castaneum* and *Ulomoides dermestoides* (Coleoptera: Tenebrionidae). *J. Econ. Entomol.* **2010**, *103*, 1094–1099. [CrossRef]
18. Pedrini, N.; Ortiz-Urquiza, A.; Huarte-Bonnet, C.; Fan, Y.; Juárez, M.P.; Keyhani, N.O. Tenebrionid secretions and a fungal benzoquinone oxidoreductase form competing components of an arms race between a host and pathogen. *Proc. Natl. Acad. Sci. USA* **2015**, *112*, E3651–E3660. [CrossRef] [PubMed]
19. Akbar, W.; Lord, J.C.; Nechols, J.R.; Howard, R.W. Diatomaceous earth increases the efficacy of *Beauveria bassiana* against *Tribolium castaneum* larvae and increases conidia attachment. *J. Econ. Entomol.* **2004**, *97*, 273–280. [CrossRef]

20. Lord, J.C.; Campbell, J.F.; Sedlacek, J.D.; Vail, P.V. Application and evaluation of entomopathogens for managing insects in stored products. In *Field Manual of Techniques in Invertebrate Pathology*; Lacey, L.A., Kaya, H.K., Eds.; Springer: Dordrecht, The Netherlands, 2007. [CrossRef]
21. Mannino, M.C.; Huarte-Bonnet, C.; Davyt-Colo, B.; Pedrini, N. Is the insect cuticle the only entry gate for fungal infection? Insights into alternative modes of action of entomopathogenic fungi. *J. Fungi* **2019**, *5*, 33. [CrossRef] [PubMed]
22. Rafaluk-Mohr, C.; Wagner, S.; Joop, G. Cryptic changes in immune response and fitness in *Tribolium castaneum* as a consequence of coevolution with *Beauveria bassiana*. *J. Invertebr. Pathol.* **2018**, *152*, 1–7. [CrossRef]
23. Peirson, H. The life history and control of the pales weevil (*Hylobius pales*). *Harvard For. Bull.* **1921**, *3*, 33.
24. Ameen, M.U.; Rahman, M.F. Larval and adult digestive tracts of *Tribolium castaneum* (Herbst) (Coleoptera: Tenebrionidae). *Int. J. Insect Morphol. Embryol.* **1973**, *2*, 137–152. [CrossRef]
25. Chapman, R. Alimentary canal, digestion and absorption. *Insect* **2012**, *6*, 36–38.
26. Toegel, S.; Salar-Behzadi, S.; Horaczek-Clausen, A.; Viernstein, H. Preservation of aerial conidia and biomasses from entomopathogenic fungi *Beauveria brongniartii* and *Metarhizium anisopliae* during lyophilization. *J. Invertebr. Pathol.* **2010**, *105*, 16–23. [CrossRef]
27. Kilaru, S.; Schuster, M.; Studholme, D.; Soanes, D.; Lin, C.; Talbot, N.J.; Steinberg, G. A codon-optimized green fluorescent protein for live cell imaging in Zymoseptoria tritici. *Fungal Genet Biol.* **2015**, *79*, 125–131. [CrossRef]
28. Bernasconi, A.; Alassimone, J.; McDonald, B.A.; Sánchez-Vallet, A. Asexual reproductive potential trumps virulence as a predictor of competitive ability in mixed infections. *Environ. Microbiol.* **2022**, *24*, 4369–4381. [CrossRef]
29. Oberti, H.; Sessa, L.; Oliveira-Rizzo, C.; Di Paolo, A.; Sánchez-Vallet, A.; Seidl, M.F.; Abreo, E. Novel genomic features in entomopathogenic fungus *Beauveria bassiana ILB308*: Accessory genomic regions and putative virulence genes involved in the infection process of soybean pest Piezodorus guildinii. *Pest. Manag. Sci.* **2025**, *81*, 2323–2336. [CrossRef]
30. Abbott, W.S. A method of computing the effectiveness of an insecticide. *J. Econ. Entomol.* **1925**, *18*, 265–267. [CrossRef]
31. Schindelin, J.; Arganda-Carreras, I.; Frise, E.; Kaynig, V.; Longair, M.; Pietzsch, T.; Preibisch, S.; Rueden, C.; Saalfeld, S.; Schmid, B.; et al. Fiji: An open-source platform for biological-image analysis. *Nat. Methods.* **2012**, *9*, 676–682. [CrossRef] [PubMed]
32. Yanagita, T. Studies on Oral Infection of Larvae of the Silkworm, *Bombyx mori*, with *Beauveria bassiana*. *J. Seric. Sci. Japan* **1987**, *56*, 279–284.
33. Rizwan, M.; Atta, B. Effect of the Entomopathogenic Fungus, *Beauveria bassiana*, Combined with Diatomaceous Earth on the Red Flour Beetle, *Tribolium castaneum* (Herbst) (Tenebrionidae: Coleoptera). *Egypt J. Biol. Pest Control* **2019**, *29*, 27. [CrossRef]
34. Vianna, F.; Pelizza, S.; Russo, L.; Ferreri, N.; Scorsetti, A.C. Colonization of Tobacco Plants by Fungal Entomopathogens and the Effect on Consumption Over *Diabrotica speciosa* (Coleoptera: Chrysomelidae). *J. Fungi.* **2021**, *7*, 1017. [CrossRef]
35. Xiao, G.; Ying, S.H.; Zheng, P.; Wang, Z.L.; Zhang, S.; Xie, X.Q.; Shang, Y.; St. Leger, R.J.; Zhao, G.P.; Wang, C.; et al. Genomic Perspectives on the Evolution of Fungal Entomopathogenicity in *Beauveria bassiana*. *Sci. Rep.* **2012**, *2*, 483. [CrossRef]
36. Butt, T.M.; Greenfield, B.P.; Greig, C.; Maffeis, T.G.; Taylor, J.W.; Piasecka, J.; Dudley, E.; Abdulla, A.; Dubovskiy, I.M.; Garrido-Jurado, I.; et al. *Metarhizium anisopliae* Pathogenesis of Mosquito Larvae: A Verdict of Accidental Death. *PLoS ONE* **2013**, *8*, e81686. [CrossRef] [PubMed]
37. Vemmer, M.; Patel, A.V. Review of Encapsulation Methods Suitable for Microbial Biological Control Agents. *Biol. Control* **2013**, *67*, 380–389. [CrossRef]
38. Yoshida, M.; Iwao, S.; Iwasaki, R.; Imura, Y.; Takano, T.; Fukui, S.; Okamoto, M. Germination and Growth of *Beauveria bassiana* in the Gut of *Bombyx mori*. *J. Insect Pathol.* **1976**, *15*, 213–225.
39. Veen, K. Oral Infection of Second-Instar Nymphs of *Schistocerca gregaria* by *Metarhizium anisopliae*. *J. Invertebr. Pathol.* **1966**, *8*, 254–256. [CrossRef]
40. Zhang, L.; Fasoyin, O.E.; Molnár, I.; Xu, Y. Secondary metabolites from hypocrealean entomopathogenic fungi: Novel bioactive compounds. *Nat. Prod. Rep.* **2020**, *37*, 1181–1206. [CrossRef]
41. Lu, H.-L.; St Leger, R.J. Insect Immunity to Entomopathogenic. *Fungi. Adv. Genet.* **2016**, *94*, 251–285. [PubMed]
42. de Bekker, C.; Smith, P.B.; Patterson, A.D.; Hughes, D.P. Metabolomics Reveals the Heterogeneous Secretome of Two Entomopathogenic Fungi to Ex Vivo Cultured Insect Tissues. *PLoS ONE* **2013**, *8*, e70609. [CrossRef] [PubMed]
43. Brewer, D.; Jen, W.-C.; Jones, G.A.; Taylor, A. The Antibacterial Activity of Some Naturally Occurring 2,5-Dihydroxy-l,4-Benzoquinones. *Can. J. Microbiol.* **1984**, *30*, 1068–1072. [CrossRef]
44. McNamara, L.; Dolan, S.K.; Walsh, J.M.D.; Stephens, J.C.; Glare, T.R.; Kavanagh, K.; Griffin, C.T. Oosporein, an Abundant Metabolite in *Beauveria caledonica*, with a Feedback Induction Mechanism and a Role in Insect Virulence. *Fungal Biol.* **2019**, *123*, 601–610. [CrossRef] [PubMed]
45. Molnár, I.; Gibson, D.M.; Krasnoff, S. Secondary Metabolites from Entomopathogenic Hypocrealean Fungi. *Nat. Prod. Rep.* **2010**, *27*, 1241–1275. [CrossRef] [PubMed]

46. Zhan, J.; Burns, A.M.; Liu, M.X.; Faeth, S.H.; Gunatilaka, A.A.L. Search for cell motility and angiogenesis inhibitors with potential anticancer activity: Beauvericin and other constituents of two endophytic strains of *Fusarium oxysporum*. *J. Nat. Prod.* **2007**, *70*, 227–232. [CrossRef] [PubMed]
47. Champlin, F.R.; Grula, E.A. Noninvolvement of beauvericin in the entomopathogenicity of *Beauveria bassiana*. *Appl. Environ. Microbiol.* **1979**, *37*, 1122–1126. [CrossRef] [PubMed]
48. Vilcinskas, A.; Matha, V.; Gotz, P.J. Inhibition of phagocytic activity of plasmatocytes isolated from *Galleria mellonella* by entomogenous fungi and their secondary metabolites. *Insect Physiol.* **1997**, *43*, 475–483. [CrossRef]
49. Vilcinskas, A.; Jegorov, A.; Landa, Z.; Gotz, P.; Matha, V. Effects of beauverolide L and cyclosporin A on humoral and cellular immune response of the greater wax moth, *Galleria mellonella*. *Comp. Biochem. Physiol. C Toxicol. Pharmacol.* **1999**, *122C*, 83–92. [CrossRef] [PubMed]
50. Erlandson, M.A.; Toprak, U.; Hegedus, D.D. Role of the peritrophic matrix in insect-pathogen interactions. *J. Ins. Phys.* **2019**, *117*, 103894. [CrossRef] [PubMed]
51. Merzendorfer, H.; Kelkenberg, M.; Muthukrishnan, S. Peritrophic Matrices. In *Encyclopedia of Insect Physiology, Biochemistry and Pharmacology*; Elsevier: Amsterdam, The Netherlands, 2016; pp. 123–145. [CrossRef]
52. Yunovitz, H.; Sneh, B.; Schuster, S.; Oron, U.; Broza, M.; Yawetz, A. A New Sensitive Method for Determining the Toxicity of a Highly Purified Fraction from δ-Endotoxin Produced by *Bacillus thuringiensis* var. *entomocidus on Isolated Larval Midgut of Spodoptera littoralis (Lepidoptera, Noctuidae)*. *J. Invertebr. Pathol.* **1986**, *48*, 223–231. [CrossRef]
53. Toprak, U.; Hegedus, D.D.; Baldwin, D.; Coutu, C.; Erlandson, M. Spatial and Temporal Synthesis of *Mamestra configurata* Peritrophic Matrix through a Larval Stadium. *Insect Biochem. Mol. Biol.* **2014**, *54*, 89–97. [CrossRef]
54. Terra, W.R. The Origin and Functions of the Insect Peritrophic Membrane and Peritrophic Gel. *Arch. Insect Biochem. Physiol.* **2001**, *47*, 47–61. [CrossRef]
55. Mondal, S.; Baksi, S.; Koris, A.; Vatai, G. Journey of Enzymes in Entomopathogenic Fungi. *Pac. Sci. Rev. A Nat. Sci. Eng.* **2016**, *18*, 85–99. [CrossRef]
56. Wang, Z.; Zhang, S.; Liu, Z.; Chang, Z.; Hu, H. Gut Bacteria Promote Phosphine Susceptibility of *Tribolium castaneum* by Aggravating Oxidative Stress and Fitness Costs. *Insects* **2023**, *14*, 815. [CrossRef]
57. Ma, M.; Luo, J.; Li, C.; Eleftherianos, I.; Zhang, W.; Xu, L. A Life-and-Death Struggle: Interaction of Insects with Entomopathogenic Fungi Across Various Infection Stages. *Front. Immunol.* **2024**, *14*, 1329843. [CrossRef]
58. Panteleev, D.; Goriacheva, I.I.; Andrianov, B.V.; Reznik, N.L.; Lazebnyĭ, O.E.; Kulikov, A.M. The endosymbiotic bacterium *Wolbachia* enhances the nonspecific resistance to insect pathogens and alters behavior of *Drosophila melanogaster*. *Genetika* **2007**, *43*, 1277–1280. [CrossRef] [PubMed]
59. Boucias, D.G.; Garcia-Maruniak, A.; Cherry, R.; Lu, H.; Maruniak, J.E.; Lietze, V.-U. Detection and characterization of bacterial symbionts in the Heteropteran, *Blissus insularis*. *FEMS Microbiol Ecol.* **2012**, *82*, 629–641. [CrossRef] [PubMed]

Journal of
Fungi

MDPI

Article

In Vitro Response of Two Strains of *Cordyceps javanica* to Six Chemical Pesticides

Ruixia Mao [1,2], Xiaoxia Cai [1], Tengyu Wang [1], Ziyang Liu [1], Peixiang Xing [1], Guisen Zhang [1], Wenwen Zhou [1], Hongliang Diao [1,2,*] and Ruiyan Ma [1,*]

[1] College of Plant Protection, Shanxi Agricultural University, Taiyuan 030031, China; 18335906241@163.com (R.M.)

[2] State Key Laboratory of Integrative Sustainable Dryland Agriculture (In Preparation), Shanxi Agricultural University, Taiyuan 030031, China

* Correspondence: 13613479481@139.com (H.D.); mary@sxau.edu.cn (R.M.)

Abstract: To determine the compatibility of two new biocontrol fungi with common chemical pesticides, this study examined the effects of three insecticides, namely, avermectin, imidacloprid, and acetamiprid, and three fungicides, namely, chlorogenonil, boscalid, and kasugamycin, on the mycelial growth and spore germination of *Cordyceps javanica* strains IF-1106 and IJ-tg19. The insecticidal effects of mixed insecticides or fungicides with good compatibility with *C. javanica* IJ-tg19 against *Myzus persicae* were tested. The results showed that the six chemical pesticides exerted different degrees of inhibition on the mycelial growth of both *C. javanica* strains, with an obvious dose-dependent effect. The inhibitory effect of chlorothalonil on the mycelial growth of IF-1106 and IJ-tg19 was greater than 75%. Different kinds and concentrations of chemical pesticides had significant effects on spore germination. Among them, acetamiprid had little inhibitory effect on *C. javanica* spores. Therefore, the two *C. javanica* strains exhibited good compatibility with the insecticide acetamiprid and had some compatibility with avermectin and imidacloprid. Among the fungicides, the compatibility of the two strains of biocontrol fungi was the best with kasugamycin, followed by boscalid, while their compatibility with chlorothalonil showed the least compatibility. The median lethal time (LT_{50}) of five concentrations of *C. javanica* IJ-tg19 (1×10^3, 1×10^4, 1×10^5, 1×10^6, and 1×10^7 spore/mL) mixed with acetamiprid against *M. persicae* were 5.28, 4.56, 3.80, 2.73, and 2.13 days, respectively, and the insecticidal rate was higher than that of fungus treatment alone (5.19, 4.59, 4.05, 3.32, and 2.94 days, respectively) or chemical pesticide treatment (5.36 days). This study provides data support and a theoretical basis for reducing the use of chemical pesticides, improving the efficiency of *C. javanica*-based insecticides, and optimizing the synergistic use of fungi and chemical pesticides.

Keywords: chemical pesticides; *Cordyceps javanica*; compatibility; growth inhibition; biological control

1. Introduction

Myzus persicae (Sulzer) (Hemiptera: Aphididae) [1], an important pest affecting peach trees, plum trees, tobacco, sugar beets, and cruciferous vegetables, is widely distributed around the world [2]. In addition to causing direct damage to crops, *M. persicae* may indirectly damage crops through the transmission of plant viral diseases. At present, chemical control is the main means used to control peach aphids, but the frequent and irregular application of chemical insecticides leads to the emergence of resistance, which increases the difficulty of control. Therefore, it is imperative to develop safe, efficient, and ecologically sustainable green prevention and control technologies.

Entomopathogenic fungi (EPF) are an important biological control resource for the control of insect pests. The wide application of fungal insecticides that utilize EPF as active substances is of great significance in the continuous control of insect pests, the protection of beneficial organisms, and the maintenance of ecological balance. As an important biocontrol fungus, *Cordyceps javanica* (Friederichs & Bally) (Hypocreales: Cordycipitaceae) can infect various instars of *Bemisia tabaci* (Gennadius) (Hemiptera: Aleyrodidae), and can effectively control agricultural pests such as *Plutella xylostella* (Linnaeus) (Lepidoptera: Plutellidae) and *M. persicae* [3–6]. Fungal insecticidal preparations with *C. javanica* as the active ingredient, such as PFR-97, which was jointly developed by the United States and Mexico, and PreFeRal, which was jointly developed by many European countries, have been successfully applied to the sustainable control of whitefly [7]. In China, *C. javanica* has also been registered for the control of *B. tabaci* and *Spodoptera litura* (Fabricius) (Lepidoptera: Noctuidae) [8].

Fungal insecticides have the advantages of non-resistance and environmental friendliness, but their wide application still faces many challenges. Fungal insecticides utilize live spores as active ingredients, and their efficacy is readily affected by synthetic chemicals applied to crops. This sensitivity to chemical pesticides leads to a limited synergistic application effect, which is an important problem in the utilization of fungal insecticides. Research has shown that the inhibition effect of different kinds of chemical pesticides on biocontrol fungi varies widely. A previous study found that the compatibility of common fungicides (difenoconazole, propiconazole, trifloxystrobin, azoxystrobin, carbendazim, and hexaconazole) with *Metarhizium anisopliae* (Sorokin) (Hypocreales: Clavicipitaceae) was low, and the spore germination rate decreased by more than 50% at 48 h, indicating that this biocontrol fungus could not be mixed with these fungicides [9,10]. Herbicides such as pendimethalin, 2,4-D, trifluralin, phenmedipham, metolachlor, and chloridazon all have an obvious inhibition effect on EPF. However, lenacil, imazapic, bentazon, and glyphosate exhibit some level of compatibility with EPF. Insecticides have good compatibility with a variety of EPF. For example, previous research demonstrated that the inhibitory rates of a sublethal concentration of fenvalerate on the spore production and mycelial growth of *M. anisopliae* were 38.63% and 25.13%, respectively, and fenvalerate had no effect on spore germination [11]. The inhibition rates of spinosad and imidacloprid on the mycelial growth of *M. anisopliae* at relatively high concentrations were only 11.41% and 14.44%, respectively. Therefore, spinosad and imidacloprid can be mixed with *M. anisopliae* at certain concentrations [9]. Bifenthrin exhibits good compatibility with *Beauveria bassiana* (Balsamo-Crivelli) (Hypocreales: Cordycipitaceae), with no significant inhibitory effect on spore germination or mycelial growth [12]. The negative effects of chemical pesticides on EPF are generally manifested as fungicides > herbicides > insecticides.

To realize the combined application of fungal insecticides and chemical pesticides to improve the efficiency of pest control, it is necessary to clarify the compatibility of EPF and common chemical pesticides. Many reported results have demonstrated the feasibility of this research direction. For example, the combination of *M. anisopliae* and chlorantraniliprole showed an apparent synergic insecticidal effect on *Locusta migratoria* (Linnaeus) (Orthoptera: Acrididae). The mortality rate of *L. migratoria* after treatment with 2.0×10^8 spores/mL + 1 mg/L chlorantraniliprole reached more than 80%, while the mortality rate following treatment with 20 mg/L chlorantraniliprole was only 63.3% [13]. The combined application of diflubenzuron with *M. anisopliae* and *B. bassiana* can improve the control efficiency of *Culex pipiens* (Linnaeus) (Diptera: Culicidae); although *M. anisopliae* is more suitable for this application than *B. bassiana* [14], there are also chemical insecticides used in conjunction with *B. bassiana*. Compared with insecticides alone, the application of the 10% recommended concentration of beta-cypermethrin combined with 1×10^7 spores/mL of *B. bassiana* PfBb resulted in a higher cumulative death rate of 93.49% for *Phauda flammans* (Walker) (Lepidoptera: Phaudidae) [15]. In addition, the application of β-cyfluthrin also significantly increased the mortality of *Alphitobius diaperinus* (Panzer) (Coleoptera: Tenebrionidae) larvae infected with *B. bassiana*. However, the combination of imidacloprid, spinosad, and *B. bassiana* showed no notable synergistic effect on the control of *A. diaperinus* larvae [16]. Although the application of fungal and chemical insecticides can be reduced to a certain extent, it is necessary to continue to find tacit partners to expand their synergies.

Various kinds and concentrations of chemical pesticides have different effects on EPF, and the same pesticide may have varying effects on different strains of the same fungus. The compatibility between pesticides and EPF follows no reliable rule or model. In this study, two potential biocontrol fungi, *C. javanica* IF-1106 and *C. javanica* IJ-tg19, were selected as the research objects, and three insecticides for the control of insects with piercing–sucking mouthparts in addition to three common fungicides were selected to conduct compatibility tests between the strains and chemical pesticides. The combined virulence of the fungi and chemical control agents was verified on the basis of compatibility. The aim of this study was to clarify the inhibitory activity and inhibitory dose of several chemical pesticides on the tested strains, and to provide a theoretical foundation and data support for the collaborative application of fungal insecticides and chemical pesticides based on the above strains.

2. Materials and Methods

2.1. Culture of Fungi

The *Cordyceps javanica* (Frieder. & Bally) Kepler, B. Shrestha & Spatafora strains IF-1106 and IJ-tg19 were provided by the Biosafety and BioControl Laboratory, Shanxi Agricultural University. *C. javanica* was obtained from the original sample using a sterile inoculation ring and inoculated onto the surface of a potato dextrose agar (PDA) medium (Qingdao Hope Bio-Technology Co., Ltd., Qingdao, China) using the streak plate method. The whole process was conducted on a super-clean workbench to ensure sterility. After inoculation, the PDA plate was placed upside down in a constant temperature incubator (Shanghai Boxun Medical Biological Instrutent Corp., Shanghai, China) at 25 °C and incubated for 10 days.

2.2. Chemical Pesticides

Six common chemical pesticides were selected, comprising three insecticides and three fungicides. The sources of the pesticides are listed in Table 1.

Table 1. The type, name, dosage form, recommended dosage, and manufacturer of the vegetable pesticide tested.

Active Ingredient		Dosage Form	Control Object	Recommended Dosage in the Field (g/L)	Manufacturer
Fungicides	75% chlorothalonil	Wettable powder	Cucumber powdery mildew	10	Shandong Xinxing Pesticide Co., Ltd., Qingzhou, China
	50% boscalid	Water-dispersible granule	Tomato early blight	2	Shaanxi Huarong Kaiwei Biological Co., Ltd., Xi'an, China
	2% kasugamycin	Water aqua	Tomato leaf mold	8	Jiangmen City Plant Protection Co., Ltd., Jiangmen, China
Insecticides	2% avermectin	Microcapsule suspension agent	Whitefly	1	Hebei Weiyuan Biochemical Pesticide Co., Ltd., Shijiazhuang, China
	25% imidacloprid	Wettable powder	Aphid	0.8	Shandong United Pesticide Industry Co., Ltd., Jinan, China
	60% acetamiprid	Wettable powder	Aphid	0.15	Shaanxi Xiannong Biotechnology Co., Ltd., Xi'an, China

2.3. Preparation of Treated Medium

Each pesticide was diluted in water according to multiples of 1, 5, 25, 125, and 625, and an equal volume of sterile water was added as a blank control. A pipette was used to inject 1 mL of pesticide from a centrifuge tube into a Petri dish, and three drops of 25% lactic acid were added to each Petri dish to inhibit bacterial growth. Finally, 9 mL of the heated melted PDA medium was added to the treated Petri dish and gently shaken for mixing, and the treated medium was obtained after solidification.

2.4. Determination of Spore Germination Rates

Spore germination assays were conducted based on Ma's method, with several modifications implemented to suit the specific requirements of this study [17]. First, 50 μL of a spore suspension was removed with a pipette, injected onto the center of the pesticide-treated medium, and then evenly spread across the medium using a triangular rod in a circular pattern. After sealing the medium, the culture was placed in an incubator at a constant temperature of 25 °C for 48 h under dark conditions, after which plates with 30–300 colonies were selected for counting. The colony diameter of each treatment was obtained as the mean of three biological replicates. The inhibition rate of spore germination was calculated as follows:

$$\text{Germination rate (\%)} = \text{Spore germination number}/300 \times 100, \quad (1)$$

2.5. Strain Growth Rate Assay

We referred to Zhang's method and made some improvements [18]. The *C. javanica* strains IF-1106 and IJ-tg19 were activated on the PDA medium using the plate scribing method and then cultured in an incubator at 25 °C in the dark for 4 d to obtain single colonies of an appropriate size. Under aseptic conditions, a sterilized 7 mm hole punch was used to obtain circular samples of an equal area from the edges of the fungal colonies, and a sterilized inoculation ring was used to inoculate the fungi in the centers of the Petri dishes containing the treated medium, with the mycelium side facing down. The dishes were placed in an incubator at a constant temperature of 25 °C under dark conditions.

After 2 d of culture, the diameter of each colony was measured twice using the cross method and averaged, and the colony diameter was recorded each day from the third day onward. The colony diameters of each treatment were obtained by averaging the results of three replicates. The inhibition ratio was calculated as follows:

$$\text{Inhibition ratio (\%)} = (\text{Control colony diameter} - \text{Colony diameter of treatment group})/\text{Control colony diameter} \times 100, \quad (2)$$

2.6. Effects of C. javanica Mixed with Low Doses of Acetamiprid and Kasugamycin on the Insecticidal Activity of M. persicae

2.6.1. Test Insect Source

The aphids, *M. persicae*, were provided by the Research Group of Biosafety and Bio-Control, the College of Plant Protection, Shanxi Agricultural University. After several generations of stable breeding on pepper leaves, the newly molted adult insects were subjected to bioassay screening. The feeding material was grown naturally without spraying pesticides.

2.6.2. Bioassay

The spore suspensions of *C. javanica* were set at five concentrations of 1×10^3, 1×10^4, 1×10^5, 1×10^6, and 1×10^7 spores/mL. The process was configured using an aqueous solution containing 0.1% Tween-80 in order to disperse the spores evenly in the liquid and avoid settling. Acetamiprid and kasugamycin were diluted 625 times, and the five concentrations of *C. javanica* were mixed with low doses of acetamiprid and kasugamycin. The test method was based on the biodetermination of the pathogenicity of *C. javanica* against *M. persicae* described by Xing [19], with appropriate modifications. The specific methods were as follows: The self-planted pepper leaves were collected and washed under running water, disinfected with 75% alcohol spray, and cut into appropriately sized rounds for placing in 60 mm Petri dishes. The Petri dishes were moisturized with a water–agar medium and the leaves were placed in the dishes. Fifteen newly molted peach aphids of the same age were added to the leaves in each dish. After a short wait for the peach aphids to begin feeding, spore suspensions of different concentrations (1×10^3, 1×10^4, 1×10^5, 1×10^6, and 1×10^7 spores/mL) were evenly sprayed into each dish using a throat sprayer (spraying 20 times). To prevent the insects from escaping, the dish cover was covered with 75% alcohol to disinfect the newspaper, and the Petri dish was sealed with a sealing film. Each dish contained 15 aphids, and the test was repeated four times. Treatment with 0.1% Tween-80 solution spray was utilized as a blank control. After spray inoculation, the insects were fed in a light incubator (25 °C, 12 h light/12 h dark). The number of *M. persicae* was recorded each day, and newborn aphids were removed and observed for 7 days. Dead *M. persicae* were cultured on PDA to determine whether they were infected by *C. javanica* IJ-tg19.

2.6.3. Data Analysis

SPSS 27 software (International Business Machines Corporation, Armonk, NY, USA) was used to test the normal distribution of the data, and a one-way analysis of variance (ANOVA) was performed to compare the difference in the spore germination rate between the two EPF strains at different concentrations. The significance of differences was analyzed using Tukey's method. The column chart of the germination rate was designed in GraphPad Prism 10 (Graphpad software, San Diego, CA, USA). The Kaplan–Meier method in Origin 2021 (OriginLab, Northampton, MA, USA) was employed to analyze the survival curve of peach aphids exposed to each fungus–drug combination.

3. Results

3.1. Effects of Three Fungicides on the Spore Germination of Two *C. javanica* Strains

Compared to the control group, each tested fungicide exerted different degrees of inhibition on the conidial germination of the two strains of biocontrol fungi. The data results conformed to a normal distribution. Among the tested fungicides, 75% chlorothalonil had the greatest effect on the inhibition of the conidial germination of *C. javanica* strains IJ-tg19 (Figure 1A) and IF-1106 (Figure 1D). The conidial germination of the two strains was completely inhibited and the conidial germination rate was close to 0%. Under normal concentrations, 50% boscalid completely inhibited the conidial germination of strains IJ-tg19 and IF-1106, but the effect gradually decreased with the dilution ratio (Figure 1B, IJ-tg19: $F = 163.856$, $df = 17$, $p < 0.001$; Figure 1E, IF-1106: $F = 726.657$, $df = 17$, $p < 0.001$). On the whole, the inhibitory effect of 2% kasugamycin decreased with the dilution ratio. At ratios greater than 625, 2% kasugamycin had little effect on the spore germination of the two strains (Figure 1C, IJ-tg19: $F = 99.998$, $df = 17$, $p < 0.001$; Figure 1F, IF-1106: $F = 90.971$, $df = 17$, $p < 0.001$).

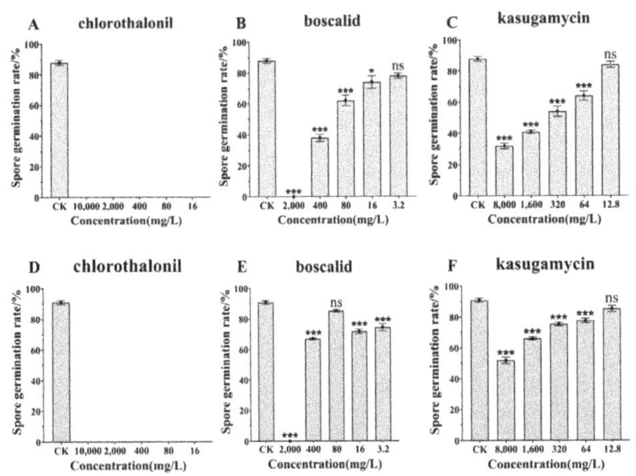

Figure 1. Effects of three fungicides on spore germination of *Cordyceps javanica* IJ-tg19 (**A–C**) and IF-1106 (**D–F**). Note: ns means no difference, * significance to $p < 0.05$, and *** significance to $p < 0.001$. CK: sterilized distilled water. Data points represent the mean ± the standard error of the mean (SEM).

3.2. Effects of Three Insecticides on the Spore Germination of Two C. javanica Strains

The three tested insecticides had significant effects on the spore germination rates of *C. javanica* strains IF-1106 and IJ-tg19 under different dilution ratios, where the lower the dilution ratio, the greater the effect. The data distribution is consistent with the normal distribution. The effects of the tested insecticides on the two strains of biocontrol fungi were in the following order: 25% imidacloprid > 2% avermectin > 60% acetamiprid. The insecticide with 25% imidacloprid was highly toxic to *C. javanica* strains IF-1106 and IJ-tg19 and inhibited spore germination even at high dilution ratios (Figure 2B, IJ-tg19: F = 49.376, df = 17, $p < 0.001$; Figure 2E, IF-1106: F = 13.103, df = 17, $p = 0.016$). The inhibition rate of 2% avermectin on the spore germination of the two strains reached more than 40% under the conventional concentration (Figure 2A, IJ-tg19: F = 140.219, df = 17, $p < 0.001$; Figure 2D, IF-1106: F = 11.639, df = 17, $p < 0.001$). The 60% acetamiprid had the least effect on the spore germination of the two strains, with the germination rates being greater than 60% under different dilutions, and the spore germination rate increased with the dilution (Figure 2C, IJ-tg19: F = 40.289, df = 17, $p < 0.001$; Figure 2F, IF-1106: F = 4.434, df = 17, $p < 0.001$).

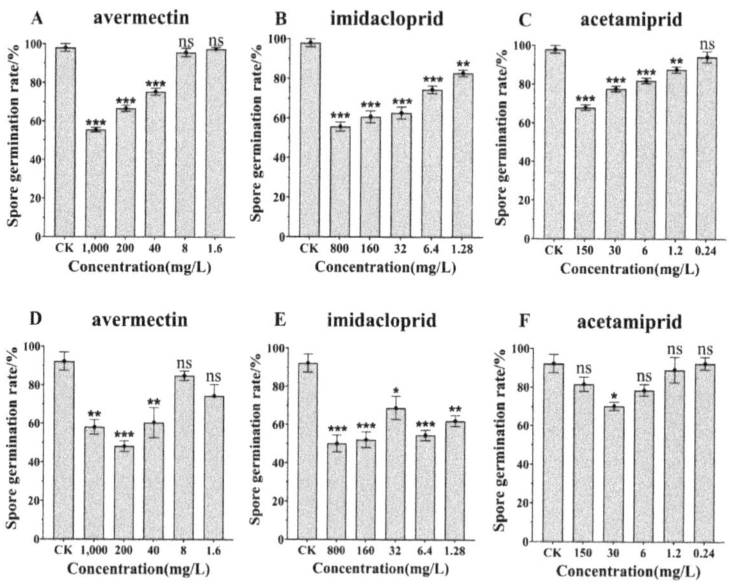

Figure 2. Effects of three insecticides on spore germination of *Cordyceps javanica* IJ-tg19 (**A–C**) and IF-1106 (**D–F**). Note: ns means no difference, * significance to $p < 0.05$, ** significance to $p < 0.01$, and *** significance to $p < 0.001$. CK: sterilized distilled water. Data points represent the mean ± the standard error of the mean (SEM).

3.3. The Effects of Pesticides on the Mycelial Growth of Two C. javanica Strains

The effects of the six pesticides on the mycelial growth of the two strains are shown in Table 2. The six tested pesticides inhibited the mycelial growth of strains IJ-tg19 and IF-1106, and the degree of inhibition varied. In general, the fungicides exerted stronger inhibitory effects on mycelial growth than the insecticides. The compatibility between 75% chlorothalonil and the two strains was the lowest, and the inhibition rate of the mycelial growth of the two strains reached more than 90%. With the increase in dilution, the inhibition rates of 50% boscalid and 2% kasugamycin on the mycelial growth of strains IF-1106 and IJ-tg19 showed decreasing trends. The inhibitory effects of the three insecticides on the mycelial growth of the two strains were weak, no more than 30%, and the inhibitory effects decreased with the increase in the dilution ratio.

Table 2. Inhibition rate of six pesticides on mycelial growth of two strains at 9th day.

Strains	Dilution Ratio	Average Rejection Rate/%					
		Fungicides			Insecticides		
		Chlorothalonil	Kasugamycin	Boscalid	Avermectin	Imidacloprid	Acetamiprid
IJ-tg19	1	100.00 ± 0.00	16.13 ± 5.59	35.48 ± 1.86	28.00 ± 1.39	29.20 ± 0.80	19.20 ± 0.80
	5	25.81 ± 1.86	15.05 ± 2.84	18.28 ± 2.84	16.80 ± 2.12	24.00 ± 2.88	11.20 ± 2.88
	25	18.28 ± 2.85	12.90 ± 1.86	16.13 ± 1.86	16.00 ± 1.38	22.40 ± 1.60	7.80 ± 2.12
	125	16.13 ± 0.52	9.68 ± 1.86	14.36 ± 1.63	5.60 ± 2.88	20.00 ± 0.80	3.73 ± 0.56
	625	6.45 ± 3.23	3.23 ± 1.86	12.90 ± 1.86	4.00 ± 1.39	7.20 ± 2.11	2.40 ± 0.80
IF-1106	1	95.10 ± 0.98	32.35 ± 3.40	32.35 ± 4.49	20.33 ± 3.12	19.39 ± 3.40	18.84 ± 1.98
	5	31.37 ± 3.53	30.39 ± 9.95	27.45 ± 4.27	15.15 ± 9.04	11.72 ± 3.38	11.78 ± 3.35
	25	23.53 ± 6.12	26.47 ± 3.39	22.55 ± 3.53	10.09 ± 1.38	10.25 ± 4.75	8.53 ± 3.81
	125	19.61 ± 4.27	14.71 ± 1.70	20.59 ± 3.40	5.54 ± 3.60	8.14 ± 2.11	3.54 ± 0.49
	625	17.65 ± 8.82	8.17 ± 2.94	8.82 ± 3.39	1.26 ± 2.65	4.51 ± 2.81	3.44 ± 1.29

3.4. Comprehensive Evaluation of Compatibility Between Six Pesticides and Two Strains of C. javanica

The effects of six commonly used pesticide preparations on the mycelial growth and spore germination of two strains of biocontrol fungi were analyzed, and the results showed that the insecticides had greater compatibility with the two strains of EPF than the fungicides. The three fungicides inhibited the growth of both strains of EPF, and the compatibility was poor. The three insecticides showed inhibitory effects on the two strains of EPF during the high-concentration stage, and the inhibitory effects gradually disappeared when the dilution ratio of the insecticides exceeded 625 times.

3.5. Pathogenicity of C. javanica Combined with Chemical Pesticides on M. persicae

The compatibility of the insecticide acetamiprid, the fungicide kasugamycin, and *C. javanica* IJ-tg19 when applied in combination or alone and their virulence against *M. persicae* were determined under indoor conditions (Table 3). The results showed that the insecticidal effect of EPF combined with chemical pesticides on *M. persicae* was obviously higher than those of EPF or insecticides alone. Among the treatments, under the condition of a dilution ratio of 625, the toxicity of acetamiprid mixed with five concentrations of *C. javanica* strain IJ-tg19 to *M. persicae* was higher than that of each agent alone, and the synergism was significant. The median lethal time (LT_{50}) values for *M. persicae* were lower than those of *C. javanica* strain IJ-tg19 alone and the pesticide alone (5.36 days), and there were significant differences in the survival analysis (Figure 3A). Under the condition of a dilution ratio of 625, there was no significant difference between the LT_{50} of kasugamycin mixed with *C. javanica* strain IJ-tg19 and *C. javanica* alone. Therefore, the mixed application of the low-dose insecticide acetamiprid and *C. javanica* IJ-tg19 has a synergistic effect, which is beneficial for sustainable pest control using *C. javanica* in the field. In addition, the low-dose fungicide kasugamycin has good compatibility with *C. javanica* and does not affect the lethal effect of *C. javanica* strain IJ-tg19 when mixed (Figure 3B).

Table 3. The toxicity to *Myzus persicae* of separately using pesticides and *C. javanica* and their combined use.

Concentration of *C. javanica* (Spores/mL)	Corrected Mortality (%) at the 7th Day (Mean ± SE)			*C. javanica*		+DCM625		+CLMS625	
	C. javanica	+DCM625	+CLMS625	LT_{50} (Days)	95% Confidence Interval	LT_{50} (Days)	95% Confidence Interval	LT_{50} (Days)	95% Confidence Interval
0	-	68.33 ± 5.69	50.00 ± 14.01	-	-	5.36	5.04~5.73	7.08	6.53~7.85
10^3	73.33 ± 17.43 b	83.33 ± 16.67 a	60.00 ± 4.71 c	5.19	4.91~5.49	5.28	4.75~5.96	5.99	5.56~6.53
10^4	86.67 ± 11.22 b	91.67 ± 8.33 a	71.67 ± 12.58 c	4.59	3.42~4.85	4.56	3.82~5.45	4.54	4.18~4.94
10^5	96.67 ± 3.33 a	98.33 ± 1.67 a	85.00 ± 5.69 b	4.05	3.44~4.69	3.80	3.33~4.26	4.40	4.12~4.69
10^6	100.00 ± 0.00 a	100.00 ± 0.00 a	100.00 ± 0.00 a	3.32	2.98~3.64	2.73	2.58~2.88	3.61	3.04~4.21
10^7	100.00 ± 0.00 a	100.00 ± 0.00 a	100.00 ± 0.00 a	2.94	2.46~3.37	2.13	1.70~2.66	2.83	2.26~3.44

Note: lowercase letters indicate differences among treatments at the 0.05 significance level.

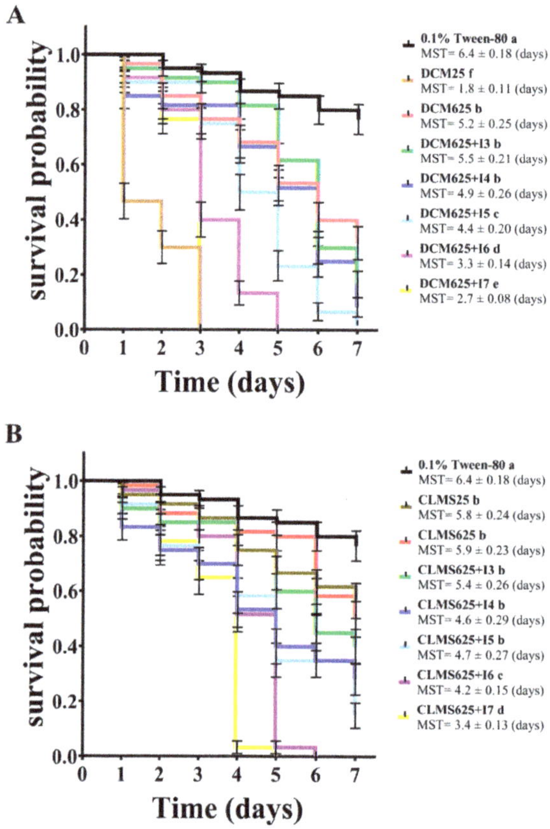

Figure 3. Survival function induced by *Cordyceps javanica* IJ-tg19 mixed with acetamiprid (**A**) and kasugamycin (**B**) in *Myzus persicae*. Note: CLMS means kasugamycin, DCM means acetamiprid, and 25 and 625 mean dilutions. I means *C. javanica* IJ-tg19; the numbers 3, 4, 5, 6, and 7 are spore concentrations of *C. javanica* IJ-tg19: 1×10^3 spores/mL, 1×10^4 spores/mL, 1×10^5 spores/mL, 1×10^6 spores/mL, and 1×10^7 spores/mL. Lowercase letters indicate differences at the 0.05 significance level in survival probabilities between groups, not differences in the median survival time (MST).

4. Discussion

In this study, the effects of six chemical insecticides and fungicides on the mycelial growth and spore germination of *C. javanica* strains IF-1106 and IJ-tg19 were measured. Compared with the control group, the three chemical fungicides exhibited strong inhibitory effects on the mycelial growth and spore germination of the two strains of biocontrol fungi, and the inhibitory effect of chlorogenonil was the most significant. Relatively speaking, the three chemical insecticides had little effect on the two strains of biocontrol fungi, and their compatibility was obviously better than that of fungicides, demonstrating the potential of these insecticides in compounding the effects of and synergistic application with EPF.

C. javanica can be employed in combination with appropriate pesticides to enhance the pest control effect and reduce the use of pesticides. In this study, the insecticidal effect of 625-fold-diluted acetamiprid combined with *C. javanica* strain IJ-tg19 was significantly better than that of acetamidine alone, and the overall dose was reduced, indicating a significant synergistic effect between the two. Acetamiprid can also be utilized in combination with *B. bassiana*, and its control effect on *B. tabaci* can reach 85.5% [20]. The combination of EPF and chemical insecticides has become an important method to reduce pesticide application and increase efficiency. For example, the combination of gamma-cyhalothrinand with *B. bassiana* LPSc 1067 can significantly increase the mortality of *Rachiplusia nu* larvae at 25% of the recommended field dose, showing a synergic effect [21]. Similarly, adding an appropriate amount of avermectin can reduce the amount of avermectin necessary and improve the control effect of *B. bassiana*. The mixed application of sublethal doses of imidacloprid and *B. bassiana* to *Leptinotarsa decemlineata* larvae also showed a synergistic effect in previous research, especially when imidacloprid was applied for 24 h followed by spraying with *B. bassiana* [22]. It is unclear why the combination of low concentrations of insecticides and EPF increases their effectiveness. Burges believes that low doses of chemical insecticides weaken insects' resistance to fungi. It has also been pointed out that low concentrations of chemical insecticides stimulate the physiological stress response of pests or change their behavior, increasing the likelihood that EPF will invade hosts [23–26].

The results of this study showed that the sensitivity of *C. javanica* to chemical pesticides was affected by different pesticide species and mechanisms of action. The three insecticides inhibited the spore germination and mycelial growth of *C. javanica* at high concentrations. Although the inhibitory activity of avermectin on fungi is unclear, avermectin can kill drug-resistant bacteria such as *Mycobacterium tuberculosis* and *M.ulcerans* [27]. High concentrations of acetamiprid inhibited the spore germination and mycelial growth of *M. anisopliae* [28] and reduced the production of its insecticidal toxin, although it did not significantly affect the infection capacity of *M. anisopliae* [29]. Imidacloprid does affect fungal spore growth [30], but the specific inhibitory mechanism remains unclear. Compared with insecticides, fungicides inhibit EPF more significantly, and the mechanism is relatively clear. Fungicides mainly inhibit fungal growth via destroying the activity of fungal metabolic enzymes, but this mechanism has little inhibitory effect on *B. bassiana* [31], which may be due to differences between strains. In contrast, *C. javanica* was less sensitive to both boscalid and kasugamycin. Boscalid inhibits the respiratory metabolism of fungal spores, while mycelial energy is mainly obtained from the outside world and does not depend on respiration [32]. Kasugamycin interferes with fungal protein synthesis, but this interference is incomplete, and normal protein synthesis still occurs [33]. These may be some of the reasons for the low inhibitory effect of kasugamycin on *C. javanica*. Although the above bactericides exerted obvious inhibitory effects on the tested strains, it is still possible to improve their compatibility. Industrial-grade boscalid inhibits fungal growth, but its preparation product Cantus does not affect the growth of *B. bassiana* or its pathogenicity against *B. tabaci* [34]. Noticeably, the inhibitory effect of fungicides on EPF can be limited to some extent by improving their formulation. In addition, to ensure the efficacy of EPF, a certain interval should be maintained between the field application of EPF and chemical fungicides.

Fungal insecticides offer environmental compatibility, safety for humans and livestock, and a reduced risk of pest resistance. However, their practical effectiveness is hampered by the prolonged infection period and high resistance in healthy pests, resulting in delayed efficacy. To address this, low concentrations of chemical insecticides can be used to weaken pest defenses, thus facilitating fungal invasion and improving performance. In addition, the combination of fungal insecticides and chemical fungicides can simultaneously prevent pests and diseases, reducing the workload of control. The collaborative application of fungal insecticides and chemical pesticides is an effective way to reduce application and increase efficiency, which can not only make up for some obvious shortcomings of fungal insecticides, but is also a reliable choice between green production and effective control. Our research and that of many scholars supports this collaborative approach, which leverages the strengths of both, showing promising progress in agricultural pest control practices.

5. Conclusions

This study determined the compatibility of *C. javanica* strains IJ-tg19 and IF-1106 with several common chemical insecticides/fungicides. For the fungus–drug combinations with poor compatibility, the application interval and method should be clearly defined during collaborative application, and for the fungus–drug combinations with good compatibility, the synergistic mechanism should be further clarified. In addition, various EPF have different sensitivities to different agents. The type and dose of agents also affect the tolerance of EPF, and the toxicological target mechanism needs to be further clarified. The development of engineered strains with high tolerance to agents and the reduction in the susceptibility of EPF through the targeted formulation of agents to increase the resistance of strains to chemical agents are also foreseeable possible approaches.

Author Contributions: Conceptualization, R.M. (Ruiyan Ma) and H.D.; methodology, R.M. (Ruixia Mao); software, P.X.; validation, R.M. (Ruixia Mao), X.C. and Z.L.; formal analysis, R.M. (Ruixia Mao); investigation, R.M. (Ruixia Mao); resources, H.D.; data curation, R.M. (Ruixia Mao); writing—original draft preparation, R.M. (Ruixia Mao); writing—review and editing, R.M. (Ruixia Mao); visualization, T.W.; supervision, G.Z. and W.Z.; project administration, H.D.; funding acquisition, R.M. (Ruiyan Ma). All authors have read and agreed to the published version of the manuscript.

Funding: This research was funded by the Key R & D Program of Shanxi Province (202302140601011), the Professor and PHD Working Station in Jinzhong Agricultural Hi-tech Industries Demonstration Zone (JZNGQBSGZZ001), the China Agricultural Research System (CARS-28-19), the special fund for Science and Technology Innovation Teams of Shanxi Province (202304051001006), the Research Program Sponsored by State Key Laboratory of Integrative Sustainable Dryland Agriculture (In Preparation), Shanxi Agricultural University (20220401091002-3-1), and the City-School Cooperative Scientific Research Program, Datong City and Shanxi Agricultural University (DT-YXHZ-202102).

Institutional Review Board Statement: Not applicable.

Informed Consent Statement: Not applicable.

Data Availability Statement: The original contributions presented in this study are included in the article. Further inquiries can be directed to the corresponding authors.

Conflicts of Interest: The authors declare no conflicts of interest.

References

1. Mohammed, A.A.; Hatcher, P.E. Combining entomopathogenic fungi and parasitoids to control the green peach aphid *Myzus persicae*. *Biol. Control* **2017**, *110*, 44–55. [CrossRef]
2. Bass, C.; Puinean, A.M.; Zimmer, C.T.; Denholm, I.; Field, L.M.; Foster, S.P.; Gutbrod, O.; Nauen, R.; Slater, R.; Williamson, M.S. The evolution of insecticide resistance in the peach potato aphid, *Myzus persicae*. *Insect Biochem. Mol. Biol.* **2014**, *51*, 41–51. [CrossRef] [PubMed]
3. Ou, D.; Zhang, L.H.; Guo, C.F.; Chen, X.S.; Ali, S.; Qiu, B.L. Identification of a new *Cordyceps javanica* fungus isolate and its toxicity evaluation against Asian citrus psyllid. *MicrobiologyOpen* **2019**, *8*, e00760. [CrossRef] [PubMed]
4. Li, Y.; Mbata, G.N.; Simmons, A.M.; Shapiro-Ilan, D.I.; Wu, S. Management of *Bemisia tabaci* on vegetable crops using entomopathogens. *Crop Prot.* **2024**, *180*, 106638. [CrossRef]
5. Wang, D.; Xing, P.X.; Diao, H.L.; Zhou, W.W.; Li, X.W.; Zhang, L.J.; Ma, R.Y. Pathogenicity characteristics of the entomopathogenic fungus *Cordyceps javanica* IJ-tg19 to *Acyrthosiphon pisum*. *BioControl* **2023**, *68*, 447–458. [CrossRef]
6. Xing, P.X.; Diao, H.L.; Wang, D.; Zhou, W.W.; Tian, J.; Ma, R.Y. Identification, pathogenicity, and culture conditions of a new isolate of *Cordyceps javanica* (Hypocreales: Cordycipitaceae) from soil. *J. Econ. Entomol.* **2022**, *116*, 98–107. [CrossRef] [PubMed]
7. Arthurs, S.P.; Bruck, D.J. Microbial control of nursery ornamental and landscape plant pests. In *Microbial Control of Insect and Mite Pests*; Elsevier: Amsterdam, The Netherlands, 2017; pp. 355–366.
8. China Pesticide Information Network. Available online: http://www.chinapesticide.org.cn/zwb/dataCenter (accessed on 27 September 2024).
9. Johnson, J.M.; Deepthy, K.B.; Chellappan, M. Tolerance of *Metarhizium anisopliae* sorokin isolates to selected insecticides and fungicides. *Entomon* **2020**, *45*, 143–148. [CrossRef]
10. Poprawski, T.J.; Majchrowicz, I. Effects of herbicides on in vitro vegetative growth and sporulation of entomopathogenic fungi. *Crop Prot.* **1995**, *14*, 81–87. [CrossRef]
11. Zhang, Y.; Nong, X.; Zhang, Z.; Wang, G. Compatibility of eighteen chemical pesticides with *Metarhizium anisopliae*. *Chin. J. Biol. Control* **2012**, *28*, 186–191.
12. Liu, F.; Zeng, M.; Wang, D.; Wang, Q.; Wu, G. Biological compatibility of six pesticides with *Beauveria bassiana* in tea plantations. *Fujian J. Agric. Sci.* **2012**, *27*, 1093–1096.

13. Jia, M.; Cao, G.; Li, Y.; Tu, X.; Wang, G.; Nong, X.; Whitman, D.W.; Zhang, Z.; Jia, M.; Cao, G.; et al. Biochemical basis of synergism between pathogenic fungus *Metarhizium anisopliae* and insecticide chlorantraniliprole in *Locusta migratoria* (Meyen). *Sci. Rep.* **2016**, *6*, 28424. [CrossRef] [PubMed]
14. Salem, H.H.A.; Mohammed, S.H.; Eltaly, R.I.; Moustafa, M.A.M.; Fónagy, A.; Farag, S.M. Co-application of entomopathogenic fungi with chemical insecticides against *Culex pipiens*. *J. Invertebr. Pathol.* **2023**, *198*, 107916. [CrossRef] [PubMed]
15. Chen, X.M.; Wang, X.Y.; Lu, W.; Zheng, X.L. Use of *Beauveria bassiana* in combination with commercial insecticides to manage *Phauda flammans* (Walker) (Lepidoptera: Phaudidae): Testing for compatibility and synergy. *J. Asia Pac. Entomol.* **2021**, *24*, 272–278. [CrossRef]
16. Rice, S.J.; Furlong, M.J. Synergistic interactions between three insecticides and *Beauveria bassiana* (Bals.-Criv.) Vuill. (Ascomycota: Hypocreales) in lesser mealworm, *Alphitobius diaperinus* Panzer (Coleoptera: Tenebrionidae), larvae. *J. Invertebr. Pathol.* **2023**, *200*, 107974. [CrossRef]
17. MA, X.; Qiao, J.; Ma, R. Effect of *Phellodendron chinense* extraction on inhibiting postharvest pathogens (*Alternatia alternata* and *Penicillium expansum*) of cherry tomato. *Food Ferment. Ind.* **2020**, *46*, 186–191.
18. Zhang, Z. Studies on Screening of Effectual Fungicides and Mixed Preparations for Controlling Potato Late Blight. Master's Thesis, Hunan Agricultural University, Changsha, China, 2017.
19. Xing, P. Discovery and Biocontrol Potential of One *Cordyceps javanica* Strain. Master's Thesis, Shanxi Agricultural University, Jinzhong, China, 2020.
20. Bevilaqua, J.G.; Padilha, G.; Pozebon, H.; Marques, R.P.; Filho, A.C.; Ramon, P.C.; Boeni, L.; Castilhos, L.B.; da Luz, G.R.; de Souza Brum, A.L.S.; et al. A sustainable approach to control whitefly on soybean: Integrating entomopathogenic fungi with insecticides. *Crop Prot.* **2023**, *164*, 106145. [CrossRef]
21. Pelizza, S.A.; Schalamuk, S.; Simón, M.R.; Stenglein, S.A.; Pacheco-Marino, S.G.; Scorsetti, A.C. Compatibility of chemical insecticides and entomopathogenic fungi for control of soybean defoliating pest, *Rachiplusia nu*. *Rev. Argent. Microbiol.* **2018**, *50*, 189–201. [CrossRef]
22. Furlong, M.J.; Groden, E. Evaluation of synergistic interactions between the colorado potato beetle (Coleoptera: Chrysomelidae) pathogen *Beauveria bassiana* and the insecticides, imidacloprid, and cyromazine. *J. Econ. Entomol.* **2001**, *94*, 344–356. [CrossRef]
23. Burges, H.D. *Formulation of Microbial Biopesticides*; Kluwer Academic Publishers: London, UK, 1998; p. 412.
24. Mohan, M.C.; Reddy, N.P.; Devi, U.K.; Kongara, R.; Sharma, H.C. Growth and insect assays of *Beauveria bassiana* with neem to test their compatibility and synergism. *Biocontrol Sci. Technol.* **2007**, *17*, 1059–1069. [CrossRef]
25. Koodalingam, A.; Dayanidhi, M.K. Studies on biochemical and synergistic effects of immunosuppressive concentration of imidacloprid with *Beauveria bassiana* and *Metarhizium anisopliae* for enhancement of virulence against vector mosquito *Culex quinquefasciatus*. *Pestic. Biochem. Physiol.* **2021**, *176*, 104882. [CrossRef]
26. Tomilova, O.G.; Kryukov, V.Y.; Duisembekov, B.A.; Yaroslavtseva, O.N.; Tyurin, M.V.; Kryukova, N.A.; Skorokhod, V.; Dubovskiy, I.M.; Glupov, V.V. Immune-physiological aspects of synergy between avermectins and the entomopathogenic fungus *Metarhizium robertsii* in Colorado potato beetle larvae. *J. Invertebr. Pathol.* **2016**, *140*, 8–15. [CrossRef] [PubMed]
27. Gao, Q.; Tan, G.Y.; Xia, X.; Zhang, L. Learn from microbial intelligence for avermectins overproduction. *Curr. Opin. Biotechnol.* **2017**, *48*, 251–257. [CrossRef] [PubMed]
28. Neves, P.; Hirose, E.; Tchujo, P.T.; Moino, A. Compatibility of entomopathogenic fungi with neonicotinoid insecticides. *Neotrop. Entomol.* **2001**, *30*, 263–268. [CrossRef]
29. Nowak, M.; Bernat, P.; Mrozińska, J.; Różalska, S. Acetamiprid affects destruxins production but its accumulation in *Metarhizium* sp. spores increases infection ability of fungi. *Toxins* **2020**, *12*, 587. [CrossRef]
30. Choudhury, R.A.; Sutherland, A.M.; Hengel, M.J.; Parrella, M.P.; Gubler, W.D. Imidacloprid movement into fungal conidia is lethal to mycophagous beetles. *Insects* **2020**, *11*, 496. [CrossRef]
31. Celar, F.A.; Kos, K. Effects of selected herbicides and fungicides on growth, sporulation and conidial germination of entomopathogenic fungus *Beauveria bassiana*. *Pest. Manag. Sci.* **2016**, *72*, 2110–2117. [CrossRef]
32. Samal, I.; Bhoi, T.K.; Vyas, V.; Majhi, P.K.; Mahanta, D.K.; Komal, J.; Singh, S.; Kumar, P.V.D.; Acharya, L.K.; Samal, I.; et al. Resistance to fungicides in entomopathogenic fungi: Underlying mechanisms, consequences, and opportunities for progress. *Tro. Plant Pathol.* **2023**, *49*, 5–17. [CrossRef]
33. Zhang, Y.; Aleksashin, N.A.; Klepacki, D.; Mankin, S.A. The context of the ribosome binding site in mRNAs defines specificity of action of kasugamycin, an inhibitor of translation initiation. *Proc. Natl. Acad. Sci. USA* **2022**, *119*, e2118553119. [CrossRef]
34. Roberti, R.; Righini, H.; Masetti, A.; Maini, S. Compatibility of *Beauveria bassiana* with fungicides in vitro and on zucchini plants infested with *Trialeurodes vaporariorum*. *Biol. Control* **2017**, *113*, 39–44. [CrossRef]

Journal of
Fungi

MDPI

Article

MaAzaR Influences Virulence of *Metarhizium acridum* against *Locusta migratoria manilensis* by Affecting Cuticle Penetration

Geng Hong [1,2,3], Siqing Wang [1,2,3], Yuxian Xia [1,2,3] and Guoxiong Peng [1,2,3,*]

1 Genetic Engineering Research Center, School of Life Sciences, Chongqing University, Chongqing 401331, China; 17602379203@163.com (G.H.); 1330878355@163.com (S.W.); yuxianxia@cqu.edu.cn (Y.X.)
2 Chongqing Engineering Research Center for Fungal Insecticide, Chongqing 401331, China
3 Key Laboratory of Gene Function and Regulation Technologies under Chongqing Municipal Education Commission, Chongqing 401331, China
* Correspondence: gxpeng@cqu.edu.cn

Abstract: The entomopathogenic fungus (EPF) *Metarhizium acridum* is a typical filamentous fungus and has been used to control migratory locusts (*Locusta migratoria manilensis*). This study examines the impact of the Zn(II)2Cys6 transcription factor, MaAzaR, in the virulence of *M. acridum*. Disruption of *MaAzaR* (Δ*MaAzaR*) diminished the fungus's ability to penetrate the insect cuticle, thereby decreasing its virulence. The median lethal time (LT_{50}) for the Δ*MaAzaR* strain increased by approximately 1.5 d compared to the wild-type (WT) strain when topically inoculated, simulating natural infection conditions. Δ*MaAzaR* compromises the formation, turgor pressure, and secretion of extracellular hydrolytic enzymes in appressoria. However, the growth ability of Δ*MaAzaR* within the hemolymph is not impaired; in fact, it grows better than the WT strain. Moreover, RNA-sequencing (RNA-Seq) analysis of Δ*MaAzaR* and WT strains grown for 20 h on locust hindwings revealed 87 upregulated and 37 downregulated differentially expressed genes (DEGs) in the mutant strain. Pathogen–host interaction database (PHI) analysis showed that about 40% of the total DEGs were associated with virulence, suggesting that MaAzaR is a crucial transcription factor that directly regulates the expression of downstream genes. This study identifies a new transcription factor involved in EPF cuticle penetration, providing theoretical support and genetic resources for the developing highly virulent strains.

Keywords: Zn(II)2Cys6 transcription factor; *Metarhizium acridum*; virulence; cuticle penetration; RNA-seq

1. Introduction

Pests like locusts, grasshoppers, termites, and cattle ticks have caused enormous economic and agricultural losses in the world [1–7]. In order to control these pests and vectors, chemical insecticides are frequently used as a solution [8]. However, the application of chemical insecticides has resulted in high environment pollution and negative effects on natural predators of pests [5,9–13]. In addition, the development of resistance in these pests has prompted in-depth research into replacing chemical insecticides with biological control agents such as fungi, bacteria, viruses, and nematodes [5,7,9,14–18]. Entomopathogenic fungi (EPF) infect insects by direct contact of the cuticle, unlike bacteria or viruses, which have to be ingested by an insect [19,20]. EPF do not cause environmental pollution and can even remove toxic contaminants such as alkylphenols, organotin compounds, pesticides, and hydrocarbons [19], so they are viewed as a viable alternative to chemical insecticides. Among them, *Metarhizium* spp. and *Beauveria* spp. are the most studied and widely used insecticidal fungi.

Metarhizium acridum has been widely studied world-wide as a model for studying insect–fungal interactions and as a genetic resource for biotechnology [21]. The key stages in the invasive and developmental processes of EPF are as follows: spore adhesion, germination, differentiation of appressoria, penetration, colonization of the hemocoel, and sporulation on insect cadavers [22]. *M. acridum* has been successfully used to control locust populations in Northern China. From 2002 to 2006, ground and aerial applications covered 6000 hectares of grasslands and the adjacent beach along the Yellow River, killing over 90% of *Locusta migratoria manilensis* (Meyen) within 11–15 days post-treatment [23]. The host's immune response to EPF results in a long period of EPF pathogenicity, limiting its practical application. During penetration, appressoria exert significant physical force and secrete proteases like subtilisin-like proteases (Pr1) and trypsin-like proteases (Pr2 and Try1) to hydrolyze the host cuticle [24]. Insects engage in cuticular sclerotization by converting polyphenol and quinone into melanin through the action of phenoloxidases (POs). In addition, POs directly attach to and melanize fungal cell walls, thereby impeding mycelial growth [25]. Once EPF invade the hemolymph, they multiply as hyphal bodies (HBs). Hosts defend against HBs via both humoral response and cellular response, including antimicrobial peptides (AMPs), phagocytosis, encapsulation, and so on. EPF produce the collagen-like protein Mcl1, which encapsulates the surface of HBs to evade the host's immune response [26]. Therefore, it is important to understand the mechanisms behind EPF infection of insects.

The Zn(II)2Cys6 binuclear cluster domain (IPR001138, PF00172) is unique to the fungal kingdom [27]. This multifunctional factor regulates numerous cellular processes, including carbon and nitrogen utilization [28,29], amino acid metabolism [27], and pyrimidine metabolism [30]. It also influences virulence in a lot of pathogenic fungi. UvZnFTF1 is involved in *Ustilaginoidea virens* vegetative growth, conidiation, pigment biosynthesis, and pathogenicity [31]. Tpc1 is an important Zn(II)2Cys6 transcriptional regulator necessary for polarized growth and virulence in the rice blast fungus *Magnaporthe oryzae* [32]. BbTpc1 in *Beauveria bassiana*, the homolog gene of Tpc1 in *M. oryzae*, is involved in chitin biosynthesis, fungal development, and virulence [33]. Azaphilone cluster-specific TF (AzaR), a Zn(II)2Cys6 fungus-specific TF, regulates the biosynthesis of polyketone secondary metabolites in *Aspergillus niger* and is involved in stress tolerance and conidiation pattern shift in *M. acridum* [34]. However, its role in the virulence of EPF remains unknown.

In this study, we report that MaAzaR is involved in the early stages of *M. acridum* infection. Disruption of *MaAzaR* delays cuticle penetration without affecting in vivo growth, thereby reducing its virulence. Further analysis revealed that Δ*MaAzaR* impedes cuticle penetration by delaying appressoria formation and impairing turgor pressure and hydrolytic enzyme secretion in appressoria. We also performed RNA-sequencing (RNA-seq) on WT and Δ*MaAzaR* strains after 20 h of locust hindwing growth to identify and analyze the key pathways in appressorium formation that may be regulated by MaAzaR. Our data provide important insights into host–pathogen interactions and are helpful to improve the practical application of biological pesticides.

2. Materials and Methods

2.1. Strains and Growth Conditions

The *M. acridum* strain CQMa102, utilized in this study, is preserved at the China General Microbial Culture Collection (strain No. CGMCC No. 0877). This strain was cultured at 28 °C for 15 d on 1/4 SDAY solid medium, consisting of 10 g glucose, 5 g yeast extract, 2.5 g tryptone, distilled water up to 1000 mL, and another 1.8% agar for solid medium (w/v). *L. migratoria* (Oriental migratory locust) were maintained in an insectary set at 28 °C with 60% relative humidity, following a 13 h light and 11 h dark cycle, and were fed bran and corn leaves. All media were sterilized using an autoclave at 121 °C for 30 min.

2.2. Bioassays

To evaluate the virulence of the three strains (WT, CP, and Δ*MaAzaR*) against *L. migratoria*, 2-day-old 5th instar locusts were subjected to topical inoculation and in vivo intra-hemocoel injection bioassays as previously described [35]. For topical inoculation,

5 μL of a conidia suspension of each strain, adjusted to a concentration of 1×10^7 conidia per mL in paraffin oil, was applied to the locust dorsal plate. For the intra-hemocoel injection bioassay, the conidia suspension of each strain was prepared with 0.05% Tween 80 to a concentration of 1×10^6 conidia/mL and injected into the third and fourth abdominal segments of the locusts using a microsyringe with a volume of 5 μL. Thirty insects were used for each treatment, with the experiment independently repeated three times. Insects were maintained at 28 °C with 70% humidity and fed every 12 h, and mortality was recorded. Conidiation on the host insect cadaver was assessed as previously described [36]. Briefly, locust carcasses were surface-sterilized, stored in humidified containers, and incubated at 28 °C. After spore emergence from the locust carcasses, the carcasses were photographed.

2.3. RNA Isolation, RT-qPCR, and RNA Sequencing

RNA extraction was performed following the manufacturer's instructions for the Ultrapure RNA kit (CWBIO, Beijing, China). Samples were ground in liquid nitrogen and extracted with Trizol. RNA quality was assessed using a 2% agarose gel. Reverse transcription was conducted with the PrimeScript™ RT Reagent Kit (Takara, Dalian, China), and qPCR analysis was performed using SYBR® Premix Ex Taq™ (Takara, Dalian, China). Relative expression levels of target genes were calculated using the $2^{-\Delta\Delta Ct}$ method [37], with *Actin* and *Gapdh* genes serving as reference genes for *L. migratoria* and *M. acridum*, respectively. For the transcriptome analysis, the washed (three times in distilled water) and sterilized (by autoclaving at 121 °C for 30 min) locust hind wings were placed in a centrifuge tube with 5 mL of conidium suspension (1×10^7 conidia/mL) and vortexed thoroughly for at least 10 min. The wings were laid flat on a clean slide, the excess water was absorbed with filter paper, and then the wings were placed in a Petri dish lined with two layers of moistened filter paper (2 mL of sterile water was evenly added in advance). The Petri dish was incubated at 28 °C for 20 h. RNA libraries were sequenced on the Illumina Novaseq 6000 platform by LC Bio Technology Co., Ltd (Hangzhou, China). The primers used in this study are listed in Table S1.

2.4. Cuticle Penetration Simulation

For the simulation of cuticle penetration, wings were obtained from adult locusts, washed three times with distilled water, and sterilized by autoclaving. The sterile locust hindwings were spread flat on 1/4 SDAY plates and inoculated with 2 μL of a conidial suspension (1×10^7 conidia/mL) of one of the three strains (WT, CP, or Δ*MaAzaR* mutant). The plates were incubated at 28 °C for 48 or 72 h. At each time point, the locust hindwings were removed and the plates were further incubated at 28 °C for five d. For the control groups, 2 μL of conidial suspension (1×10^7 conidia/mL) of each strain was directly spotted onto 1/4 SDAY plates. Clone sizes of the strains on the plates were then compared [38].

2.5. Appressorium Formation and Turgor Pressure Measurement

The method for inducing appressorium formation on locust hindwings, as described in Section 2.3, was used for transcriptome sample collection. Conidia were incubated on the wings for 16, 18, 20, 24, 28, 32, or 36 h, and the rate of appressorium formation was counted in 300 conidia at each time point. Turgor pressure was estimated by adding different concentrations (0.2 to 1.2 g/mL) of PEG-8000 (polyethylene glycol) to the appressoria; after 10 min, the collapse rate of the appressoria was calculated.

2.6. Phenoloxidase Assay

Locusts were inoculated as previously described in Methods Section 2.4. Hemocoel was extracted from the locusts (10 μL per locust, five insects per treatment) at the appropriate times, with a total of 50 μL of hemocoel being prepared. After centrifugation at $30\times g$ for 10 min, the supernatant (hemolymph) was aspirated from the pellet and placed on ice. Subsequently, 300 μL of 0.1 M potassium dihydrogen phosphate buffer (pH 6.0, 28 °C), 10 μL of 0.01 M levodopa, and 10 μL of hemolymph were added to individual wells of a

96-well plate, which was then shaken to ensure uniform mixing of the solution. One unit of PO enzyme activity is defined as an increase of one OD_{485} unit per mg of protein per min. Three biological replicates were performed for each sample [35].

2.7. In Vitro *Culture of Hyphal Bodies and its DNA Content Quantification*

Conidia from WT were inoculated in 1/4 SDAY liquid medium and incubated in a shaker at 28 °C for 3 d. Mycelium was ground in liquid nitrogen after filtration, and the genome of the mycelium was extracted following the instructions on the OMEGA Genome Extraction Kit (Omega, GA, USA). A gradient dilution (10–100 pg) of the genome of WT was then used as a template for real-time quantitative PCR (qPCR) using *M. acridum*-specific ITS primers to make a standard curve.

Hemolymph was extracted from healthy fifth instar *L. migratoria* (10 μL per insect, 50 insects per group, a total of 500 μL hemolymph). Centrifugation at 600 rpm for 10 min was performed to separate the hemocytes from the hemolymph. A total of 2 μL of conidia suspension (1×10^6) of each strain was added to the supernatant. The mixture was then incubated at 28 °C with shaking. HBs DNA was extracted 1 to 4 days post-inoculation (dpi) according to the instructions of the OMEGA Genome Extraction Kit (Omega, GA, USA) and was used as a template for the detection of HBs Ct values using *M. acridum*-specific ITS primers, which were then substituted into a standard curve to derive the genomic concentration of HBs. There were three replicates per group, and the experiment was repeated three times independently [39].

2.8. *Data Analysis*

All experiments were carried out with three independent replicates. Statistical analysis was performed using SPSS (SPSS Inc., Chicago, IL, USA, version 26.0 for Windows). Data were expressed as mean ± standard error of the mean (SEM). One-way ANOVA was used to compare more than two samples, and the Tukey test was conducted as the multiple pairwise comparison test once ANOVA was shown to be significant.

3. Results

3.1. *Disruption of MaAzaR Reduces the Virulence of M. acridum by Impeding Cuticle Penetration*

To investigate the function of MaAzaR in *M. acridum*, the *MaAzaR* knockout (Δ*MaAzaR*) and complementation (CP) strains were constructed in previous research [34]. To ask whether MaAzaR affects *M. acridum* virulence, we used *L. migratoria* as the host to perform a topical inoculation bioassay to simulate natural infection. The results revealed a significantly reduced mortality rate in locusts inoculated with Δ*MaAzaR* compared to the WT and CP strains. At 8 dpi, all locusts in the other treatments had succumbed, while 20% of locusts treated with Δ*MaAzaR* remained alive, with complete mortality occurring only at 10 dpi. Further analysis of median lethal time (LT_{50}) showed an average increase of 1.5 d in LT_{50} for the Δ*MaAzaR* knockout strain compared to the WT (Figure 1A,B). Visual inspection of locust cadavers killed by the WT and CP strains exhibited robust conidiation 7 days post-death (Figure 1C), disruption of *MaAzaR* further impairing *M. acridum* prevalence in the locust population.

The insect integument is the first obstacle against fungal invasion [21]. Cuticle penetration is the initial step for the successful infection of the host by the EPF. Therefore, we simulated the process of *M. acridum* infecting locusts using locust hind wings, as described in Methods 2.4. At 48 h post-inoculation (hpi), both the WT and CP strains successfully penetrated the hind wings of locusts. However, the knockout strain could only achieve penetration at 72 hpi. This suggests that MaAzaR may influence the virulence of *M. acridum* by affecting the process of host cuticle penetration (Figure 1D).

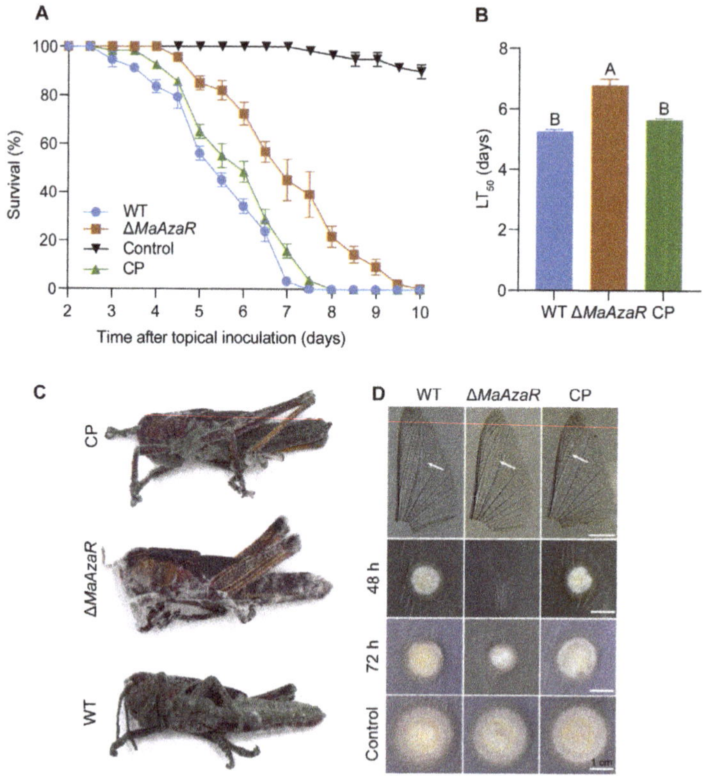

Figure 1. Insect bioassay and simulation of *Metarhizium acridum* cuticle penetration. (**A**) Survival of *Locusta migratoria manilensis* topically inoculated with each strain. (**B**) Calculated LT_{50} (median lethal time, i.e., time until death) values of *L. migratoria* in the topical inoculation bioassay. (**C**) Image of locust cadaver 7 days post-death in topical inoculation bioassay. (**D**) Cuticle penetration simulation assay; the white arrow points to the location of the conidia vaccinate. Data points represent the mean ± standard error of the mean (SEM). The same capital letters above two bars indicate no significant difference between the samples by one-way ANOVA and Tukey test ($p > 0.01$).

3.2. *MaAzaR* Influences Cuticle Penetration Process by Affecting Appressoria

We firstly quantified the expression levels of adhesion-related genes, including *Mad1* and *Mad2*, and observed that the expression levels of these genes in Δ*MaAzaR* were significantly lower than those in the WT and CP strains, with reductions of at least 50% (Figure 2A). Differentiation of appressoria is a prerequisite for infection by most EPF [22]. So, we then quantified the appressoria formation rates of different strains and found that Δ*MaAzaR* formed less appressoria than WT (Figure 2B). The AFT_{50} (the time required for a 50% appressoria formation rate) for Δ*MaAzaR* was 4.6 h longer than that for WT (Figure 2C). Additionally, the turgor pressure of the appressoria formed by Δ*MaAzaR* strain was lower than that of the WT strain (Figure 2D,E). Then, we stained appressoria with Nile Red and observed that the appressoria formed by Δ*MaAzaR* had fewer lipid droplets compared to those formed by WT (Figure 2F,G). Also, the relative expression of proteases *MaPR1*, *MaPR1C*, and *MaPR1E* and the chitinase *MaCHIT1* in Δ*MaAzaR* during appressoria formation is also significantly lower than those in the WT and CP strains (Figure 2A). These results suggest that the absence of *MaAzaR* leads to slowing down of the cuticle penetration process by reducing the formation of appressoria as well as impairing turgor pressure and hydrolytic enzymes of appressoria.

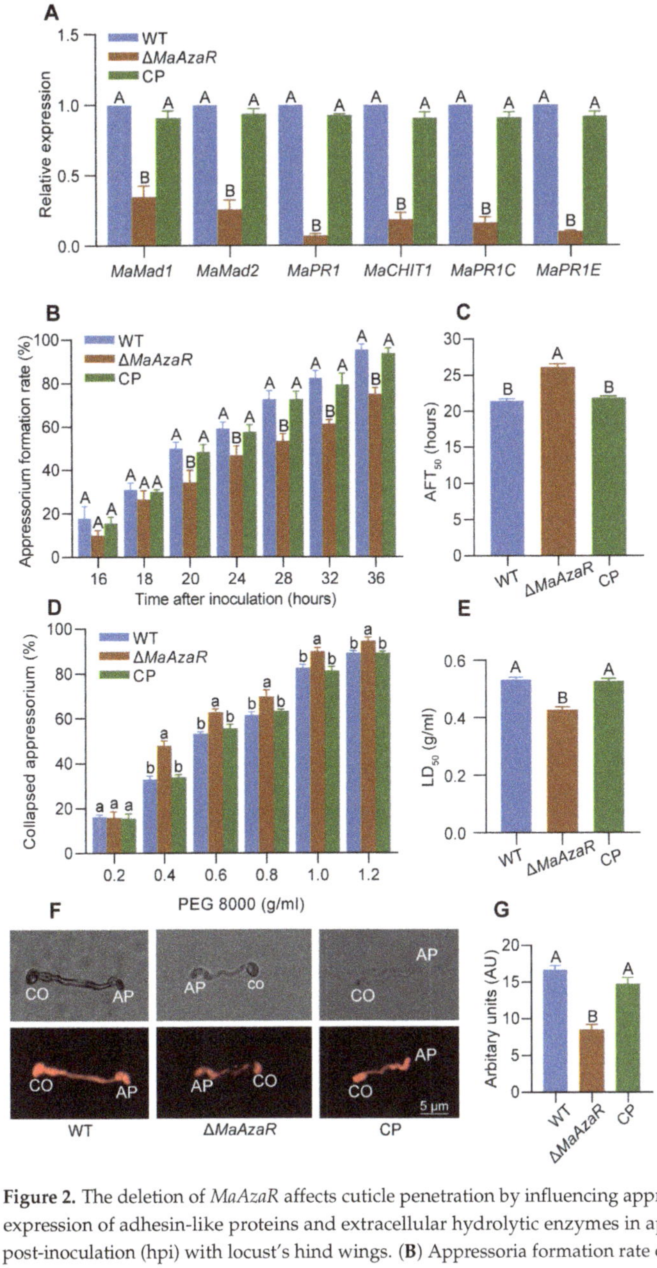

Figure 2. The deletion of *MaAzaR* affects cuticle penetration by influencing appressoria. (**A**) Relative expression of adhesin-like proteins and extracellular hydrolytic enzymes in appressoria after 24 h post-inoculation (hpi) with locust's hind wings. (**B**) Appressoria formation rate on locust wings from 16 to 36 hpi of each strain. (**C**) The time required for 50% appressoria formation (AFT_{50}). (**D**) The collapsed rate of appressoria treated with different concentrations of PEG-8000. (**E**) Concentration of PEG-8000 to make 50% of appressorium collapse (LD_{50}). (**F**) Nile Red staining of lipid droplets in the appressorium (AP) and conidium (CO). (**G**) Fluorescence of lipid droplet staining with Nile Red was quantified. Data points represent the mean ± standard error of the mean (SEM). The same lowercase letters and capital letters above two bars indicate no significant difference between the samples by one-way ANOVA and Tukey test ($p > 0.05$ and $p > 0.01$, respectively).

3.3. Disruption of *MaAzaR* Induces Higher Host Immune Responses During Cuticle Penetration

During the process of EPF infection, the host protects itself from fungal development through the collaboration between cellular immunity and humoral immunity [21,22]. We assessed the host's immune response by measuring the PO enzyme activity, as well as the expression levels of Toll pathway genes, *spaetzle* and *myd88*, and antimicrobial peptide, *defensin*. The results showed that the PO activity in locusts inoculated with the Δ*MaAzaR* strain was significantly higher than that in locusts inoculated with the WT and CP strains from 8 hpi to 36 hpi (Figure 3A). Additionally, the relative expression levels of Toll pathway genes and the antimicrobial peptide *defensin* induced by Δ*MaAzaR* were higher compared to those induced by WT (Figure 3B–D). Those suggest that the absence of *MaAzaR* may induce a stronger host immune response during the process of cuticle penetration.

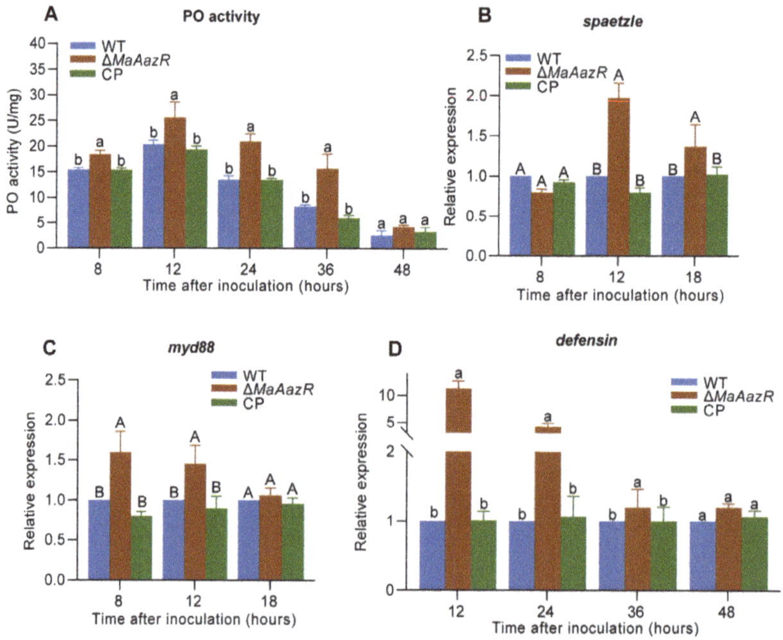

Figure 3. Disruption of *MaAzaR* induces higher host immune responses during cuticle penetration. (**A**) Phenoloxidases (PO) enzyme activity of locust hemolymph after topical inoculation with each strain. (**B**) Relative expression of *spaetzle* at 8, 12, and 18 h post-inoculation (hpi) after topical inoculation. (**C**) Relative expression of *myd88* at 8, 12, and 18 hpi after topical inoculation. (**D**) Relative expression of *defensin* at 12, 24, 36, and 48 hpi after topical inoculation. Data points represent the mean ± standard error of the mean (SEM). The same lowercase letters and capital letters above two bars indicate no significant difference between the samples by one-way ANOVA and Tukey test ($p > 0.05$ and $p > 0.01$, respectively).

3.4. The Absence of *MaAzaR* Does Not Impair In Vivo Growth Stage of *M. acridum*

Upon successful penetration of the host insect's cuticle and entry into the hemocoel, fungal hyphae converted to yeast-like HBs to adapt to the host hemocoel. Hemolymph samples were collected during the topical inoculation bioassay to observe HBs. At 4.5 dpi, HBs were observed in the hemolymph of locusts treated with the WT and CP strains. However, in locusts treated with the Δ*MaAzaR* strain, HBs were observed at 5.5 dpi, approximately 1 d later than those treated with WT. Furthermore, the number of HBs observed in locusts inoculated with Δ*MaAzaR* was lower compared to those formed by the WT strain (Figure 4A). The reduction in HBs, is it due to an extended duration of penetration, or is it attributed to growth inhibition within the hemolymph? We directly injected conidia into the hemocoel of locusts, bypassing the process of cuticle penetration, to simulate the growth of different fungal strains within the hemolymph. The results showed that the absence of *MaAzaR* resulted in accelerated locust mortality (Figure 4B,C). This revealed that the absence of *MaAzaR* does not impair the in vivo growth capability, and it even led to an enhancement of growth within the hemolymph. Next, we collected the healthy locust hemolymph for conidia cultivation in vitro, simulating the growth conditions of different fungal strains within the hemolymph. We then quantified the growth ability of different strains by measuring fungal DNA concentrations in the locust hemolymph. Similar results were observed (Figure 4D).

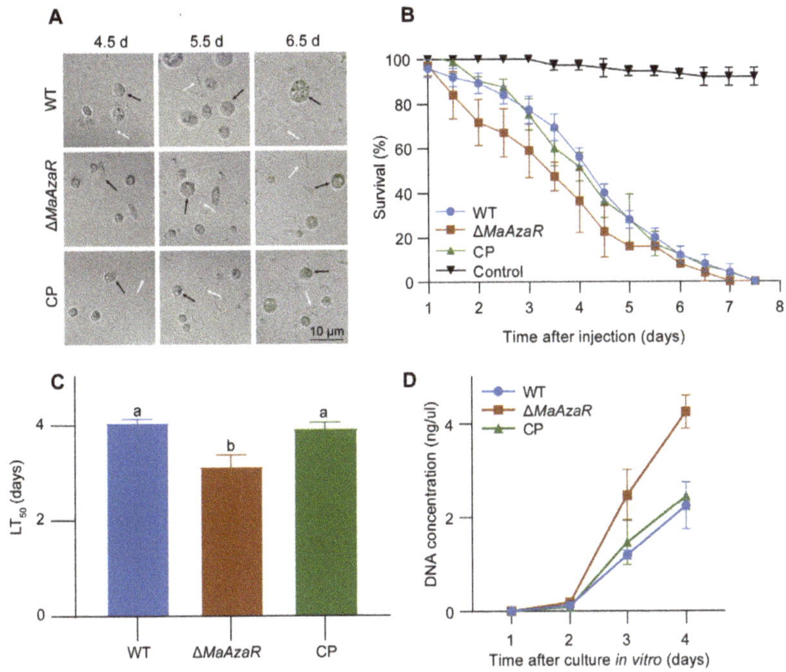

Figure 4. In vivo growth stage does not cause the decreased virulence of Δ*MaAzaR* strain. (**A**) Microscopic images of hyphal bodies and insect hemocytes in hemolymph after topical inoculation. Black arrow: insect hemocytes. White arrow: hyphal bodies. (**B**) Survival of *Locusta migratoria manilensis* following intrahemocoel injection with different strains. (**C**) Calculated LT_{50} (median lethal time) values of *L. migratoria* in the intrahemocoel injection bioassay. (**D**) Quantification of DNA concentration of in vitro cultured hyphal bodies. Data points represent the mean ± standard error of the mean (SEM). The same lowercase letters above two bars indicate no significant difference between the samples by one-way ANOVA and Tukey test ($p > 0.05$).

3.5. Transcriptomic Insights into the Roles of MaAzaR in Appressorium Formation

As mentioned above, the absence of *MaAzaR* resulted in a reduction in the virulence of *M. acridum* by decreasing the turgor pressure, relative expression of hydrolytic enzymes, and formation rate of appressoria. Transcriptomic analysis was performed to reveal differentially expressed genes (DEGs) in Δ*MaAzaR* and WT strains grown on wings for 20 hpi. There were 124 DEGs in Δ*MaAzaR* compared to WT ($|\log2FC| > 1$, where FC = fold-change, $q < 0.05$): 87 upregulated and 37 downregulated (Figure 5A, Table S2). The Kyoto Encyclopedia of Genes and Genomes (KEGG) pathways of total DEGs were mainly enriched in various metabolic pathways, including fructose and mannose metabolism, glycerolipid metabolism, glycerophospholipid metabolism, and amino sugar and nucleotide sugar metabolism (Figure 5B). Gene ontology (GO) annotation was performed to reveal the function of DEGs. The results showed significant enrichment of GO terms such as oxidation–reduction process, metabolic process, membrane, oxidoreductase activity, and zinc ion binding (Figure 5C). To mine important genes associated with the pathogenicity of fungal infestations affected by MaAzaR, the DEGs were subjected to BLAST alignment in the Pathogen–Host Interactions (PHI) database (http://www.phi-base.org/index.jsp), which contains the reported pathogenicity, virulence, and effector genes of pathogenic microorganisms. A total of 49 DEGs (E-value: 1×10^{-5}) were successfully matched to homologs in the database, implying that approximately 40% of the total DEGs were associated with virulence (Figure 5D, Table S3). This suggests that MaAzaR may be a crucial transcription factor associated with virulence by regulating metabolite synthesis.

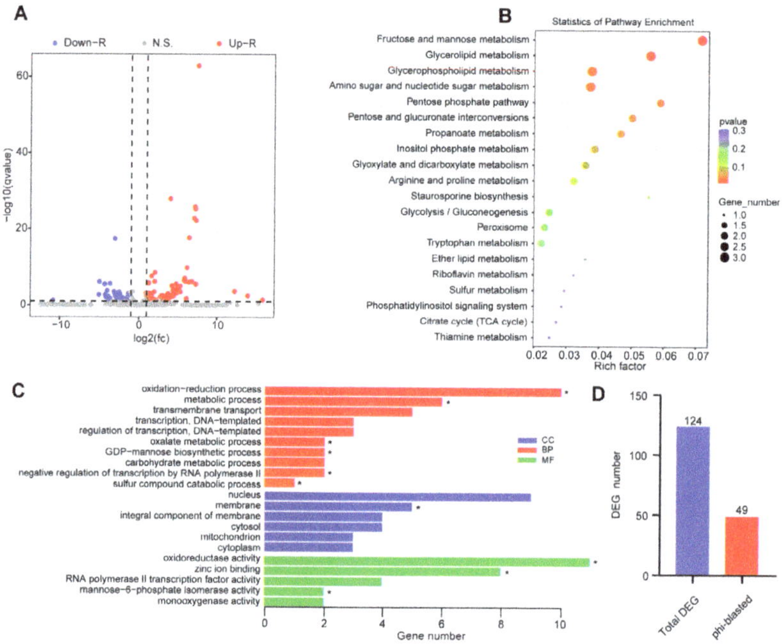

Figure 5. Transcriptomic analysis of MaAzaR in *Metarhizium acridum*. (**A**) Volcano plot of differentially expressed genes (DEGs) between the knockout mutant Δ*MaAzaR* and the wild-type (WT) strain grown 20 h in locust hindwings. (**B**) Enrichment of total DEGs to Kyoto Encyclopedia of Genes and Genomes (KEGG) pathways. (**C**) Enrichment of total DEGs to gene ontology (GO) item. BP, biological process; CC, cellular component; MF, molecular function. * indicates the significantly enriched terms ($q < 0.05$). (**D**) Quantification of total DEGs and homolog genes identified in the Pathogen–Host Interactions (PHI) database (phi-blasted DEGs).

4. Discussion

Zn(II)2Cys6 transcription factors are unique to the fungal kingdom and have diverse functions. It has contributed to virulence in various pathogenic fungi. VpFSTF1, a fungal-specific TF with a Zn(II)2Cys6 binuclear cluster, is essential for fruiting body formation and virulence in *Valsa pyri* [40]. Zcf15 and Zcf29 are required for the virulence of human pathogen *Candida albicans* [41]. In our study, deleting *MaAzaR* decreased its virulence to *L. migratoria* under natural infection and resulted in fewer conidia on cadavers, further influencing its prevalence in the locust population.

The insect integument serves as the primary barrier against fungal invasion [21], and cuticle penetration is the initial step essential for successful host infection by the pathogenic fungus. Our results indicate that the absence of *MaAzaR* impairs *M. acridum*'s ability to penetrate the cuticle. The appressorium is the main infective structure of a pathogenic fungus, facilitating invasion of the host from the cuticle [21]. Previous research has shown that disruption of *GPF1, PCF1, CNF1,* and *CCA1* affects appressorium formation and is required for virulence in *M. oryzae* [42]. In *Metarhizium*, the absence of *Mamk1, Makatg1,* and *Mrark1* decreased the rate of appressorium formation and thus reduced the virulence of knockout strains [43–45]. In our research, the absence of *MaAzaR* delayed appressorium formation. Appressoria establish adequate turgor pressure through the transport of lipids to the developing appressoria, which are subsequently degraded to glycerol. This process increases turgor pressure, serving as the driving force for mechanical penetration [22]. This mechanism enables the upward growth of penetration pegs and facilitates the breach of the host cuticle [21]. During cuticle penetration, appressoria produce a variety of extracellular hydrolytic enzymes, including lipases, proteases, and chitinases, to facilitate penetration [21,22]. We found that the loss of *MaAzaR* reduced lipid droplet content and turgor pressure in appressoria. The relative expression of the extracellular hydrolase gene in Δ*MaAzaR* was also reduced compared to the WT strain. Those results suggest that MaAzaR is an important transcription factor during appressoria formation.

Transcriptome analysis of appressorium formation revealed that the KEGG pathways of the DEGs primarily highlighted various metabolic pathways, including fructose and mannose metabolism, glycerolipid metabolism, glycerophospholipid metabolism, and amino sugar and nucleotide sugar metabolism. DEGs involved in fructose and mannose metabolism, such as MAC_03237, MAC_05390, and MAC_07720, are associated with the integrity of the fungal cell wall or polysaccharide capsule, transportation, and catalysis, influencing nutrient absorption, morphological growth, and virulence [46–49]. Approximately 40% of the total DEGs have homolog genes in the PHI database. Among these, MAC_06847 (log2FC = −10.8) encodes a serine protease, its homologous gene, *Bcser2*, contributes to hyphal growth, sclerotial formation, and conidiation in *Botrytis cinerea*. When *Bcser2* is disrupted, *B. cinerea* loses the ability to infect intact plant leaves of *Arabidopsis* and *Solanum lycopersicum* L. [50]. MAC_05902 (log2FC = −4.11) has a homologous gene in the PHI database, *Ss-oas1*, which encodes an oxaloacetate acetylhydrolase. The Δ*Ss-oah1* mutant of *Sclerotinia sclerotiorum* cannot infect efficiently due to a defect in appressorium development [51]. This reveals the crucial role of MaAzaR in regulating appressorium formation, but the specific pathways it regulates require further research. A previous study found that MaAzaR plays a role in conidiation pattern shifts and stress tolerance in *M. acridum* [34]. Transcriptomic data from conidia production and appressoria formation revealed three identical DEGs: MAC_00247, MAC_01894, MAC_08357. These DEGs are likely downstream targets of MaAzaR, although further investigation is needed to confirm their direct regulation.

The host protects itself against fungal development through cellular immunity and humoral immunity [21,22]. During cuticle penetration, the host activates the prophenoloxidase (PPO) system and antimicrobial compounds in the cuticle to inhibit fungal development. PO directly attaches to and melanizes the fungal cell wall to impede hyphal growth [25]. Following conidial attachment to locusts, increased expression of Toll pathway genes, such as *Spatzle, Toll9,* and *myd88*, is observed in the fat bodies during *M. acridum* germination [52]. We observed increased immune responses, including higher PO activity and increased expression levels of *Spatzle, myd88,* and *defensin,* induced by Δ*MaAzaR* during cuticle penetration. During the in vivo growth stage, fungal hyphae convert to yeast-like hyphal bodies to adapt to the host hemocoel [21,22]. The fungal cell wall remodels to evade the host immune response [21,52]. Furthermore, hyphal bodies release toxic substances while consuming host nutrients, ultimately leading to host death [21]. During this phase, the absence of *MaAzaR* does not impair the growth of *M. acridum* in the hemolymph of locusts, and it even enhances growth within the hemolymph.

5. Conclusions

MaAzaR impacts the virulence of *M. acridum* by affecting the cuticular penetration process rather than the intracellular growth within the host. During cuticle penetration, disruption of *MaAzaR* slows the formation rate of appressoria and impairs their turgor pressure and the expression of hydrolytic enzymes. The present study provides an important theoretical basis for selecting EPF strains and for promoting and utilizing EPF.

Supplementary Materials: The following supporting information can be downloaded at: https://www.mdpi.com/article/10.3390/jof10080564/s1, Table S1. Primers used in the study. Table S2. The DEGs in appressoria formation of Δ*MaAzaR* compared to WT. Table S3. DEGs which identified homolog genes in the Pathogen–Host Interactions (PHI) Database.

Author Contributions: Conceptualization, Y.X. and G.P.; investigation, G.H. and S.W.; methodology, Y.X. and G.P.; project administration, G.P.; funding acquisition, G.P.; writing—original draft, G.H.; software, S.W.; writing—review and editing, G.P. All authors have read and agreed to the published version of the manuscript.

Funding: This work was supported by International Cooperation and Exchange Program of the National Natural Science Foundation of China (Nos. 82261128004 and 2022YMFL1002), the technology innovation and application development project of Chongqing (No. CSTB2023TIAD-KPX0045), and the fundamental research funds for the central universities (No. 2024CDJXY016 and 2023CDJKYJH043).

Institutional Review Board Statement: Not applicable.

Informed Consent Statement: Not applicable.

Data Availability Statement: Clean data were deposited in the NCBI Sequence Read Archive database (accession number: PRJNA883628).

Conflicts of Interest: The authors declare no conflicts of interest.

References

1. Long, Z.; Hunter, D.M. Laboratory and field trials of Green Guard® (*Metarhizium anisopliae* var. *acridum*) (Deuteromycotina: Hyphomycetes) against the oriental migratory locust (*Locusta migratoria manilensis*) (Orthoptera: Acrididae) in China. *J. Orthoptera Res.* **2005**, *14*, 27–30.
2. Yanagawa, A.; Shimizu, S. Resistance of the termite, *Coptotermes formosanus* Shiraki to *Metarhizium anisopliae* due to grooming. *BioControl* **2007**, *52*, 75–85. [CrossRef]
3. Leemon, D.M.; Jonsson, N.N. Laboratory studies on Australian isolates of *Metarhizium anisopliae* as a biopesticide for the cattle tick *Boophilus microplus*. *J. Invertebr. Pathol.* **2008**, *97*, 40–49. [CrossRef] [PubMed]
4. Hoe, P.K.; Bong, C.F.J.; Jugah, K.; Rajan, A. Evaluation of *Metarhizium anisopliae* var. *anisopliae* (Deuteromycotina: Hyphomycete) Isolates and their Effects on Subterranean Termite *Coptotermes curvignathus* (Isoptera: Rhinotermitidae). *Am. J. Agric. Biol. Sci.* **2009**, *4*, 289–297. [CrossRef]
5. Niassy, S.; Diarra, K. Pathogenicity of local *Metarhizium anisopliae* var. *acridum* strains on *Locusta migratoria migratorioides* Reiche and Farmaire and *Zonocerus variegatus* Linnaeus in Senegal. *Afr. J. Biotechnol.* **2011**, *10*, 28–33. [CrossRef]
6. Guerrero, C.; Reyes-Montes, M.d.R.; Toriello, C.; Velázquez, V.M.; Santiago-López, I.; Palomino, L.; Calderón-Segura, M.; Docampo, S.; Ezquerro, C. Study of the persistence and viability of *Metarhizium acridum* in Mexico's agricultural area. *Aerobiologia* **2013**, *29*, 249–261. [CrossRef]
7. Camargo, M.; Marciano, A.; Sá, F.; Perinotto, W.; Quinelato, S.; Golo, P.; Angelo, I.; Prata, M.; Bittencourt, V. Commercial formulation of *Metarhizium anisopliae* for the control of *Rhipicephalus microplus* in a pen study. *Vet. Parasitol.* **2014**, *205*, 271–276. [CrossRef] [PubMed]
8. Zhang, W.; Ren, H.; Sun, F.; Shen, T.; Yuan, S.; Gao, X.; Tan, Y. Evaluation of the Toxicity of Chemical and Biogenic Insecticides to Three Outbreaking Insects in Desert Steppes of Northern China. *Toxins* **2022**, *14*, 546. [CrossRef]
9. Rezende-Teixeira, P.; Dusi, R.G.; Jimenez, P.C.; Espindola, L.S.; Costa-Lotufo, L.V. What can we learn from commercial insecticides? Efficacy, toxicity, environmental impacts, and future developments. *Environ. Pollut.* **2022**, *300*, 118983. [CrossRef]
10. Machate, O.; Schmeller, D.S.; Loyau, A.; Paschke, A.; Krauss, M.; Carmona, E.; Schulze, T.; Moyer, A.; Lutz, K.; Brack, W. Complex chemical cocktail, containing insecticides diazinon and permethrin, drives acute toxicity to crustaceans in mountain lakes. *Sci. Total Environ.* **2022**, *828*, 154456. [CrossRef]
11. de Oliveira Barbosa Bitencourt, R.; Corrêa, T.A.; Santos-Mallet, J.; Santos, H.A.; Lowenberger, C.; Moreira, H.V.S.; Gôlo, P.S.; Bittencourt, V.R.E.P.; da Costa Angelo, I. *Beauveria bassiana* interacts with gut and hemocytes to manipulate *Aedes aegypti* immunity. *Parasites Vectors* **2023**, *16*, 17. [CrossRef] [PubMed]

12. Kandel, Y.; Vulcan, J.; Rodriguez, S.D.; Moore, E.; Chung, H.-N.; Mitra, S.; Cordova, J.J.; Martinez, K.J.L.; Moon, A.S.; Kulkarni, A.; et al. Widespread insecticide resistance in *Aedes aegypti L.* from New Mexico, U.S.A. *PLoS ONE* **2019**, *14*, e0212693. [CrossRef]
13. Sharma, R. Analytical Concept of Fungicide Resistance: A Review. *Int. J. Curr. Microbiol. Appl. Sci.* **2019**, *8*, 1672–1684. [CrossRef]
14. Paula, A.R.; Carolino, A.T.; Silva, C.P.; Samuels, R.I. Susceptibility of adult female *Aedes aegypti* (Diptera: Culicidae) to the entomopathogenic fungus *Metarhizium anisopliae* is modified following blood feeding. *Parasites Vectors* **2011**, *4*, 91. [CrossRef] [PubMed]
15. Greenfield, B.P.J.; Peace, A.; Evans, H.; Dudley, E.; Ansari, M.A.; Butt, T.M. Identification of *Metarhizium* strains highly efficacious against *Aedes*, *Anopheles* and *Culex* larvae. *Biocontrol Sci. Technol.* **2015**, *25*, 487–502. [CrossRef]
16. Chouvenc, T.; Su, N.-Y.; Kenneth Grace, J. Fifty years of attempted biological control of termites—Analysis of a failure. *Biol. Control.* **2011**, *59*, 69–82. [CrossRef]
17. Cheraghi, A.; Habibpour, B.; Mossadegh, M. Application of Bait Treated with the Entomopathogenic Fungus *Metarhizium anisopliae* (Metsch.) Sorokin for the Control of *Microcerotermes diversus* Silv. *Psyche A J. Entomol.* **2013**, *2013*, 865102. [CrossRef]
18. Atheimine, M.O.; Bashir, M.O.; Ely, S.O.; Kane, C.M.H.; Mohamed, S.A.O.; Babah, M.A.O.; Benchekroun, M. Efficacy and persistence of *Metarhizium acridum* (Hypocreales: Clavicipitaceae) used against desert locust larvae, *Schistocerca gregaria* (Orthoptera: Acrididae), under different vegetation cover types. *Int. J. Trop. Insect Sci.* **2014**, *34*, 106–114. [CrossRef]
19. Litwin, A.; Nowak, M.; Różalska, S. Entomopathogenic fungi: Unconventional applications. *Rev. Environ. Sci. Bio/Technol.* **2020**, *19*, 23–42. [CrossRef]
20. Bilgo, E.; Lovett, B.; St. Leger, R.J.; Sanon, A.; Dabiré, R.K.; Diabaté, A. Native entomopathogenic *Metarhizium* spp. from Burkina Faso and their virulence against the malaria vector *Anopheles coluzzii* and non-target insects. *Parasites Vectors* **2018**, *11*, 209. [CrossRef]
21. Li, J.; Xia, Y. Host–Pathogen Interactions between *Metarhizium* spp. and Locusts. *J. Fungi* **2022**, *8*, 602. [CrossRef] [PubMed]
22. Butt, T.M.; Coates, C.J.; Dubovskiy, I.M.; Ratcliffe, N.A. Chapter Nine—Entomopathogenic Fungi: New Insights into Host–Pathogen Interactions. *Adv. Genet.* **2016**, *94*, 307–364. [CrossRef]
23. Peng, G.; Wang, Z.; Yin, Y.; Zeng, D.; Xia, Y. Field trials of *Metarhizium anisopliae* var. *acridum* (Ascomycota: Hypocreales) against oriental migratory locusts, *Locusta migratoria manilensis* (Meyen) in Northern China. *Crop Prot.* **2008**, *27*, 1244–1250. [CrossRef]
24. St. Leger, R.J.; Joshi, L.; Roberts, D.W. Ambient pH Is a Major Determinant in the Expression of Cuticle-Degrading Enzymes and Hydrophobin by *Metarhizium anisopliae*. *Appl. Environ. Microbiol.* **1998**, *64*, 709–713. [CrossRef] [PubMed]
25. Söderhäll, K.; Häll, L.; Unestam, T.; Nyhlén, L. Attachment of phenoloxidase to fungal cell walls in arthropod immunity. *J. Invertebr. Pathol.* **1979**, *34*, 285–294. [CrossRef]
26. Wang, C.; St Leger, R.J. A collagenous protective coat enables *Metarhizium anisopliae* to evade insect immune responses. *Proc. Natl. Acad. Sci. USA* **2006**, *103*, 6647–6652. [CrossRef]
27. MacPherson, S.; Larochelle, M.; Turcotte, B. A fungal family of transcriptional regulators: The zinc cluster proteins. *Microbiol. Mol. Biol. Rev.* **2006**, *70*, 583–604. [CrossRef]
28. Turcotte, B.; Liang, X.B.; Robert, F.; Soontorngun, N. Transcriptional regulation of nonfermentable carbon utilization in budding yeast. *FEMS Yeast Res.* **2009**, *10*, 2–13. [CrossRef]
29. Chang, P.-K.; Ehrlich, K.C. Genome-wide analysis of the Zn(II)2Cys6 zinc cluster-encoding gene family in *Aspergillus flavus*. *Appl. Microbiol. Biotechnol.* **2013**, *97*, 4289–4300. [CrossRef]
30. Flynn, P.J.; Reece, R.J. Activation of transcription by metabolic intermediates of the pyrimidine biosynthetic pathway. *Mol. Cell. Biol.* **1999**, *19*, 882–888. [CrossRef]
31. Tianqiao, S.; Xiong, Z.; You, Z.; Dong, L.; Jiaoling, Y.; Junjie, Y.; Mina, Y.; Huijuan, C.; Mingli, Y.; Xiayan, P.; et al. Genome-Wide Identification of Zn2Cys6 Class Fungal-Specific Transcription Factors (ZnFTFs) and Functional Analysis of UvZnFTF1 in *Ustilaginoidea virens*. *Rice Sci.* **2021**, *28*, 567–578. [CrossRef]
32. Galhano, R.; Illana, A.; Ryder, L.S.; Rodríguez-Romero, J.; Demuez, M.; Badaruddin, M.; Martinez-Rocha, A.L.; Soanes, D.M.; Studholme, D.J.; Talbot, N.J.; et al. Tpc1 is an important Zn(II)2Cys6 transcriptional regulator required for polarized growth and virulence in the rice blast fungus. *PLoS Pathog.* **2017**, *13*, e1006516. [CrossRef] [PubMed]
33. Qiu, L.; Zhang, J.; Song, J.-Z.; Hu, S.-J.; Zhang, T.-S.; Li, Z.; Wang, J.-J.; Cheng, W. Involvement of BbTpc1, an important Zn(II)2Cys6 transcriptional regulator, in chitin biosynthesis, fungal development and virulence of an insect mycopathogen. *Int. J. Biol. Macromol.* **2021**, *166*, 1162–1172. [CrossRef] [PubMed]
34. Zhou, J.; Wang, S.; Xia, Y.; Peng, G. MaAzaR, a Zn2Cys6/Fungus-Specific Transcriptional Factor, Is Involved in Stress Tolerance and Conidiation Pattern Shift in *Metarhizium acridum*. *J. Fungi* **2024**, *10*, 468. [CrossRef] [PubMed]
35. Song, D.; Jin, Y.; Shi, Y.; Xia, Y.; Peng, G. The carbon catabolite repressor CreA is an essential virulence factor of *Metarhizium acridum* against *Locusta migratoria*. *Pest Manag. Sci.* **2022**, *78*, 3676–3684. [CrossRef] [PubMed]
36. Peng, G.; Xia, Y. Expression of scorpion toxin LqhIT2 increases the virulence of *Metarhizium acridum* towards *Locusta migratoria manilensis*. *J. Ind. Microbiol. Biotechnol.* **2014**, *41*, 1659–1666. [CrossRef] [PubMed]
37. Livak, K.J.; Schmittgen, T.D. Analysis of Relative Gene Expression Data Using Real-Time Quantitative PCR and the $2^{-\Delta\Delta Ct}$ Method. *Methods* **2001**, *25*, 402–408. [CrossRef] [PubMed]
38. Li, C.; Xia, Y.; Jin, K. N-terminal zinc fingers of MaNCP1 contribute to growth, stress tolerance, and virulence in *Metarhizium acridum*. *Int. J. Biol. Macromol.* **2022**, *216*, 426–436. [CrossRef] [PubMed]

39. Wei, Q.; Du, Y.; Jin, K.; Xia, Y. The Ste12-like transcription factor MaSte12 is involved in pathogenicity by regulating the appressorium formation in the entomopathogenic fungus, *Metarhizium acridum*. *Appl. Microbiol. Biotechnol.* **2017**, *101*, 8571–8584. [CrossRef] [PubMed]
40. Kange, A.M.; Xia, A.; Si, J.; Li, B.; Zhang, X.; Ai, G.; He, F.; Dou, D. The Fungal-Specific Transcription Factor VpFSTF1 Is Required for Virulence in *Valsa pyri*. *Front. Microbiol.* **2020**, *10*, 2945. [CrossRef]
41. Issi, L.; Farrer, R.A.; Pastor, K.; Landry, B.; Delorey, T.; Bell, G.W.; Thompson, D.A.; Cuomo, C.A.; Rao, R.P. Zinc Cluster Transcription Factors Alter Virulence in *Candida albicans*. *Genetics* **2017**, *205*, 559–576. [CrossRef] [PubMed]
42. Lu, J.; Cao, H.; Zhang, L.; Huang, P.; Lin, F. Systematic Analysis of Zn2Cys6 Transcription Factors Required for Development and Pathogenicity by High-Throughput Gene Knockout in the Rice Blast Fungus. *PLoS Pathog.* **2014**, *10*, e1004432. [CrossRef] [PubMed]
43. Li, G.; Fan, A.; Peng, G.; Keyhani, N.; Xin, J.; Cao, Y.; Xia, Y. A bifunctional catalase-peroxidase, MakatG1, contributes to virulence of *Metarhizium acridum* by overcoming oxidative stress on the host insect cuticle. *Environ. Microbiol.* **2017**, *19*, 4365–4378. [CrossRef]
44. Wang, Z.; Jiang, Y.; Li, Y.; Feng, J.; Huang, B. MrArk1, an actin-regulating kinase gene, is required for endocytosis and involved in sustaining conidiation capacity and virulence in *Metarhizium robertsii*. *Appl. Microbiol. Biotechnol.* **2019**, *103*, 4859–4868. [CrossRef] [PubMed]
45. Jin, K.; Han, L.; Xia, Y. MaMk1, a FUS3/KSS1-type mitogen-activated protein kinase gene, is required for appressorium formation, and insect cuticle penetration of the entomopathogenic fungus *Metarhizium acridum*. *J. Invertebr. Pathol.* **2014**, *115*, 68–75. [CrossRef] [PubMed]
46. Pitkänen, J.-P.; Törmä, A.; Alff, S.; Huopaniemi, L.; Mattila, P.; Renkonen, R. Excess Mannose Limits the Growth of Phosphomannose Isomerase PMI40 Deletion Strain of *Saccharomyces cerevisiae**. *J. Biol. Chem.* **2004**, *279*, 55737–55743. [CrossRef] [PubMed]
47. Scott, B.; Eaton, C.J. Role of reactive oxygen species in fungal cellular differentiations. *Curr. Opin. Microbiol.* **2008**, *11*, 488–493. [CrossRef] [PubMed]
48. Galamba, A.; Soetaert, K.; Buyssens, P.; Monnaie, D.; Jacobs, P.; Content, J. Molecular and biochemical characterisation of *Mycobacterium smegmatis* alcohol dehydrogenase C. *FEMS Microbiol. Lett.* **2001**, *196*, 51–56. [CrossRef]
49. Wills, E.A.; Roberts, I.S.; Del Poeta, M.; Rivera, J.; Casadevall, A.; Cox, G.M.; Perfect, J.R. Identification and characterization of the *Cryptococcus neoformans* phosphomannose isomerase-encoding gene, *MAN1*, and its impact on pathogenicity. *Mol. Microbiol.* **2001**, *40*, 610–620. [CrossRef]
50. Liu, X.; Xie, J.; Fu, Y.; Jiang, D.; Chen, T.; Cheng, J. The Subtilisin-Like Protease Bcser2 Affects the Sclerotial Formation, Conidiation and Virulence of *Botrytis cinerea*. *Int. J. Mol. Sci.* **2020**, *21*, 603. [CrossRef]
51. Liang, X.; Liberti, D.; Li, M.; Kim, Y.-T.; Hutchens, A.; Wilson, R.; Rollins, J.A. Oxaloacetate acetylhydrolase gene mutants of *Sclerotinia sclerotiorum* do not accumulate oxalic acid, but do produce limited lesions on host plants. *Mol. Plant Pathol.* **2015**, *16*, 559–571. [CrossRef] [PubMed]
52. Zheng, X.; Li, S.; Si, Y.; Hu, J.; Xia, Y. Locust can detect β-1, 3-glucan of the fungal pathogen before penetration and defend infection via the Toll signaling pathway. *Dev. Comp. Immunol.* **2020**, *106*, 103636. [CrossRef] [PubMed]

Journal of
Fungi

MDPI

Article

The Forkhead Box Gene, *MaSep1*, Negatively Regulates UV- and Thermo-Tolerances and Is Required for Microcycle Conidiation in *Metarhizium acridum*

Tiantian Song [1,2,3,4,†], Chan Li [1,2,3,4,†], Kai Jin [1,2,3,4,*] and Yuxian Xia [1,2,3,4,*]

1 Genetic Engineering Research Center, School of Life Sciences, Chongqing University, Chongqing 401331, China; 202226021001@cqu.edu.cn (T.S.); 202026021001@cqu.edu.cn (C.L.)
2 Chongqing Engineering Research Center for Fungal Insecticide, Chongqing 401331, China
3 Key Laboratory of Gene Function and Regulation Technologies Under Chongqing Municipal Education Commission, Chongqing 401331, China
4 National Engineering Research Center of Microbial Pesticides, Chongqing 401331, China
* Correspondence: jinkai@cqu.edu.cn (K.J.); yuxianxia@cqu.edu.cn (Y.X.)
† These authors contributed equally to this work.

Abstract: Insect pathogenic fungi have shown great potential in agricultural pest control. Conidiation is crucial for the survival of filamentous fungi, and dispersal occurs through two methods: normal conidiation, where conidia differentiate from mycelium, and microcycle conidiation, which involves conidial budding. The conidiation process is related to cell separation. The forkhead box gene *Sep1* in *Schizosaccharomyces pombe* plays a crucial role in cell separation. Nevertheless, the function of Sep1 has not been clarified in filamentous fungi. Here, MaSep1, the homolog of Sep1 in *Metarhizium acridum*, was identified and subjected to functional analysis. The findings revealed that conidial germination of the *MaSep1*-deletion strain (Δ*MaSep1*) was accelerated and the time for 50% germination rate of conidial was shortened by 1 h, while the conidial production of Δ*MaSep1* was considerably reduced. The resistances to heat shock and UV-B irradiation of Δ*MaSep1* were enhanced, and the expression of some genes involved in DNA damage repair and heat shock response was significantly increased in Δ*MaSep1*. The disruption of *MaSep1* had no effect on the virulence of *M. acridum*. Interestingly, Δ*MaSep1* conducted the normal conidiation on the microcycle conidiation medium, SYA. Furthermore, 127 DEGs were identified by RNA-Seq between the wild-type and Δ*MaSep1* strains during microcycle conidiation, proving that MaSep1 mediated the conidiation pattern shift by governing some genes associated with conidiation, cell division, and cell wall formation.

Keywords: entomopathogenic fungus; *Metarhizium acridum*; MaSep1; stress tolerances; conidiation pattern shift

1. Introduction

Insect pathogenic fungi have shown great potential in agricultural pest control [1,2]. *Metarhizium acridum* has been successfully applied as a mycoinsecticide to control locusts in Asia, Africa, and Australia [3,4]. Conidia are the infective unit and reproductive basis of entomopathogenic fungi, and are also the main active components of mycopesticides [5,6]. In general, conidia are mainly produced through solid-state fermentation, since they display more resistance to abiotic stress in the field [7]. However, substrate aggregation seriously affects production efficiency due to the rapid growth of hyphae during solid fermentation [8]. The two modes of asexual development are normal conidiation and microcycle conidiation in filamentous fungi [9]. Microcycle conidiation is the direct production of conidia by conidia, with almost no hyphae formation [10], which can make up for the lack of a solid-state fermentation process [11]. In addition, the quality of conidia produced by microcycle conidiation was obviously improved, with higher conidial production and higher resistance to abiotic stress, which are beneficial to the industrial production of entomopathogenic fungi [12]. Consequently, exploring the mechanism of conidiation pattern shift is of great significance in entomopathogenic fungi.

In specific circumstances, for instance, variations in nutrition or temperature, normal conidiation can be shifted to microcycle conidiation [13]. Some genes related to microcycle conidiation have been identified in *Metarhizium acridum*, such as *MaNCP1*, *MaCreA*, *MaMsn2*, *Mavib-1*, *MaSln1*, *MaSho1*, *MaPEPDA*, *MaPpt1*, *MaOpy2*, and *MaH1* [14–22]. In addition, *MaMsn2* and *MaNsdD* regulate microcycle conidiation by negatively regulating *MaAbaA*, in which *MaMsn2* is regulated by *MaH1* [23]. Nevertheless, the mechanisms that regulate conidiation pattern shift are still largely unknown.

The conidiation process is related to cell separation. The regulatory mechanism of cell separation was mainly focused on yeast [24]. In *Schizosaccharomyces pombe*, 16 cell separation-related genes Sep1-16 were identified. When these genes are deleted, the cells are unable to separate properly [25,26]. Sep1, as a transcription factor, belongs to the HNF-3/forkhead family, and *Sep1* deletion can hinder septum division [25]. The expression of genes involved in cell division is regulated by Sep1, Ace2, and Fkh2 [27]. Sep1 and Fkh2 share the same forkhead domain. These transcription factors are believed to play opposing roles in regulating transcription during mitosis [28]. The forkhead protein Sep1p triggers the ZnF protein Ace2p. The main role of the Sep1p-Ace2p transcription pathway is to activate gene expression necessary for completing cell separation and cytokinesis upon completion of the cell cycle [29]. The forkhead transcription factors contain a DNA-binding domain with a winged helix structure [30] and are conserved in fungi. The forkhead transcription factor Mo-HCM1, the homolog of Sep1 in *M. oryzae*, is important in fungal development [31]. CaFkh2p, the homolog of Sep1 in *C. albicans*, is crucial for morphogenesis and virulence [32]. FkhB, the homolog of Sep1 in *A. nidulans*, is essential for fungal growth, development, and stress tolerance [33]. However, the functions of Sep1 have not been clarified in insect pathogenic fungi.

In this research, MaSep1, the homolog of Sep1 in *M. acridum*, was identified and subjected to functional analysis. Disruption of *MaSep1* led to accelerated conidial germination, reduced conidial production, and increased resistance to heat shock and UV-B irradiation, with no impact on fungal virulence. Interestingly, the *MaSep1*-disruption strain could not perform microcycle conidiation on SYA medium. RNA-Seq analysis further revealed that MaSep1 regulates this conidiation pattern shift by influencing genes related to conidiation, cell separation, cell cycle, cell polarity, and cell wall formation.

2. Materials and Methods

2.1. Strains and Cultivations

The wild-type strain *M. acridum* CQMa102 (CGMCC 0877) and mutants derived from the wild-type strain were cultivated as described previously [14]. Routine DNA manipulation was conducted using *Escherichia coli* DH5α competent cells (TransGen Biotech, Beijing, China). Fungal transformations were carried out using *Agrobacterium tumefaciens* AGL-1 (Weidi Biotech, Shanghai, China). To observe conidiation processes, 100 μL conidial suspensions (1×10^7 conidia/mL) from different fungal strains were evenly spread onto 1/4 SDAY and SYA media, and grown at 28 °C. Starting from 8 h, the conidiation patterns were checked under a light microscope and photographed every 2 h.

2.2. Bioinformatic Analyses

Online analyses were conducted to retrieve the sequences of the *MaSep1* gene and the MaSep1 protein (https://www.ncbi.nlm.nih.gov/, accessed on 15 December 2021) and predict isoelectric point (pI) and molecular weight (https://web.expasy.org/protparam/, accessed on 15 December 2021), conserved domains (https://www.ncbi.nlm.nih.gov/Structure/cdd/wrpsb.cgi, accessed on 15 December 2021), and the putative target DNA motif of MaSep1 (https://jaspar.genereg.net/, accessed on 10 December 2022). The multiple sequence alignment analysis was performed using the DNAMAN program 8. The program

MEGA v7.0 (www.megasoftware.net, accessed on 20 December 2021) was used to construct the neighbor-joining tree under 1000 bootstrap replicates.

2.3. Gene Disruption and Complementation

About 1000 bp of 5′ and 3′ flanking fragments of *MaSep1* (MAC_03141) or *MaFkh2* (MAC_06478) genes were inserted into the pK2-PB vector, respectively. The completed vectors were introduced into CQMa102 using the *Agrobacterium*-mediated method, respectively. Potential mutants were selected using Czapek–Dox medium supplemented with 500 μg/mL glyphosate (Sigma, St. Louis, MO, USA), which were subsequently validated by PCR. The disruption strain of *MaAce2* (MAC_04654) has been constructed previously [34]. To rescue *MaSep1* in the Δ*MaSep1* strain, the native promoter and open reading frame (ORF) of *MaSep1* were incorporated into the pK2-Sur vector [35]. The resultant vector was introduced into the Δ*MaSep1* strain using the method described above. Complemented transformants (CP) of *MaSep1* were screened on Czapek–Dox medium with 20 μg/mL chlorimuron ethyl (Sigma, Bellefonte, PA, USA). Quantitative reverse transcription PCR (qRT-PCR) was adopted to further verify the *MaSep1* disruption and complementation strains. The above-mentioned primers are recorded in Table S1.

2.4. Conidial Germination and Conidiation Capacity Assays

Conidial germination was evaluated following previously established methods on 1/4 SDAY medium [36]. One hundred microliters of the suspension (1×10^7 conidia/mL) were evenly spread onto 1/4 SDAY medium and incubated at 28 °C. The conidial germination rates of the WT, Δ*MaSep1*, and CP strains were examined every 2 h by taking samples using a previously described method [37]. For conidiation capacity assays, two microliters of conidial suspensions (1×10^6 conidia/mL) from the WT, Δ*MaSep1*, and CP strains were placed onto 1/4 SDAY medium in a 12-well plate and grown at 28 °C. Fungal samples incubated for 3, 6, 9, 12, and 15 days were harvested in each well with 0.05% Tween 80 to disperse the conidia evenly by vortexing. Conidia were examined using a hemocytometer.

2.5. Stress Tolerance Assays

Fungal resistance to ultraviolet (UV-B) irradiation and heat shock were evaluated as previously described [38]. In short, 100 μL of the suspensions (1×10^7 conidia/mL) were spread onto 1/4 SDAY medium and treated by UV-B irradiation for 0.5, 1.0, 1.5, 2.0, and 2.5 h, or heat shock at 42.5 °C for 2.0, 4.0, 6.0, 8.0, and 10.0 h, followed by incubation at 28 °C for 20 h. The germination of conidia at each time point after different treatments was recorded and statistically analyzed. Three replicates for each time point were performed [35]. For spot assays, mature conidia from WT, Δ*MaSep1*, and CP strains were prepared at a concentration of 1×10^6 conidia/mL. Then, 2 μL of conidial suspensions were inoculated onto 1/4 SDAY medium added with sorbitol (SOR), NaCl, H_2O_2, SDS, and Congo red (CR), respectively. The inoculated plates were allowed to dry and then incubated upside down at 28 °C. After a 6-day incubation, the colony sizes were measured.

2.6. DNA Damage and Repair

One hundred microliters of the suspension (5×10^7 conidia/mL) from each fungal strain was spread onto 1/4 SDAY medium and subjected to 1350 mW/m^2 UV light for 1.5 h or to a 42.5 °C water bath for 8 h. DAPI was employed to stain the DNA of each strain and the conidia were photographed under laser confocal scanning microscope (LSCM) to observe the nucleic acid damage. The remaining UV/heat shock-treated conidia were incubated at 28 °C for 20 h, and then taken out for DAPI staining and photographing as stated above.

2.7. Pathogenicity Assays

Locusts, *Locusta migratoria manilensis*, were reared at 28 °C with a 16:8 h (light–dark) cycle in our lab. The bioassays with 5th instar nymphs were conducted by a previous method [39]. Conidial suspensions (1×10^7 conidia/mL) were prepared with paraffin oil using mature conidia from WT, Δ*MaSep1*, and CP strains, and 5 μL of conidial suspensions were dripped onto the locusts with paraffin oil as the blank control. Thirty locusts were dripped in each group. Locusts were raised on fresh corn leaves and counted every 12 h.

2.8. qRT-PCR and RNA-Seq

Total RNA was obtained from fungal cultures on SYA medium using the Ultrapure RNA Kit (CWBIO, Beijing, China). RNA quality and quantity were examined using a Nano Drop spectrophotometry and an Agilent 2100 Bioanalyzer (Thermo Fisher Scientific, Waltham, MA, USA). qRT-PCR was conducted using SYBRPrime qPCR Set (BAOGUANG, Chongqing, China). The transcriptional levels of target genes were calculated using the $2^{-\Delta\Delta Ct}$ method [40]. The *gpdh* gene (MAC_09584) was employed as the internal control. For the transcriptome analysis, samples of Δ*MaSep1* and WT grown on SYA medium for 8, 10, and 12 h were harvested to extract RNA for RNA-Seq with three biological replicates, respectively. RNA library construction and RNA sequencing were conducted on a BGIseq500 platform by BGI (Wuhan, China). Clean reads were obtained using SOAPnuke software (v1.5.2) [41] and mapped to the reference genome of *M. acridum* (SUB13250283) using HISAT2 software (v2.0.4) [42]. Gene expression levels were calculated using RSEM software (v1.2.12) [43]. Differentially expressed genes (DEGs) were defined by the threshold of $|\text{Log2 ratio}| \geq 1$ and Qvalue ≤ 0.05. DEGs were annotated through Gene Ontology (GO) enrichment analysis (http://www.geneontology.org/, accessed on 16 March 2022) and Kyoto Encyclopedia of Genes and Genomes (KEGG) enrichment analysis (https://www.genome.jp/kegg, accessed on 16 March 2022).

2.9. Data Analyses

The primers were designed using Primer Premier 5 software. The specific primers used in qRT-PCR were designed using Beacon Designer 2.0 software (Bio-Rad, Hercules, CA, USA) (Table S1). The Shapiro–Wilk test and Levene's test were applied to check the normality and homogeneity of variances. Statistical analysis of the experimental data was conducted using one-way analysis of variance (ANOVA) in SPSS 22.0. Graphpad Prism 5 and Adobe Photoshop 2022 software were used for image processing.

3. Results

3.1. Features of MaSep1

MaSep1, the *Sep1* homologous gene in *M. acridum*, was obtained from the CQMa102 genome (SUB13250283). MaSep1 comprises a 2271 bp ORF, encoding a predicted 756 amino acids (82 kDa) with a pI of 9.61. Domain predictions using SMART revealed the presence of a forkhead domain in MaSep1, and the sequences of the forkhead domain displayed highly conserved in fungi (Figure 1A). Phylogenetic analysis of MaSep1 and its homologs in other fungi indicated that MaSep1 is closely related to *M. robertsii* and *M. anisopliae* (Figure 1B).

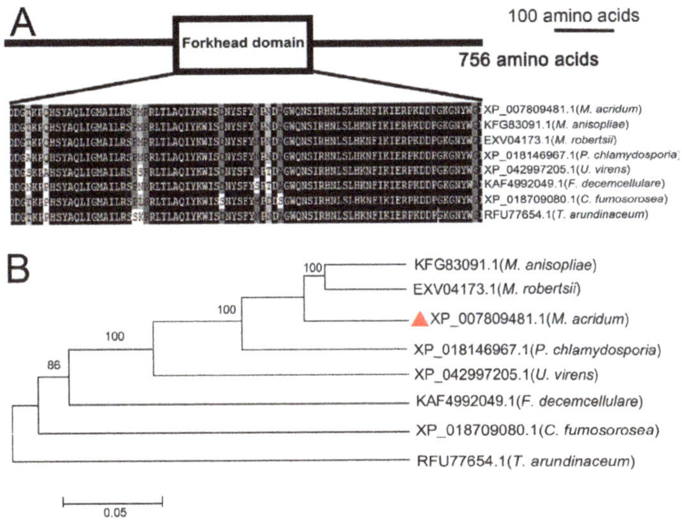

Figure 1. Features of MaSep1. (**A**) The location and multiple sequence alignment analysis of Forkhead domain. (**B**) Phylogenetic tree of Sep1 in different filamentous fungi. The phylogenetic tree was performed using MEGA software. *M. acridum* (XP_007809481.1), *Ustilaginoidea virens* (XP_042997205.1), *M. anisopliae* (KFG83091.1), *M. robertsii* (EXV04173.1), *Pochonia chlamydosporia* (XP_018146967.1), *Fusarium decemcellulare* (KAF4992049.1), *Cordyceps fumosorosea* (XP_018709080.1), and *Trichoderma arundinaceum* (RFU77654.1). Red triangle represented the Sep1 homologous protein in *M. acridum.*

3.2. Deletion of MaSep1 Affected Conidial Germination and Conidial Production, but Not Virulence

To reveal the roles of *MaSep1* in conidial germination and conidial production in *M. acridum*, the conidial germination rates and yields of WT, Δ*MaSep1*, and CP strains were assessed on 1/4 SDAY medium. The conidial germination rate of Δ*MaSep1* was markedly higher compared to the WT and CP strains. Approximately 60% of conidia from Δ*MaSep1* had germinated, whereas less than 40% of conidia from the WT and CP strains had germinated at 6 h after incubation (Figure 2A). The average 50% germination time (GT_{50}) of Δ*MaSep1* was remarkably lower relative to the WT and CP strains (Figure 2A). On 1/4 SDAY medium, the conidial yield of Δ*MaSep1* was markedly lower compared to the WT and CP strains. (Figure 2B). Bioassays conducted by topical inoculation showed that the half-lethal time (LT_{50}) of Δ*MaSep1* did not differ significantly from the WT and CP strains (Figure 2C), indicating that disruption of *MaSep1* did not affect the virulence of *M. acridum*.

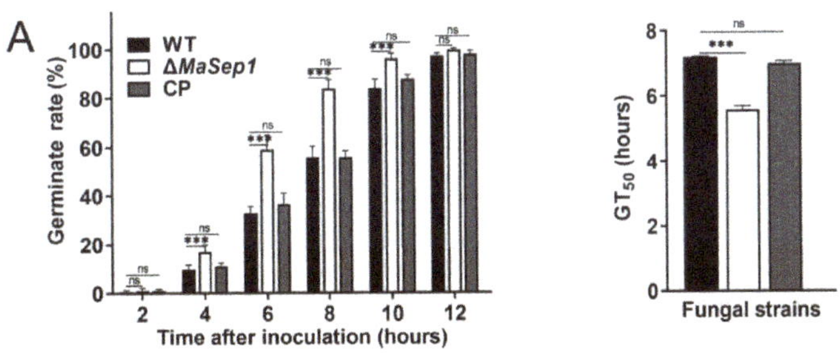

Figure 2. *Cont.*

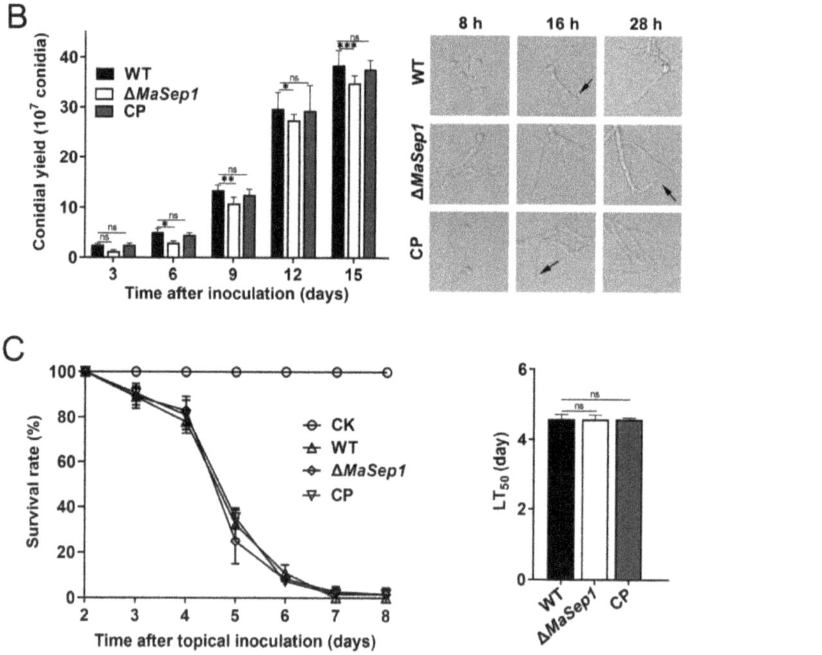

Figure 2. Conidial germination and conidial production. (**A**) Conidial germinations. (**B**) Conidiation. Black arrows represent conidia on the conidiophores. (**C**) The survival rates of the locusts, *L. migratoria manilensis*, and LT_{50} of the fungal strains were evaluated following topical inoculation with 5 μL conidial suspensions. Error bars indicate the standard error of the mean (SEM) from triplicated assays. ns, not significant, $p > 0.05$. *, $p < 0.05$; **, $p < 0.01$; ***, $p < 0.001$.

3.3. Deletion of MaSep1 Affects Multiple Stress Tolerances in M. acridum

To reveal the function of *MaSep1* in stress tolerances, we examined the germination rates of conidia following heat shock and UV-B irradiation. The findings showed that Δ*MaSep1* demonstrated notably higher conidial germination rates after UV-B irradiation for 1.5, 2.0, and 2.5 h. Additionally, the half-inhibition time of germination (IT_{50}) for the Δ*MaSep1* strain was considerably higher in comparison with those of the WT and CP (Figure 3A). Similarly, Δ*MaSep1* exhibited significantly increased conidial germination rates compared to the WT and CP strains after 4, 6, 8, and 10 h of treatments at 42.5 °C (Figure 3B).

To assess DNA damage in conidia from different fungal strains after UV-B irradiation, conidia treated with 1.5 h of UV-B irradiation and conidia incubated on 1/4 SDAY medium for 20 h after 1.5 h of UV-B irradiation were stained with DAPI, respectively. The results indicated that the DNA in fungal strains was severely damaged to a comparable extent and exhibited a diffuse pattern after 1.5 h of treatment with UV-B irradiation (Figure 4A). However, the conidia of Δ*MaSep1* germinated to form long mycelium, and the damaged DNA was repaired and became aggregated again after 20 h of incubation, while the conidia of WT and CP still had the ability to germinate but their DNA was still in a diffuse state (Figure 4B). qRT-PCR showed that the expression of most of the genes (*Rad3*, *Rad4*, *Rad23*) in the NER pathway was significantly upregulated in Δ*MaSep1* (Figure 4C). To understand the differences in the ROS scavenging ability of the WT, Δ*MaSep1*, and CP strains, conidia incubated for 20 h after UV-B irradiation were collected to detect the transcriptional levels of some genes related to ROS scavenging by qRT-PCR. The findings displayed that most of the oxidoreductase-encoding genes (*Gpx*, *Sod*, and *Cat*) involved in ROS scavenging were significantly upregulated in Δ*MaSep1* (Figure 4D).

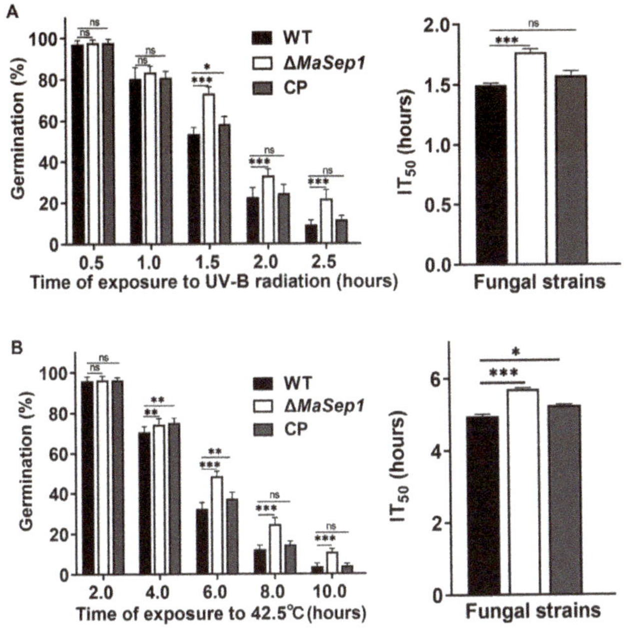

Figure 3. Deletion of *MaSep1* increased the fungal tolerances to UV-B and heat shock. (**A**) Conidial germination rates and IT_{50} of fungal strains after UV-B irradiation at 1350 mW/m^2 for 0.5, 1.0, 1.5, 2.0, and 2.5 h. (**B**) Conidial germination rates and IT_{50} of fungal strains after heat shock at 42.5 °C for 2, 4, 6, 8, and 10 h. Error bars indicate the SEM from triplicated assays. ns, not significant, $p > 0.05$. *, $p < 0.05$; **, $p < 0.01$; ***, $p < 0.001$.

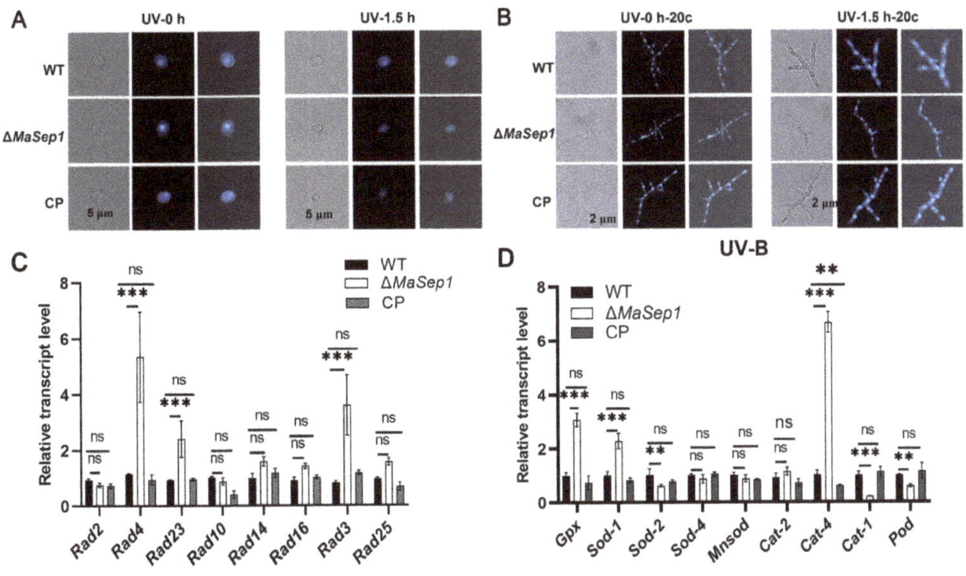

Figure 4. Analysis of DNA damage and repair under UV-B irradiation. (**A**) DNA damage was observed after UV-B irradiation. "UV-0 h" means no UV-B irradiation; "UV-1.5 h" means UV-B irradiation for 1.5 h. (**B**) DNA repair was observed after UV-B irradiation. "UV-0 h-20c" means grown

for 20 h at 28 °C without UV-B irradiation; "UV-1.5 h-20c" means grown for 20 h at 28 °C after UV-B irradiation for 1.5 h. (**C**) Transcriptional expression of key genes in the NER pathway in different strains after UV-B irradiation. (**D**) The relative transcriptional expression of ROS scavenging genes in different strains after UV treatment. Error bars indicate the SEM from triplicated assays. ns, not significant, $p > 0.05$. **, $p < 0.01$; ***, $p < 0.001$.

To explore the effect of heat shock stress on the DNA of WT, Δ*MaSep1*, and CP strains, conidia exposed to heat shock for 6 h and conidia incubated for 20 h after 6 h of heat shock were, respectively, collected and stained using DAPI to observe the DNA damage. The results showed that after 6 h of exposure to heat shock, DNA in all fungal strains was severely damaged in a diffuse manner (Figure 5A). However, after 20 h of incubation, the conidia of Δ*MaSep1* germinated to form long mycelium and the damaged DNA was repaired and became aggregated again, while the conidia of WT and CP still had the ability to germinate but their DNA was still in a diffuse state (Figure 5B). qRT-PCR showed that *Hsp40-2* in Δ*MaSep1* was significantly upregulated (Figure 5C). Similar to the UV-B treatments, conidia from the WT, Δ*MaSep1*, and CP strains incubated for 20 h after heat shock were collected to detect the transcriptional levels of some genes related to ROS scavenging by qRT-PCR. The results displayed that most of the oxidoreductase-encoding genes (*Gpx*, *Sod*, *Cat*) involved in ROS scavenging were significantly upregulated in Δ*MaSep1* (Figure 5D).

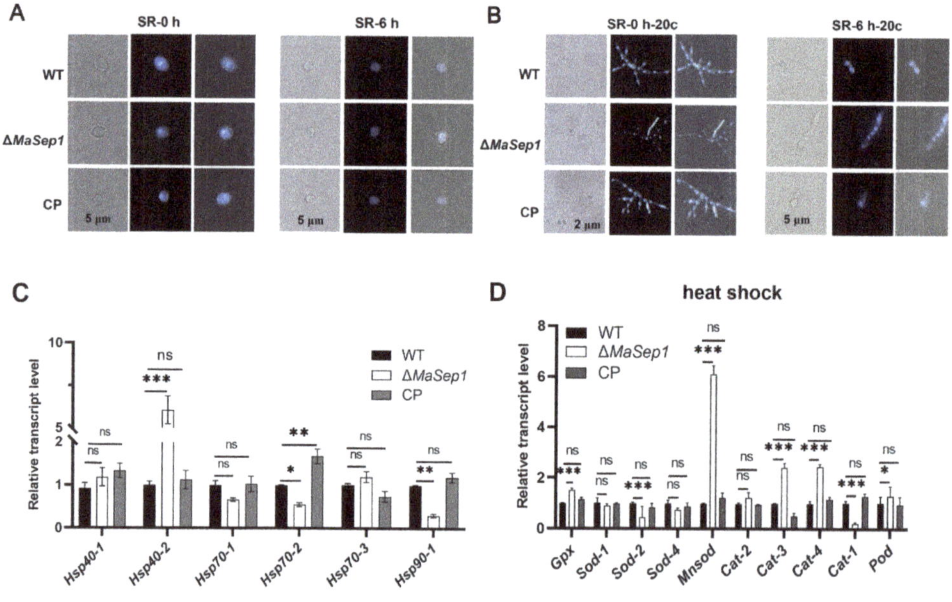

Figure 5. Analysis of DNA damage and repair under heat shock. (**A**) DNA damage was observed after thermo treatment. "SR-0 h" means no heat shock; "SR-6 h" means heat shock for 6 h. (**B**) DNA repair was observed after heat shock. "SR-0 h-20c" means grown for 20 h at 28 °C without heat shock; "SR-6 h-20c" means grown for 20 h at 28 °C after heat shock for 6 h. (**C**) The relative transcriptional expression levels of heat shock protein-encoding genes among different strains after heat shock. (**D**) The relative transcriptional expression of ROS scavenging genes in different strains after heat shock. Error bars indicate the SEM from triplicated assays. ns, not significant, $p > 0.05$. *, $p < 0.05$; **, $p < 0.01$; ***, $p < 0.001$.

Additionally, fungal responses to oxidative stress (H_2O_2), cell wall stress (CR), cell membrane stress (SDS), and osmotic stress (SOR and NaCl) were evaluated. The findings demonstrated that the colony morphology of Δ*MaSep1* exhibited insignificant differences compared to the WT and CP strains, except when grown on 1/4 SDAY medium added with CR (Figure 6A). The disruption of *MaSep1* significantly reduced the fungal growth rate on 1/4 SDAY containing the CR (Figure 6B). For the relative inhibition rate, Δ*MaSep1* was more sensitive to CR than WT and CP (Figure 6C).

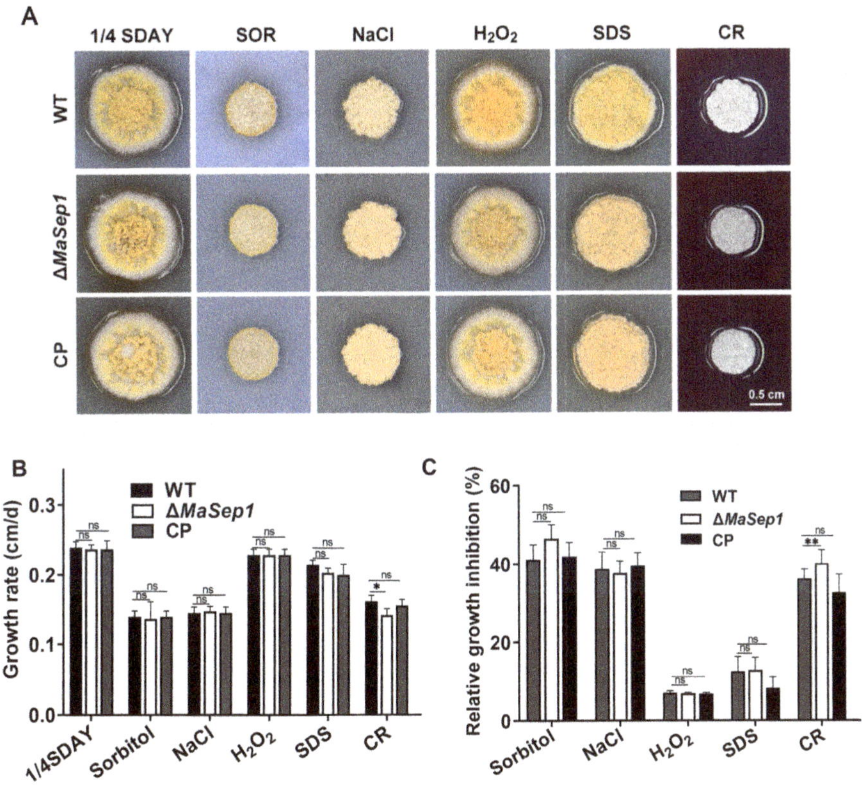

Figure 6. Stress tolerance assays. (**A**) Colonies of different fungal strains on 1/4 SDAY medium with SOR, NaCl, H_2O_2, SDS, and CR for 6 days. (**B**) Growth rates of different fungal strains. (**C**) Relative growth inhibition rates of different fungal strains. Error bars indicate the SEM from triplicated assays. ns, not significant, $p > 0.05$. *, $p < 0.05$; **, $p < 0.01$.

3.4. MaSep1 Is Required for Microcycle Conidiation of M. acridum

To examine the impact of MaSep1 on the conidiation pattern of *M. acridum*, the conidiation process of Δ*MaSep1* was monitored on the SYA medium. On the SYA medium, Δ*MaSep1* displayed microcycle conidiation, while WT and CP exhibited normal conidiation (Figure 7A). Sep1, Ace2, and Fkh2 jointly regulate cell division in *S. pombe* [27]. To reveal the roles of Ace2 and Fkh2 on microcycle conidiation in *M. acridum*, *MaFkh2* and *MaAce2* were deleted, respectively, in *M. acridum*, and the conidiation of the *MaFkh2*-disruption strain (Δ*MaFkh2*) and the *MaAce2*-disruption strain (Δ*MaAce2*) was observed on the SYA medium. The results showed that WT, Δ*MaFkh2*, and Δ*MaAce2* performed microcycle conidiation on SYA medium, while Δ*MaSep1* showed normal conidiation on SYA medium (Figure 7B). These results suggest that MaSep1 contributed to the microcycle conidiation independent of MaFkh2 and MaAce2 in *M. acridum*.

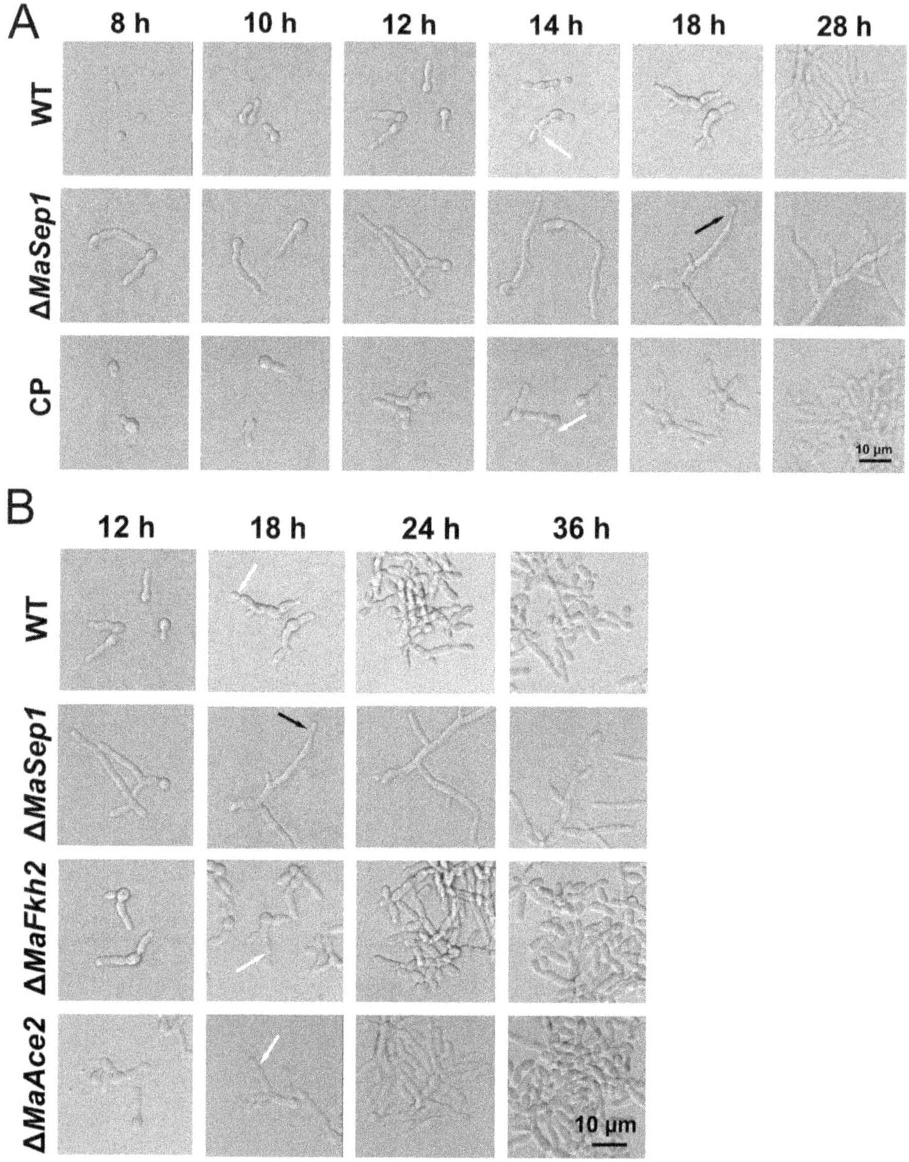

Figure 7. Conidiation pattern analyses of fungal strains grown on SYA medium. (**A**) The conidiation patterns of WT, Δ*MaSep1*, and CP strains. (**B**) The conidiation patterns of WT, Δ*MaSep1*, Δ*MaFkh2*, and Δ*MaAce2* strains. Black arrows represent normal conidiation. White arrows represent microcycle conidiation.

3.5. Identification of the DEGs Regulated by MaSep1 Using RNA-Seq

RNA-seq was performed to recognize DEGs between the Δ*MaSep1* and WT strains cultivated on SYA medium for 8, 10, and 12 h. In total, 127 DEGs were found between the Δ*MaSep1* and WT strains (Table S2). Of these DEGs, 112 DEGs were upregulated and 15 DEGs were down-regulated in Δ*MaSep1* (Figure 8A). At 8 h, 29 DEGs showed upregulation, while 4 DEGs were down-regulated in Δ*MaSep1*. The KEGG pathways of these DEGs were significantly enriched in DNA replication, purine metabolism, glutathione metabolism, nucleotide excision repair, pentose and glucuronate interconversions, carbon metabolism, cyanoamino acid metabolism, base excision repair, and TCA cycle (Figure 8B). At 10 h, 80 DEGs were upregulated and 11 DEGs were down-regulated in Δ*MaSep1*. The KEGG pathways of these DEGs were significantly enriched in fatty acid degradation, glycerolipid metabolism, β-alanine metabolism, phenylalanine metabolism, isoquinoline alkaloid biosynthesis, starch and sucrose metabolism, limonene and pinene degradation, tyrosine metabolism, glycine, serine and threonine metabolism, and tryptophan metabolism (Figure 8C). At 12 h, 59 DEGs were upregulated and 4 DEGs were down-regulated in Δ*MaSep1*. The KEGG pathways of these DEGs were significantly enriched in limonene and pinene degradation, pyruvate metabolism, arginine and proline metabolism, ascorbate and aldarate metabolism, betalain biosynthesis, isoquinoline alkaloid biosynthesis, cyanoamino acid metabolism, histidine metabolism, fatty acid degradation, and lysine degradation (Figure 8D).

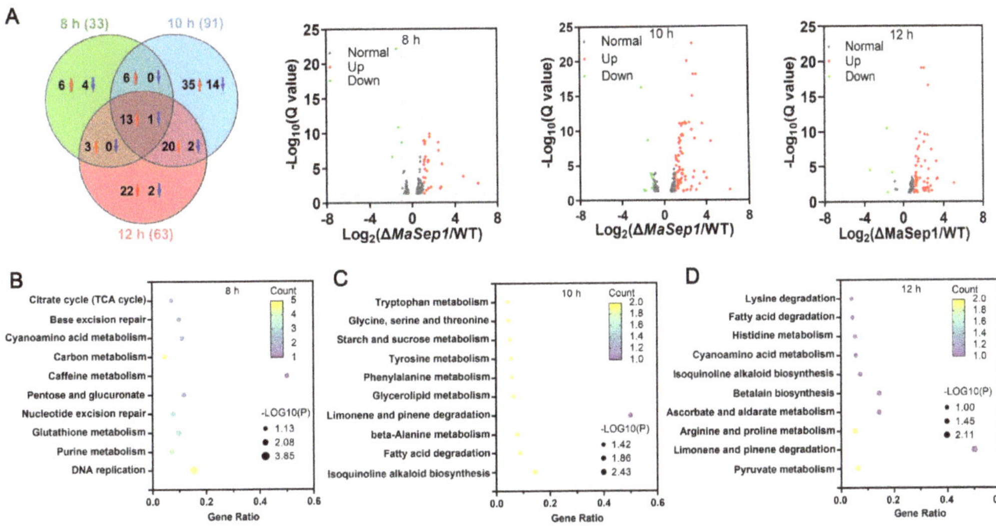

Figure 8. The analyses of RNA-seq data. (**A**) The number of DEGs. Red arrows represented up-regultaion, blue arrows represented down-regulation. The KEGG pathway enrichment at 8 (**B**), 10 (**C**), and 12 h (**D**).

In GO function enrichment analysis, DEGs were mainly enriched in biological processes, cellular components, and molecular functions. At 8 h, DEGs were mainly enriched in MCM complex, oxidoreductase activity, FAD binding, oxidation-reduction processes, and DNA replication (Figure 9A). At 10 h, DEGs were mainly enriched in the nucleus, preribosome, nucleolus, oxidoreductase activity, oxidation-reduction processes, rRNA processing, and cellular amino acid metabolic processes (Figure 9B). At 12 h, DEGs were mainly enriched in integral components of membrane, membrane, nucleus, oxidoreductase activity, transmembrane transporter activity, oxidation-reduction processes, transmembrane transport, regulation of transcription, and DNA-templated processes (Figure 9C).

Bioinformatics analysis showed that there were 105 putative target genes of MaSep1 in the 127 DEGs identified by RNA-seq during the conidiation pattern shift of *M. acridum* (Table S3). Of these putative target DEGs, 26 genes were related to conidiation in Δ*MaSep1* (Table S4). Fourteen of those participated in cell wall formation, comprising cell wall protein (MAC_06850 and MAC_06624), hydrophobic protein (MAC_04376 and MAC_07330), glucosidase (MAC_00623, MAC_04525, and MAC_08796) and NLPC/P60-like cell wall peptidase (MAC_06296). Nine of those participated in hyphal development and conidiation, comprising transcription factors (MAC_00186), ammonium osmotic enzyme MEPC (MAC_03001), and MFS family transporters (MAC_06625 and MAC_08197). Three of those were involved in the cell cycle, RNA binding protein MSSP-2 (MAC_01348), and reverse transcriptase protein (MAC_09371). These results indicate that MaSep1 governs the conidiation pattern shift by regulating the expression of genes associated with cell wall formation, conidiation, and cell cycle in *M. acridum*.

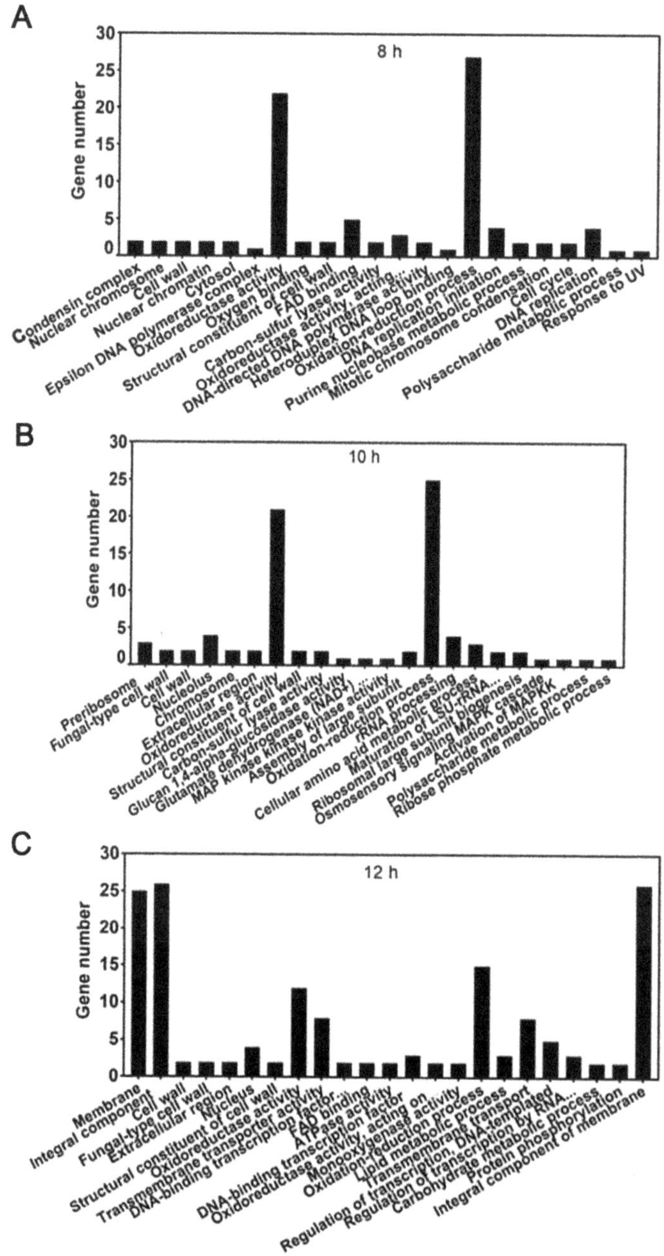

Figure 9. Significant enrichment of DEGs on GO terms at 8 (**A**), 10 (**B**), and 12 h (**C**).

4. Discussion

In this investigation, we identified the forkhead transcription factor MaSep1 in *M. acridum* and explored its functions. The disruption of *MaSep1* led to accelerated conidial germination, reduced conidial production, and enhanced resistance to UV-B irradiation and heat shock while fungal virulence remained unaffected. More importantly, *M. acridum* could not perform microcycle conidiation when *MaSep1* was disrupted. Furthermore, some DEGs involved in microcycle conidiation were identified in Δ*MaSep1* by RNA-seq.

For insect pathogenic fungi, stress tolerance is a crucial factor for survival and infectivity in hosts [44]. In this study, Δ*MaSep1* showed significantly increased tolerances to UV-B irradiation and heat shock, and the tolerance of Δ*MaSep1* to the cell wall disruptor Congo red was decreased. The stress tolerance of fungi is closely related to cell wall integrity [45]. Congo red can bind to β-1,3-glucan specifically, thus hindering the normal formation of cell walls [46]. NER is the pathway to remove bulky lesions from UV-B irradiation, and NER plays a crucial role in fungal tolerance to UV-B irradiation [47]. The *Rad25* gene of *S. cerevisiae* is essential for NER after UV-B irradiation [48]. RAD4 exhibits very high activity in NER after UV irradiation in *M. robertsii* [49]. Thus, the enhanced ability of DNA repair increased the tolerance of *M. acridum* to UV-B irradiation. HSPs are a class of proteins induced by high temperatures in fungi, which can help proteins fold correctly and restore the original spatial conformation and biological activity, which plays a crucial role in fungal heat tolerance [50]. Hsp70 is widely present in fungi, and its expression is influenced by temperature [51]. The high expression of genes encoded HSPs after heat shock treatment may be responsible for the heat shock tolerance of fungi [52].

Large amounts of reactive oxygen species (ROS) were induced after UV-B irradiation and heat shock treatments. The ability to clean ROS reflects the tolerance to abiotic stress [53]. ROS are usually produced as by-products of metabolic processes in living organisms [54]. Although ROS play an important role in various cellular processes, excessive ROS accumulation can cause irreversible damage to cellular biomacromolecules [55]. Therefore, ROS scavenging capacity is a major determinant of fungal stress response. The enzyme system consisting of POD, CAT, SOD, and GPX is an important weapon for ROS scavenging [47]. In this study, the upregulated expression of genes related to DNA excision repair, HSPs, and ROS scavenging in Δ*MaSep1* may be related to the enhanced tolerances of Δ*MaSep1* to UV irradiation and heat shock.

Sep1 involved in cell separation was first reported in *S. pombe*. The disruption of *Sep1* hindered the separation of the cell septum and resulted in impaired cell separation in *S. pombe* [25,56]. In this study, MaSep1, the homolog of Sep1 in *S. pombe*, was found to be required for microcycle conidiation in *M. acridum*. The expression of *Ace2* is regulated by a transcriptional complex composed of Sep1 and Fkh2 [27]. Sep1p and Ace2p transcriptionally regulate the initiation of the division of daughter cells following cytoplasmic division [29]. Sep1 is required for Fkh2 activity, and stimulation of forkhead transcription factors, Fkh2p and Sep1p, triggers the mitotic gene expression in *S. pombe* [57]. In this research, we discovered that Δ*MaSep1* displayed normal conidiation on SYA medium, but the Δ*MaAce2* and Δ*MaFkh2* strains performed microcycle conidiation, which is similar to WT, demonstrating that MaSep1 contributed to microcycle conidiation in *M. acridum* independent of MaAce2 and MaFkh2. RNA-seq data show that MaSep1 regulates conidiation pattern shift mainly by governing the expression of genes associated with cell wall synthesis, the cell cycle, and cell separation. Among the DEGs we studied, some are involved in cell development. A gene for hydrophobin (MAC_04376) was upregulated in Δ*MaSep1*. In *M. brunneum*, the growth of the deletion strain is slowed and conidiation is delayed [58]. The MFS transporter gene (MAC_06625), critical for fungal hyphal morphology, exhibited upregulation [59]. Moreover, some DEGs relevant to cell wall synthesis were identified. A gene for cell wall protein (MAC_06850) was upregulated in Δ*MaSep1* and was critical for the integrity of fungal cell walls [60]. A GNAT family acetyltransferase (MAC_04676), which is involved in chitin metabolism and cell wall rearrangement in fungi, was upregulated [61]. In addition, several genes related to the cell cycle and cell division showed differential expression. Upregulated genes included those encoding integral membrane proteins (MAC_07739, MAC_06568, and MAC_01217), which play crucial roles in linking the plasma membrane to the actomyosin ring and the assembly of the segmentation septum machinery in fission yeast [62].

In conclusion, MaSep1 negatively regulates the UV- and thermo-tolerances of *M. acridum* by influencing the expression of some genes associated with DNA damage repair and heat shock response, and mediates the conidiation pattern shift of *M. acridum* by governing some genes associated with conidiation, cell division and cell wall formation. These results will provide theoretical insights for further elucidating the molecular mechanism of stress tolerances and microcycle conidiation in filamentous fungi.

Supplementary Materials: The following supporting information can be downloaded at: https://www.mdpi.com/article/10.3390/jof10080544/s1, Figure S1: The disruption and complementation of *MaSep1*. (A) The knockout and complement schematic diagram of *MaSep1*. (B) Transcription level analysis of *MaSep1* in WT, Δ*MaSep1*, and CP strains. *MaSep1* expression in WT as control. Figure S2: PCR was used to verify the disruption of *MaFkh2*. (A) Fkh2-VF/Pt-R was used to verify the left arm of Δ*MaFkh2*. (B) Fkh2-VR/Bar-F was used to verify the right arm of Δ*MaFkh2*. M: DNA marker. Lanes 25–32: different mutants. Δ*MaFkh2*-27# was used for observation of conidiation pattern. Table S1: Primers used in this work. Table S2: Description of these DEGs. Table S3: 105 putative target genes of MaSep1. Table S4: DEGs involved in conidiation pattern shift. References [58–60,62–81] are cited in the Supplementary Materials.

Author Contributions: Conceptualization, K.J. and Y.X.; Data curation, T.S. and C.L.; Funding acquisition, K.J. and Y.X.; Investigation, T.S. and C.L.; Methodology, K.J. and Y.X.; Project administration, K.J. and Y.X.; Software, T.S. and C.L.; Writing—original draft, T.S. and C.L.; Writing—review and editing, K.J. and Y.X. All authors have read and agreed to the published version of the manuscript.

Funding: This work was supported by the National Natural Science Foundation of China (31972330), the Technology Innovation and Application Development Project of Chongqing (CSTB2023TIAD-KPX0045), the Natural Science Foundation Project of Chongqing (CSTB2022NSCQ-MSX1185), and Fundamental Research Funds for the Central Universities (2024CDJXY016).

Institutional Review Board Statement: Not applicable.

Informed Consent Statement: Not applicable.

Data Availability Statement: RNA-seq data have been deposited in the NCBI BioProject database (accession number: PRJNA974819).

Conflicts of Interest: The authors declare no conflicts of interest.

References

1. Li, J.; Xie, J.; Zeng, D.; Xia, Y.; Peng, G. Effective control of *Frankliniella occidentalis* by *Metarhizium anisopliae* CQMa421 under field conditions. *J. Pest Sci.* **2020**, *94*, 111–117. [CrossRef]
2. Peng, G.; Xie, J.; Guo, R.; Keyhani, N.O.; Zeng, D.; Yang, P.; Xia, Y. Long-term field evaluation and large-scale application of a *Metarhizium anisopliae* strain for controlling major rice pests. *J. Pest Sci.* **2021**, *94*, 969–980. [CrossRef]
3. Lomer, C.J.; Bateman, R.P.; Johnson, D.L.; Langewald, J.; Thomas, M. Biological control of locusts and grasshoppers. *Annu. Rev. Ėntomol.* **2001**, *46*, 667–702. [CrossRef]
4. Peng, G.; Wang, Z.; Yin, Y.; Zeng, D.; Xia, Y. Field trials of *Metarhizium anisopliae* var. *acridum* (Ascomycota: Hypocreales) against oriental migratory locusts, Locusta migratoria manilensis (Meyen) in Northern China. *Crop. Prot.* **2008**, *27*, 1244–1250. [CrossRef]
5. Mascarin, G.M.; Jaronski, S.T. The production and uses of *Beauveria bassiana* as a microbial insecticide. *World J. Microbiol. Biotechnol.* **2016**, *32*, 177. [CrossRef]
6. Muñiz-Paredes, F.; Miranda-Hernández, F.; Loera, O. Production of conidia by entomopathogenic fungi: From inoculants to final quality tests. *World J. Microbiol. Biotechnol.* **2017**, *33*, 57. [CrossRef]
7. Faria, M.R.D.; Wraight, S.P. Mycoinsecticides and mycoacaricides: A comprehensive list with worldwide coverage and international classification of formulation types. *Biol. Control* **2007**, *43*, 237–256. [CrossRef]
8. Weber, F.J.; Oostra, J.; Tramper, J.; Rinzema, A. Validation of a model for process development and scale-up of packed-bed solid-state bioreactors. *Biotechnol. Bioeng.* **2002**, *77*, 381–393. [CrossRef]
9. Park, H.-S.; Yu, J.-H. Genetic control of asexual sporulation in filamentous fungi. *Curr. Opin. Microbiol.* **2012**, *15*, 669–677. [CrossRef]

10. Anderson, J.G.; Smith, J.E. The Production of conidiophores and conidia by newly germinated conidia of *Aspergillus niger* (Microcycle Conidiation). *J. Gen. Microbiol.* **1971**, *69*, 185–197. [CrossRef]
11. van Heerden, A.; Mouton, M.; Postma, F.; van Wyk, P.W.J.; Lerm, B.; Van Zyl, W.H.; Borstlap, C.J.; Botha, A. The microcyclic conidial stage of *Coniochaeta pulveracea* and its effect on selected biological interactions. *Folia Microbiol.* **2016**, *61*, 319–328. [CrossRef] [PubMed]
12. Zhang, S.Z.; Peng, G.X.; Xia, Y.X. Microcycle conidiation and the conidial properties in the entomopathogenic fungus *Metarhizium acridum* on agar medium. *Biocontrol Sci. Technol.* **2010**, *20*, 809–819. [CrossRef]
13. Wang, Z.; Jin, K.; Xia, Y. Transcriptional analysis of the conidiation pattern shift of the entomopathogenic fungus *Metarhizium acridum* in response to different nutrients. *BMC Genom.* **2016**, *17*, 586. [CrossRef] [PubMed]
14. Li, C.; Xia, Y.; Jin, K. The C2H2 zinc finger protein mancp1 contributes to conidiation through governing the nitrate assimilation pathway in the entomopathogenic fungus *Metarhizium acridum*. *J. Fungi* **2022**, *8*, 942. [CrossRef] [PubMed]
15. Li, C.C.; Zhang, Q.P.; Xia, Y.X.; Jin, K. MaAreB, a GATA transcription Factor, is involved in nitrogen source utilization, stress tolerances and virulence in *Metarhizium acridum*. *J. Fungi* **2021**, *7*, 512. [CrossRef] [PubMed]
16. Li, J.; Su, X.; Cao, Y.; Xia, Y. Dipeptidase PEPDA is required for the conidiation pattern shift in *Metarhizium acridum*. *Appl. Environ. Microbiol.* **2021**, *87*, e0090821. [CrossRef] [PubMed]
17. Song, D.; Shi, Y.; Ji, H.; Xia, Y.; Peng, G. The *MaCreA* gene regulates normal conidiation and microcycle conidiation in *Me-tarhizium acridum*. *Front. Microbiol.* **2019**, *10*, 1946. [CrossRef] [PubMed]
18. Song, D.; Cao, Y.; Xia, Y. Transcription Factor MaMsn2 regulates conidiation pattern shift under the control of MaH1 through homeobox domain in *Metarhizium acridum*. *J. Fungi* **2021**, *7*, 840. [CrossRef] [PubMed]
19. Su, X.; Jiao, R.; Liu, Z.; Xia, Y.; Cao, Y. Functional and characteristic analysis of an appressorium-specific promoter *PMagas1* in *Metarhizium acridum*. *J. Invertebr. Pathol.* **2021**, *182*, 107565. [CrossRef]
20. Su, X.; Liu, H.; Xia, Y.; Cao, Y. Transcription factor *Mavib-1* negatively regulates conidiation by affecting utilization of carbon and nitrogen source in *Metarhizium acridum*. *J. Fungi* **2022**, *8*, 594. [CrossRef]
21. Wen, Z.Q.; Fan, Y.; Xia, Y.X.; Jin, K. MaOpy2, a transmembrane protein, is involved in stress tolerances and pathogenicity and negatively regulates conidial yield by shifting the conidiation pattern in *Metarhizium acridum*. *J. Fungi* **2022**, *8*, 587. [CrossRef] [PubMed]
22. Zhang, J.; Wang, Z.L.; Keyhani, N.O.; Peng, G.X.; Jin, K.; Xia, Y.X. The protein phosphatase gene MaPpt1 acts as a pro-grammer of microcycle conidiation and a negative regulator of UV-B tolerance in *Metarhizium acridum*. *Appl. Microbiol. Bio-Technol.* **2019**, *103*, 1351–1362. [CrossRef] [PubMed]
23. Song, D.; Cao, Y.; Xia, Y. *MaNsdD* regulates conidiation negatively by inhibiting the *AbaA* expression required for normal conidiation in *Metarhizium acridum*. *Environ. Microbiol.* **2022**, *24*, 2951–2961. [CrossRef] [PubMed]
24. Roncero, C.; Sánchez, Y. Cell separation and the maintenance of cell integrity during cytokinesis in yeast: The assembly of a septum. *Yeast* **2010**, *27*, 521–530. [CrossRef] [PubMed]
25. Ribar, B.; Banrevi, A.; Sipiczki, M. *sep1*$^+$ encodes a transcription-factor homologue of the HNF-3/forkhead DNA-binding-domain family in *Schizosaccharomyces pombe*. *Gene* **1997**, *202*, 1–5. [CrossRef]
26. Ribár, B.; Grallert, A.; Oláh, E.; Szállási, Z. Deletion of the *sep1*$^+$ forkhead transcription factor homologue is not lethal but causes hyphal growth in *Schizosaccharomyces pombe*. *Biochem. Biophys. Res. Commun.* **1999**, *263*, 465–474. [CrossRef] [PubMed]
27. Suárez, M.B.; Alonso-Nuñez, M.L.; del Rey, F.; McInerny, C.J.; de Aldana, C.R.V. Regulation of Ace2-dependent genes requires components of the PBF complex in *Schizosaccharomyces pombe*. *Cell Cycle* **2015**, *14*, 3124–3137. [CrossRef] [PubMed]
28. Voth, W.P.; Yu, Y.; Takahata, S.; Kretschmann, K.L.; Lieb, J.D.; Parker, R.L.; Milash, B.; Stillman, D.J. Forkhead proteins control the outcome of transcription factor binding by antiactivation. *EMBO J.* **2007**, *26*, 4324–4334. [CrossRef] [PubMed]
29. Bähler, J. A Transcriptional pathway for cell separation in fission yeast. *Cell Cycle* **2005**, *4*, 39–41. [CrossRef]
30. Kaestner, K.H.; Knöchel, W.; Martínez, D.E. Unified nomenclature for the winged helix/forkhead transcription factors. *Genes Dev.* **2000**, *14*, 142–146. [CrossRef]
31. Park, J.; Kong, S.; Kim, S.; Kang, S.; Lee, Y.-H. Roles of forkhead-box transcription factors in controlling development, pathogenicity, and stress response in *Magnaporthe oryzae*. *Plant Pathol. J.* **2014**, *30*, 136–150. [CrossRef] [PubMed]
32. Bensen, E.S.; Filler, S.G.; Berman, J. A forkhead transcription factor is important for true hyphal as well as yeast morphogenesis in *Candida albicans*. *Eukaryot. Cell* **2002**, *1*, 787–798. [CrossRef] [PubMed]
33. Jang, S.-Y.; Son, Y.-E.; Oh, D.-S.; Han, K.-H.; Yu, J.-H.; Park, H.-S. The Forkhead gene *fkhB* is necessary for proper development in *Aspergillus nidulans*. *J. Microbiol. Biotechnol.* **2023**, *33*, 1420–1427. [CrossRef] [PubMed]
34. Zou, Y.; Li, C.; Wang, S.; Xia, Y.; Jin, K. MaCts1, an endochitinase, is involved in conidial germination, conidial yield, stress tolerances and microcycle conidiation in *Metarhizium acridum*. *Biology* **2022**, *11*, 1730. [CrossRef] [PubMed]
35. Jin, K.; Ming, Y.; Xia, Y.X. MaHog1, a Hog1-type mitogen-activated protein kinase gene, contributes to stress tolerance and virulence of the entomopathogenic fungus *Metarhizium acridum*. *Microbiology* **2012**, *158*, 2987–2996. [CrossRef] [PubMed]
36. Du, Y.; Jin, K.; Xia, Y. Involvement of MaSom1, a downstream transcriptional factor of cAMP/PKA pathway, in conidial yield, stress tolerances, and virulence in *Metarhizium acridum*. *Appl. Microbiol. Biotechnol.* **2018**, *102*, 5611–5623. [CrossRef] [PubMed]
37. He, M.; Xia, Y. Construction and analysis of a normalized cDNA library from *Metarhizium anisopliae* var. *acridum germinating and differentiating on Locusta migratoria wings. FEMS Microbiol. Lett.* **2009**, *291*, 127–135. [PubMed]

38. Ma, Q.; Jin, K.; Peng, G.; Xia, Y. An ENA ATPase, MaENA1, of *Metarhizium acridum* influences the Na + -, thermo- and UV-tolerances of conidia and is involved in multiple mechanisms of stress tolerance. *Fungal Genet. Biol.* **2015**, *83*, 68–77. [CrossRef] [PubMed]
39. Zhang, M.; Wei, Q.; Xia, Y.; Jin, K. MaPacC, a pH-responsive transcription factor, negatively regulates thermotolerance and contributes to conidiation and virulence in *Metarhizium acridum*. *Curr. Genet.* **2019**, *66*, 397–408. [CrossRef]
40. Livak, K.J.; Schmittgen, T.D. Analysis of relative gene expression data using real-time quantitative PCR and the $2^{-\Delta\Delta CT}$ method. *Methods* **2001**, *25*, 402–408. [CrossRef]
41. Li, R.; Li, Y.; Kristiansen, K.; Wang, J. SOAP: Short oligonucleotide alignment program. *Bioinformatics* **2008**, *24*, 713–714. [CrossRef]
42. Kim, D.; Langmead, B.; Salzberg, S.L. HISAT: A fast spliced aligner with low memory requirements. *Nat. Methods* **2015**, *12*, 357–360. [CrossRef]
43. Li, B.; Dewey, C.N. RSEM: Accurate transcript quantification from RNA-Seq data with or without a reference genome. *BMC Bioinform.* **2011**, *12*, 323. [CrossRef] [PubMed]
44. Ortiz-Urquiza, A.; Luo, Z.; Keyhani, N.O. Improving mycoinsecticides for insect biological control. *Appl. Microbiol. Biotechnol.* **2015**, *99*, 1057–1068. [CrossRef] [PubMed]
45. Gow, N.A.R.; Latge, J.-P.; Munro, C.A. The fungal cell wall: Structure, biosynthesis, and function. *Microbiol. Spectr.* **2017**, *5*, 28513415. [CrossRef]
46. Ortiz-Ramírez, J.A.; Cuéllar-Cruz, M.; López-Romero, E. Cell compensatory responses of fungi to damage of the cell wall induced by Calcofluor White and Congo Red with emphasis on *Sporothrix schenckii* and *Sporothrix globosa*. *A review. Front. Cell. Infect. Microbiol.* **2022**, *12*, 976924. [CrossRef] [PubMed]
47. Chatterjee, N.; Walker, G.C. Mechanisms of DNA damage, repair, and mutagenesis. *Environ. Mol. Mutagen.* **2017**, *58*, 235–263. [CrossRef]
48. Qiu, H.; Park, E.; Prakash, L.; Prakash, S. The Saccharomyces cerevisiae DNA repair gene RAD25 is required for transcription by RNA polymerase II. *Genes Dev.* **1993**, *7*, 2161–2171. [CrossRef]
49. Zhang, Y.L.; Peng, H.; Zhang, K.; Ying, S.H.; Feng, M.G. Divergent roles of Rad4 and Rad23 homologs in *Metarhizium rob-ertsii*'s resistance to solar ultraviolet damage. *Appl. Environ. Microbiol.* **2023**, *89*, e0099423. [CrossRef]
50. Sottile, M.L.; Nadin, S.B. Heat shock proteins and DNA repair mechanisms: An updated overview. *Cell Stress Chaperon* **2018**, *23*, 303–315. [CrossRef]
51. Plesofsky-Vig, N.; Brambl, R. Heat shock response of *Neurospora crassa*: Protein synthesis and induced thermotolerance. *J. Bacteriol.* **1985**, *162*, 1083–1091. [CrossRef] [PubMed]
52. Angelidis, C.E.; Lazaridis, I.; Pagoulatos, G.N. Constitutive expression of heat-shock protein 70 in mammalian cells confers thermoresistance. *Eur. J. Biochem.* **1991**, *199*, 35–39. [CrossRef] [PubMed]
53. Vanhaelewyn, L.; Van Der Straeten, D.; De Coninck, B.; Vandenbussche, F. Ultraviolet radiation from a plant perspective: The plant-microorganism context. *Front. Plant Sci.* **2020**, *11*, 597642. [CrossRef] [PubMed]
54. Zhang, Z.; Chen, Y.; Li, B.; Chen, T.; Tian, S. Reactive oxygen species: A generalist in regulating development and pathogenicity of phytopathogenic fungi. *Comput. Struct. Biotechnol. J.* **2020**, *18*, 3344–3349. [CrossRef] [PubMed]
55. Lennicke, C.; Cochemé, H.M. Redox metabolism: ROS as specific molecular regulators of cell signaling and function. *Mol. Cell* **2021**, *81*, 3691–3707. [CrossRef] [PubMed]
56. Grallert, A.; Grallert, B.; Ribar, B.; Sipiczki, M. Coordination of initiation of nuclear division and initiation of cell division in *Schizosaccharomyces pombe*: Genetic interactions of mutations. *J. Bacteriol.* **1998**, *180*, 892–900. [CrossRef] [PubMed]
57. Buck, V.; Ng, S.S.; Ruiz-Garcia, A.B.; Papadopoulou, K.; Bhatti, S.; Samuel, J.M.; Anderson, M.; Millar, J.B.A.; McInerny, C.J. Fkh2p and Sep1p regulate mitotic gene transcription in fission yeast. *J. Cell Sci.* **2004**, *117*, 5623–5632. [CrossRef] [PubMed]
58. Sevim, A.; Donzelli, B.G.G.; Wu, D.; Demirbag, Z.; Gibson, D.M.; Turgeon, B.G. Hydrophobin genes of the entomopathogenic fungus, *Metarhizium brunneum*, are differentially expressed and corresponding mutants are decreased in virulence. *Curr. Genet.* **2012**, *58*, 79–92. [CrossRef]
59. Liu, L.; Yan, Y.; Huang, J.; Hsiang, T.; Wei, Y.; Li, Y.; Gao, J.; Zheng, L. A novel MFS transporter gene *ChMfs1* is important for hyphal morphology, conidiation, and pathogenicity in *Colletotrichum higginsianum*. *Front. Microbiol.* **2017**, *8*, 1953. [CrossRef]
60. Free, S. Fungal cell wall organization and biosynthesis. *Adv. Genet.* **2013**, *81*, 33–82.
61. Qin, Z.; Xiao, Y.; Yang, X.; Mesters, J.R.; Yang, S.; Jiang, Z. A unique GCN5-related glucosamine N-acetyltransferase region exist in the fungal multi-domain glycoside hydrolase family 3 β-N-acetylglucosaminidase. *Sci. Rep.* **2015**, *5*, 18292. [CrossRef] [PubMed]
62. Sethi, K.; Palani, S.; Cortés, J.C.G.; Sato, M.; Sevugan, M.; Ramos, M.; Vijaykumar, S.; Osumi, M.; Naqvi, N.I.; Ribas, J.C.; et al. A new membrane protein Sbg1 links the contractile ring apparatus and septum synthesis machinery in fission yeast. *PLoS Genet.* **2016**, *12*, e1006383. [CrossRef] [PubMed]
63. Zhao, J.; Shi, D.; Yang, S.; Lin, H.; Chen, H. Identification of an intracellular β-glucosidase in *Aspergillus niger* with transglycosylation activity. *Appl. Microbiol. Biotechnol.* **2020**, *104*, 8367–8380. [CrossRef] [PubMed]
64. Tamano, K.; Satoh, Y.; Ishii, T.; Terabayashi, Y.; Ohtaki, S.; Sano, M.; Takahashi, T.; Koyama, Y.; Mizutani, O.; Abe, K.; et al. The beta-1,3-exoglucanase gene *exgA* (*exg1*) of *Aspergillus oryzae* is required to catabolize extracellular glucan, and is induced in growth on a solid surface. *Biosci. Biotechnol. Biochem.* **2007**, *71*, 926–934. [CrossRef] [PubMed]
65. Hou, S.; Chen, G.; Wang, W.; Xia, L.; Wang, Z.; Lu, Y. Identification of a cell-wall peptidase (NlpC/P60) from Nocardia seriolae which induces apoptosis in fathead minnow cells. *J. Fish Dis.* **2020**, *43*, 571–581. [CrossRef] [PubMed]

66. Lah, L.; Podobnik, B.; Novak, M.; Korošec, B.; Berne, S.; Vogelsang, M.; Kraševec, N.; Zupanec, N.; Stojan, J.; Bohlmann, J.; et al. The versatility of the fungal cytochrome P450 monooxygenase system is instrumental in xenobiotic detoxification. *Mol. Microbiol.* **2011**, *81*, 1374–1389. [CrossRef] [PubMed]
67. Tong, S.M.; Chen, Y.; Ying, S.H.; Feng, M.G. Three DUF1996 proteins localize in vacuoles and function in fungal responses to multiple stresses and metal Ions. *Sci. Rep.* **2016**, *6*, 20566–20576. [CrossRef] [PubMed]
68. Schmidt, F.; Thywißen, A.; Goldmann, M.; Cunha, C.; Cseresnyés, Z.; Schmidt, H.; Rafiq, M.; Galiani, S.; Gräler, M.H.; Chamilos, G.; et al. Flotillin-dependent membrane microdomains are required for functional phagolysosomes against fungal infections. *Cell Rep.* **2020**, *32*, 108017–108039. [CrossRef] [PubMed]
69. Bailey, L.A.; Ebbole, D.J. The fluffy gene of *Neurospora crassa* encodes a Gal4p-type C6 zinc cluster protein required for conidial development. *Genetics* **1998**, *148*, 1813–1820. [CrossRef]
70. Teichert, S.; Rutherford, J.C.; Wottawa, M.; Heitman, J.; Tudzynski, B. Impact of ammonium permeases mepA, mepB, and mepC on nitrogen-regulated secondary metabolism in *Fusarium fujikuroi*. *Eukaryot Cell.* **2008**, *7*, 187–201. [CrossRef]
71. Semighini, C.P.; Goldman, M.H.; Goldman, G.H. Multi-copy suppression of an *Aspergillus nidulans* mutant sensitive to camptothecin by a putative monocarboxylate transporter. *Curr. Microbiol.* **2004**, *49*, 229–233. [CrossRef]
72. Fujita, Y.; Takegawa, K. Characterization of two genes encoding putative cysteine synthase required for cysteine biosynthesis in *Schizosaccharomyces pombe*. *J. Agric. Chem. Soc. Jpn.* **2004**, *68*, 306–311. [CrossRef]
73. Yan, Y. The screening and identification of A lysine 6-aminotransferase-producing fungi. *Pharm. Biotechnol.* **2007**, *14*, 259–263. [CrossRef]
74. Iida, M.; Taira, T.; Ariga, H.; Iguchi-Ariga, S.M. Nduction of apoptosis in HeLa cells by MSSP, c-myc binding proteins. *Biol. Pharm. Bull.* **1997**, *20*, 10–15. [CrossRef] [PubMed]
75. Liu, X.; Pan, X.; Chen, D.; Yin, C.; Peng, J.; Shi, W.; Qi, L.; Wang, R.; Zhao, W.; Zhang, Z.; et al. Prp19-associated splicing factor Cwf15 regulates fungal virulence and development in the rice blast fungus. *Environ. Microbiol.* **2021**, *23*, 5901–5916. [CrossRef]
76. Singh, P.K.; Shrivastava, A.K.; Singh, S.; Rai, R.; Chatterjee, A.; Rai, L.C. Alr2954 of *Anabaena sp.* PCC 7120 with ADP-ribose pyrophosphatase activity bestows abiotic stress tolerance in *Escherichia coli*. *Funct. Integr. Genom.* **2017**, *17*, 39–52. [CrossRef] [PubMed]
77. Yao, S.H.; Guo, Y.; Wang, Y.Z.; Zhang, D.; Xu, L.; Tang, W.H. A cytoplasmic Cu-Zn superoxide dismutase SOD1 contributes to hyphal growth and virulence of *Fusarium graminearum*. *Fungal Genet. Biol.* **2016**, *91*, 32–42. [CrossRef] [PubMed]
78. Mingot, J.M.; Peñalva, M.A.; Fernández-Cañón, J.M. Disruption of *phacA*, an *Aspergillus nidulans* gene encoding a novel cytochrome P450 monooxygenase catalyzing phenylacetate 2-hydroxylation, results in penicillin overproduction. *J. Biol. Chem.* **1999**, *274*, 14545–14550. [CrossRef]
79. Ďurišová, K.; Šimkovič, M.; Varečka, L'.; Kaliňák, M.; Šoltys, K.; Mosná, B.; Kryštofová, S. Functional dissection of γ-aminobutyric acid metabolism in *Neurospora crassa*. *Gen. Physiol. Biophys.* **2020**, *39*, 205–218. [CrossRef]
80. Wang, H.; Sun, S.; Ge, W.; Zhao, L.; Hou, B.; Wang, K.; Lyu, Z.; Chen, L.; Xu, S.; Guo, J.; et al. Horizontal gene transfer of *Fhb7* from fungus underlies *Fusarium* head blight resistance in wheat. *Science* **2020**, *368*, 1–7. [CrossRef]
81. Norvienyeku, J.; Zhong, Z.; Lin, L.; Dang, X.; Chen, M.; Lin, X.; Zhang, H.; Anjago, W.M.; Lin, L.; Abdul, W.; et al. Methylmalonate-semialdehyde dehydrogenase mediated metabolite homeostasis essentially regulate conidiation, polarized germination and pathogenesis in *Magnaporthe oryzae*. *Environ. Microbiol.* **2017**, *19*, 4256–4277. [CrossRef] [PubMed]

Journal of *Fungi*

MDPI

Article

The Catalase Gene *MrCat1* Contributes to Oxidative Stress Tolerance, Microsclerotia Formation, and Virulence in the Entomopathogenic Fungus *Metarhizium rileyi*

Yu Su [1,2], Xuyi Wang [2], Yuanli Luo [2], Huan Jiang [2], Guiting Tang [2] and Huai Liu [1,*]

[1] College of Plant Protection, Southwest University, Chongqing 400716, China; 15978967939@163.com
[2] Southeast Chongqing Academy of Agricultural Sciences, Chongqing 408000, China; wxyi135@163.com (X.W.); luoyuanli163@163.com (Y.L.); jhhyw521@163.com (H.J.); guitingtang616@163.com (G.T.)
* Correspondence: liuhuai@swu.edu.cn; Tel.: +86-023-68251269

Abstract: Catalases play a crucial role in the metabolism of reactive oxygen species (ROS) by converting H_2O_2 into molecular oxygen and water. They also contribute to virulence and fungal responses to various stresses. Previously, the *MrCat1*-deletion mutant (Δ*MrCat1*) was generated using the split-marker method in *Metarhizium rileyi*. In this study, the *Cat1* gene was identified, and its function was evaluated. Under normal culture conditions, there were no significant differences in colony growth or dimorphic switching between Δ*MrCat1* and the wild-type (WT) strains. However, under oxidative stress, the colony growth was inhibited, and the yeast–hyphal transition was suppressed in the Δ*MrCat1* strain. Hyperosmotic stress did not differ significantly between the two strains. In the Δ*MrCat1* strain, microsclerotia (MS) formation was delayed, resulting in less uniform MS size and a 76% decrease in MS yield compared to the WT strain. Moreover, the Δ*MrCat1* strain exhibited diminished virulence. Gene expression analysis revealed up-regulation of Δ*MrCat1, MrCat2, MrCat4*, and *MrAox* in the Δ*MrCat1* strain. These findings indicate that the *MrCat1* gene in *M. rileyi* is essential for oxidative stress tolerance, MS formation, and virulence.

Keywords: *Metarhizium rileyi*; catalase; oxidative stress tolerance; microsclerotium; virulence

1. Introduction

Metarhizium rileyi is a valuable entomopathogenic fungus known for its ability to infect lepidopterous pests, particularly *Noctuidae* spp., making it a potential candidate for insect biocontrol [1–5]. However, the sporulation of *M. rileyi* is dependent on specific conditions such as light stimulation and a maltose carbon source, which limits its commercial potential. To overcome this limitation, researchers have successfully induced microsclerotia (MS) formation in *M. rileyi* using liquid-amended media (AM) [6]. These MS are specialized, hyperpigmented hyphal aggregations with a diameter of 50–600 μm. Compared to *M. rileyi* conidia, MS have higher production efficiency and exhibit longer persistence in the field. Other entomopathogenic fungi like *Metarhizium brunneum* [7] and *Metarhizium anisopliae* [8] also produce MS and have been used in insect biocontrol.

In filamentous fungi, studies have shown that reactive oxygen species (ROS) play a critical role in various aspects of cell physiology, cellular differentiation, cell signaling, and defense against pathogens [9]. In the case of *M. rileyi*, comparative transcriptome analysis has been conducted to investigate the molecular mechanisms underlying MS development. This analysis revealed that oxidative stress is a key factor in MS development [10]. Several genes involved in MS development have been identified in *M. rileyi*. For example, the small GTPase *RacA* and *Cdc42*, along with their regulatory factors *Cdc24* and *Bem1*, regulate MS formation by controlling ROS generation [11,12]. Additionally, the NADH:flavin oxidoreductase/NADH oxidase gene (*Nox*) and alternative oxidase (*Aox*) genes are involved in MS formation by regulating intracellular H_2O_2 concentrations [13–15]. Defects in the transmembrane sensor genes *Sho1* and *Sln1* [16], as well as the deletion of two MAPK genes, *Hog1* and *Slt2*, have also been found to inhibit MS formation [17].

Catalases are crucial enzymes in the metabolism of ROS. They, along with superoxide dismutases (SODs) and catalases, convert superoxide and H_2O_2 into water and molecular oxygen [18–21]. Previous studies have highlighted the important adaptive role of catalases in response to environmental stress. For example, in *Beauveria bassiana*, five catalase genes (*CatA*, *CatB*, *CatP*, *CatC*, and *CatD*) have been implicated in the regulation of virulence and tolerance to oxidative stress, high temperatures, and UV-B radiation [22]. Overexpression of the *Cat1* gene in *M. anisopliae* has been shown to enhance resistance to exogenous H_2O_2, reduce germination time, and increase virulence [23]. In contrast, *Neurospora crassa* exhibited defects in the survival of conidia under oxidative and light-induced stress [24], while *Cat3* mutants exhibited growth and differentiation defects [25].

In *Aspergillus oryzae*, the *CatB* gene has been implicated in the detoxification of oxidative stress [26]. In the phytopathogenic fungus *Claviceps purpurea*, catalase has been shown to suppress the host defense system [27]. Furthermore, peroxisomal catalases may be involved in insect hydrocarbon catabolism [28]. These results collectively demonstrate the importance of fungal catalase genes in detoxification, cellular differentiation, and catabolism. However, there have been no reports on the involvement of catalase genes in MS formation. In this study, we isolated genomic DNA and cDNA of the *Cat1* gene in *M. rileyi* and investigated its role in oxidative stress tolerance, MS formation, and virulence through gene deletion experiments.

2. Materials and Methods

2.1. Strains and Growth Conditions

The *M. rileyi* strain Nr01 utilized in this study was obtained from the Engineering Research Center of Fungal Insecticides located in Chongqing, China. The mutant strain Δ*MrCat1*, lacking the *MrCat1* gene, was previously generated and confirmed by a study [29]. Both the wild-type (WT) and Δ*MrCat1* strains were cultured on solid SMAY media containing 40 g L^{-1} maltose, 15 g L^{-1} yeast extract powder, and 10 g L^{-1} peptone. Blastospores were collected during the early stages of growth, while conidia were collected after spore production by suspending the colonies in a sterile 0.05% Tween-80 solution and filtering them through lens wiping paper. The resulting blastospores of the WT and Δ*MrCat1* strains were washed and suspended in sterile phosphate-buffered saline (PBS) using a process involving three rounds of centrifugation. The *M. rileyi* spores were then cultured in liquid AM medium at 28 °C with continuous shaking at 250 rpm, following a described protocol [6].

2.2. Cloning the MrCat1 Gene of M. rileyi

The partial *MrCat1* sequence in this study was obtained from transcriptome data reported by Song [10]. To obtain the full cDNA and genomic sequences, the utilization of the fusion primer and nested integrated PCR (FPNI-PCR) method described by Wang [30]. The amino acid sequence was deduced by performing BlastX searches in GenBank. Signal peptide prediction was carried out by signalP 6.0, a web-based tool available at http://www.cbs.dtu.dk/serbices.signal/ (accessed on 21 February 2023). The conserved domain of *MrCat1* was predicted using the SMART web resource, available at http://smart.embl.de/ (accessed on 21 February 2023). The obtained sequences were aligned using MUSCLE with default settings, and an unrooted phylogenetic tree was generated using the Maximum Likelihood method in MEGA-X (https://www.megasoftware.net/) (accessed on 25 February 2023) with a 1000 replicates bootstrap test, as detailed by Kumar [31].

2.3. Gene Expression Patterns of M. rileyi WT and ΔMrCat1 Strains during MS Development

The transcription levels of *MrCat1*, *MrCat2*, *MrCat4*, and *MrAox* were quantified using real-time quantitative PCR (RT-qPCR) at different time points during the development of MS. The WT and Δ*MrCat1* strains were introduced into flasks containing 100 mL of liquid AM, along with 0.5 mL of conidia suspension (1×10^8 sp mL^{-1}). The flasks were then incubated with agitation at 28 °C and 250 rpm for 1.5–7 days. Samples were collected at specific time intervals corresponding to different stages of development: germinating spores (1.5–2 days), yeast-like cells (2.5–3 days), MS initiation and hyphal period (3.5–4 days), MS formation (4.5–5 days), MS maturation (5.5–6 days), and secondary mycelial growth (6.5–7 days). After centrifugation and two washes with sterile distilled water, total RNA was extracted using TRIzol® reagent (Invitrogen, Carlsbad, CA, USA). First-strand cDNA was then synthesized following the manufacturer's protocol using SuperScript II Reverse Transcriptase (Invitrogen, Carlsbad, CA, USA). RT-qPCR was performed using SYBR® Green II mix (TaKaRa, Shiga, Japan) according to the manufacturer's instructions. The primers used to assess the expression levels of *MrCat1*, *MrCat2*, *MrCat4*, and *MrAox* are detailed in Table A1. Each sample was prepared in triplicate, and each reaction was carried out three times. The *MrTef* and *MrTub* genes were employed as reference genes for normalization.

2.4. H_2O_2 Sensitivity of WT and Δ*MrCat1* Strains

To assess the sensitivity to H_2O_2 in the WT and Δ*MrCat1* strains, the filter paper method was employed. Conidial suspensions (1×10^6 sp mL^{-1}) of the WT and Δ*MrCat1* strains were spread onto plates. On the center of each plate, a 4-mm filter paper containing 800, 1600, or 3200 mM H_2O_2 was placed. For colony morphology evaluation, conidial suspensions (1×10^7, 1×10^6, and 1×10^5 sp mL^{-1}) of the WT and Δ*MrCat1* strains were pipetted onto SMAY plates and SMAY plates supplemented with 2 mM H_2O_2. All treatments were then incubated at 25 °C.

2.5. Hyperosmolarity Tolerance of WT and Δ*MrCat1* Strains

To evaluate the tolerance of the WT and Δ*MrCat1* strains to NaCl, KCl, LiCl, and sorbitol, solid SMAY media was used. Three concentrations of conidial suspensions (1×10^7, 1×10^6, and 1×10^5 sp mL^{-1}) were pipetted onto plates for the evaluation of colony morphology.

2.6. MS Formation of WT and Δ*MrCat1* Strains

To examine the MS formation capacities of the WT and Δ*MrCat1* strains, a conidial suspension (1×10^8 sp mL^{-1}) was inoculated into 100 mL of liquid AM. The inoculated cultures were then incubated at 28 °C with shaking at 250 rpm for 3.5–6.0 days. During the incubation period, the formation of MS yield was monitored, and the yield of MS was assessed. Additionally, the morphologies of the MS structures were observed using a microscope.

2.7. Virulence Assays

To determine the virulence of the WT and Δ*MrCat1* strains, third-instar larvae of *Prodenia litura* were used in the experiment. The larvae were immersed in three different concentrations of conidial suspensions: 2.5×10^7 sp mL^{-1}, 5×10^7 sp mL^{-1}, and 1×10^8 sp mL^{-1}. Control experiments were also conducted using a 0.05% Tween-80 solution without the fungus. After treatment, the insects were incubated at 26 °C, and mortality was recorded every 24 h. The median lethal time (LT_{50}) values were calculated using the minimum squares method, and the median lethal concentration (LC_{50}) values were calculated using the linearized regression model (LRM) method. Each treatment was repeated three times, with 15 larvae per replicate, to ensure reliable and statistically significant results.

2.8. Data Statistical Analysis

To assess variations in the relative expression levels of the genes among the samples, a variance analysis was conducted. Following the variance analysis, multiple comparisons were performed using Duncan's least significant range (LSR) tests in the SPSS statistical software (SPSS 16.0, SPSS Inc., Armonk, NY, USA). If a column lacks a common superscript letter, it indicates a significant difference compared to other columns ($p < 0.05$). On the other hand, columns labeled with the same letter were not found to be significantly different at the 5% level, based on Duncan's multiple range test.

3. Results

3.1. Features of *MrCat1* in *M. rileyi*

The *MrCat1* gene was identified from the expressed sequence tag (EST) of the *Cat1* gene isolated from the transcriptome library of *M. rileyi*. It has a full-length sequence of 2561 base pairs (bp) and encodes 716 amino acid residues. Comparison of cDNA and genomic sequences revealed the presence of four introns in *MrCat1*. Similar to other fungal genes, it contains a signal peptide for secretion and exhibits a typical catalase structural domain (Figure 1A), indicating that it is likely an exocrine protein. The predication molecular weight of the *MrCat1* is 79.085 KDa, and its isoelectric point (pI) is 5.92. A phylogenetic analysis using MEGA-X software revealed that *MrCat1* is closely associated with *Metarhizium* spp. (Figure 1B). The deduced amino acid sequence of *MrCat1* showed similarities with catalase in *Metarhizium acridum* (78.85% identity) [32] and catalaseB in *Purpureocillium lilacinum* (69.18% identity) [33], suggesting functional similarities with these enzymes.

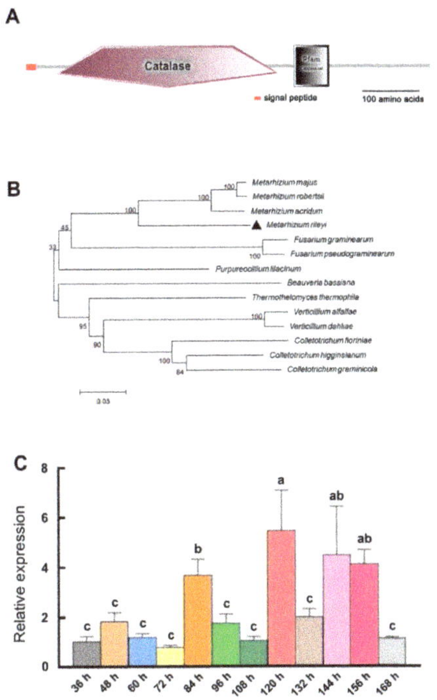

Figure 1. Features of *MrCat1* in *Metarhizium rileyi*. (**A**) The schematic of the protein primary structure encoded by the *MrCat1* gene, with the small red box indicating the signal peptide structure. (**B**) A phylogenetic tree inferred from *Cat1* DNA sequence alignment using the neighbor-joining (NJ). The aligned sequences of *Cat1* are from *Metarhizium rileyi* (OAA38085.1) *Metarhizium acridum* CQMa 102 (XM_007811861.1), *Metarhizium majus* ARSEF 297 (XM_014726471.1), *Purpureocillium lilacinum* (XM_018322662.1), *Fusarium graminearum* PH-1 (XM_011328072.1), *Fusarium pseudograminearum* CS3096 (XM_009262196.1), *Thermothelomyces thermophila* ATCC 42464 (XM_003662984.1), *Colletotrichum higginsianum* IMI 349063 (XM_018298034.1), *Colletotrichum fioriniae* PJ7 (XM_007598278.1), *Colletotrichum graminicola* M1.001 (XM_008092892.1), *Verticillium alfalfae* VaMs.102, (XM_003001071.1), *Verticillium dahliae* VdLs.17 (XM_009658953.1), *Metarhizium robertsii* ARSEF 23 (XM_007823877.1), and *Beauveria bassiana* (JX050139.1). (**C**) RT-qPCR analysis to assess the expression of *MrCat1*. Total RNA was isolated during different periods of MS development from the WT strain. Results are mean relative expression ± SD. The letters (a, b, c) indicated a significant difference compared to a column lacking a common superscript letter ($p < 0.05$). Conversely, columns labeled with the same letter were not significantly different at the 5% level.

To assess the expression of *MrCat1* during MS development, RT-qPCR analysis was performed. The results showed that *MrCat1* was expressed at all stages of MS development. It exhibited significantly up-regulated during MS initiation (84 h, 3.6-fold), MS formation (120 h, 5.4-fold), MS maturation (144 h, 4.4-fold), and the early stage of secondary mycelial growth (156 h, 4.1-fold) compared to the germinating spore period (36 h). These findings suggest that the *MrCat1* may play a potential role in hyphal growth and MS formation (Figure 1C).

3.2. Deletion of MrCat1 Impaired the Tolerance to H_2O_2 in M. rileyi

To investigate the role of *MrCat1* in H_2O_2 metabolism, a deletion mutant of *MrCat1* was created in *M. rileyi* using the split-marker method [29]. Both the WT and Δ*MrCat1* conidia were grown on SMAY plates supplemented with 2 mM H_2O_2 at a temperature of 25 °C for 7 days. Notably, there were no noticeable differences in morphology between the colonies of the WT and Δ*MrCat1* strains. However, a delay in the dimorphic switch was observed in the Δ*MrCat1* strain grown on SMAY plates with 2 mM H_2O_2. Specifically, the dimorphic switch was delayed by approximately 1 and 2 days compared to the WT strain (Figure 2A). In H_2O_2 inhibition assays, the Δ*MrCat1* strain exhibited larger inhibition zones compared to the WT strain. The size of the inhibition zones for the Δ*MrCat1* strain grown at H_2O_2 concentrations of 800, 1600, and 3200 mM were 1.7, 1.3, and 1.1 times larger, respectively, than those of the WT strain under the same conditions (Figure 2B). Variance analysis revealed that the diameter of the inhibition zones for the Δ*MrCat1* strain grown at an H_2O_2 concentration of 800 mM was significantly different from the WT strain ($p < 0.01$).

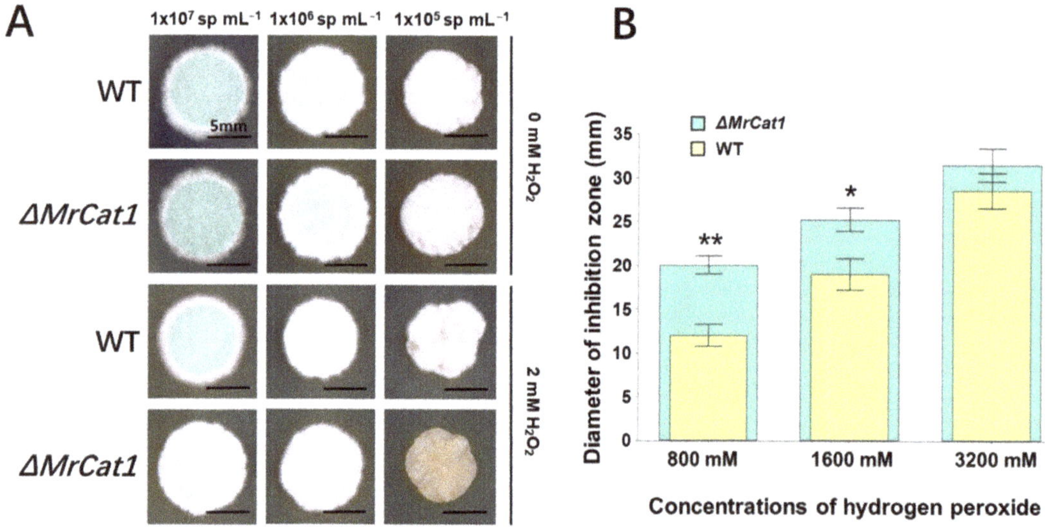

Figure 2. Deletion of *MrCat1* impaired the tolerance to H_2O_2 in *M. rileyi*. (**A**) The colony morphology of the wild-type (WT) strain and the Δ*MrCat1* mutant strain observed on SMAY plates or SMAY plates supplemented with 2 mM H_2O_2 for 7 days. Three concentrations of conidial suspensions (1×10^7, 1×10^6, and 1×10^5 sp mL^{-1}) were pipetted onto the plates. (**B**) The diameter of H_2O_2 inhibition zones measured at H_2O_2 concentrations of 800, 1600, and 3200 mM. Values are presented as the relative mean ± SD from five independent assays. Standard error bars indicate variation in measurements. * $p < 0.05$, ** $p < 0.01$, compared with WT grown at the same concentration of H_2O_2.

3.3. Deletion of MrCat1 Did Not Affect the Hyperosmotic Tolerance of M. rileyi

Both the WT and Δ*MrCat1* conidia were cultivated on SMAY plates supplemented with different hyperosmotic conditions, including 1 M NaCl, 1 M KCl, 0.04 M LiCl, and 1 M sorbitol. The cultivation period lasted for 18 days at a temperature of 25 °C. The growth rate of both strains, WT and Δ*MrCat1*, was significantly hindered by the presence of hyperosmotic conditions, except for the presence of sorbitol. The growth of Δ*MrCat1* strain exhibited a promotion at the beginning of the inoculation period (6d and 9d) compared with the WT strain (Figure 3A–D). However, despite the differences in growth rate, there were no discernible differences in colony morphologies between the two strains (Figure 3A–D).

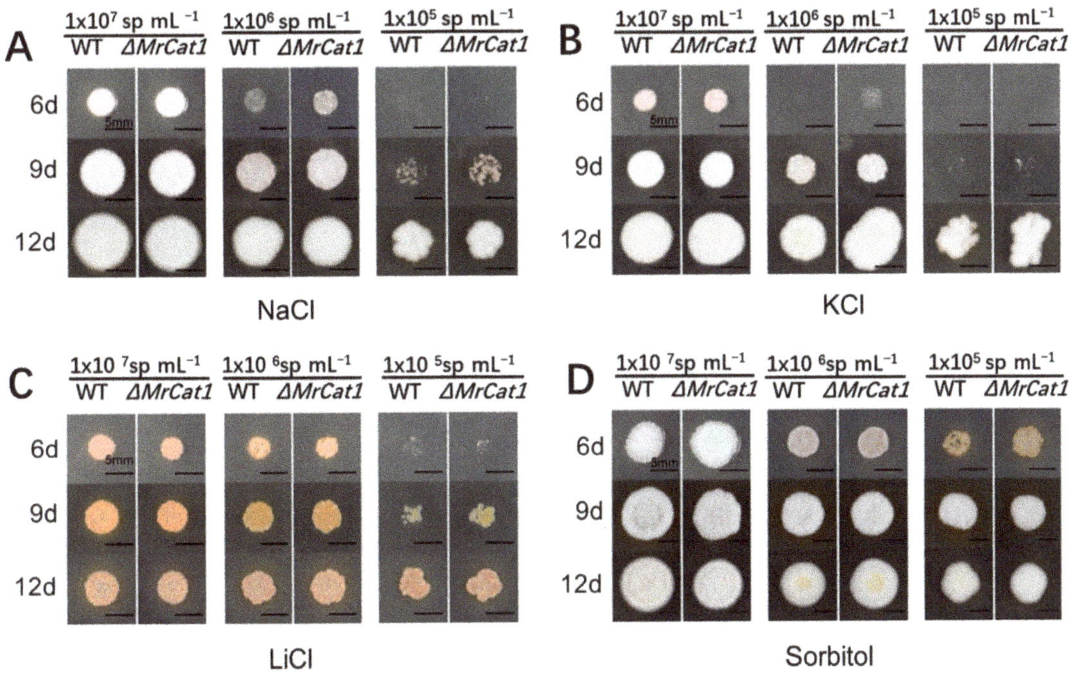

Figure 3. Deletion of *MrCat1* did not affect the hyperosmotic tolerance of *M. rileyi*. The colony morphology of the WT and Δ*MrCat1* strains observed on SMAY plates supplemented with (**A**) 1 M NaCl, (**B**) 1 M KCl, (**C**) 0.04 M LiCl, and (**D**) 1 M sorbitol for 6, 9, and 12 days, respectively.

3.4. Deletion of MrCat1 Affected MS Development of M. rileyi

Both the WT and Δ*MrCat1* conidia were introduced into a liquid AM medium and subjected to agitation. The WT strains exhibited visible MS formation after a period of 3.5–4 days, while the Δ*MrCat1* strain experienced a delay in MS formation, occurring after 4.5–5 days. Furthermore, the fermentation broth of the Δ*MrCat1* strain exhibited lower viscosity and a lighter pigment compared to the WT strain (Figure 4A,B). Additionally, the size distribution of WT MS displayed greater uniformity, ranging from 70 to 300 μm, compared to the Δ*MrCat1* MS, which ranged from 50 to 1500 μm. Moreover, the MS yield of the Δ*MrCat1* strain was diminished by approximately 76% compared to the WT strain (Figure 4C).

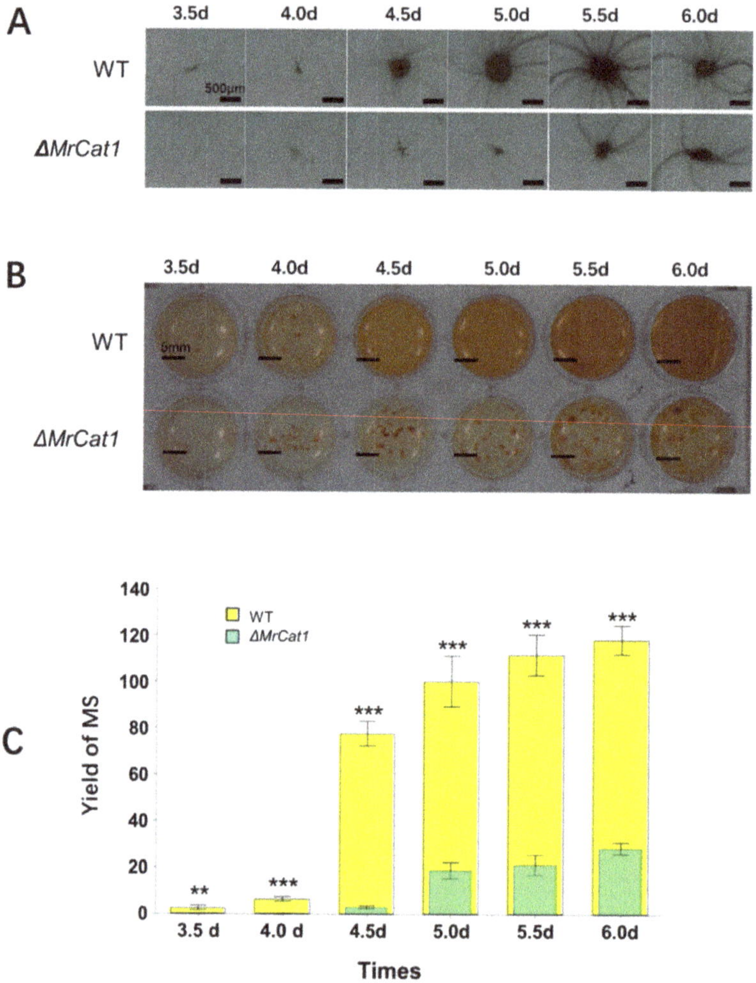

Figure 4. Deletion of *MrCat1* affected the MS development of *M. rileyi*. Microscopic images (**A**) and morphology (**B**) of the WT and Δ*MrCat1* strains grown in AM media from day 3.5–6. (**C**) Numbers of MS generated by the WT and Δ*MrCat1* strains from day 3.5–6. Values are presented as relative mean ± SD from three independent assays. Standard error bars indicate variation in measurements. ** $p < 0.01$, *** $p < 0.001$, compared with the WT strain at the same time point.

3.5. Deletion of MrCat1 Resulted in Enhancing the Expression of MrCat2, MrCat4, and MrAox during MS Development of M. rileyi

The expressions of *MrCat2*, *MrCat4*, and *MrAox* genes were analyzed during different stages of MS development, including spore germinating, MS initiation, formation stage, and secondary mycelial growth phase, in both the WT and Δ*MrCat1* strains of *M. rileyi*. Remarkably, the transcription levels of the *MrCat2* and *MrAox* were consistently higher in the Δ*MrCat1* strain compared to the WT strain (Figure 5A,C). Specifically, for *MrCat2*, the transcription levels were higher during the spore germinating to MS initiation period (36 h–84 h) and exhibited higher expression in the early period of MS formation (108 h) and early period of secondary mycelial growth (156 h). The most significant compensation occurred at 60 h, where *MrCat2* was up-regulated by 10.2-fold (Figure 5A).

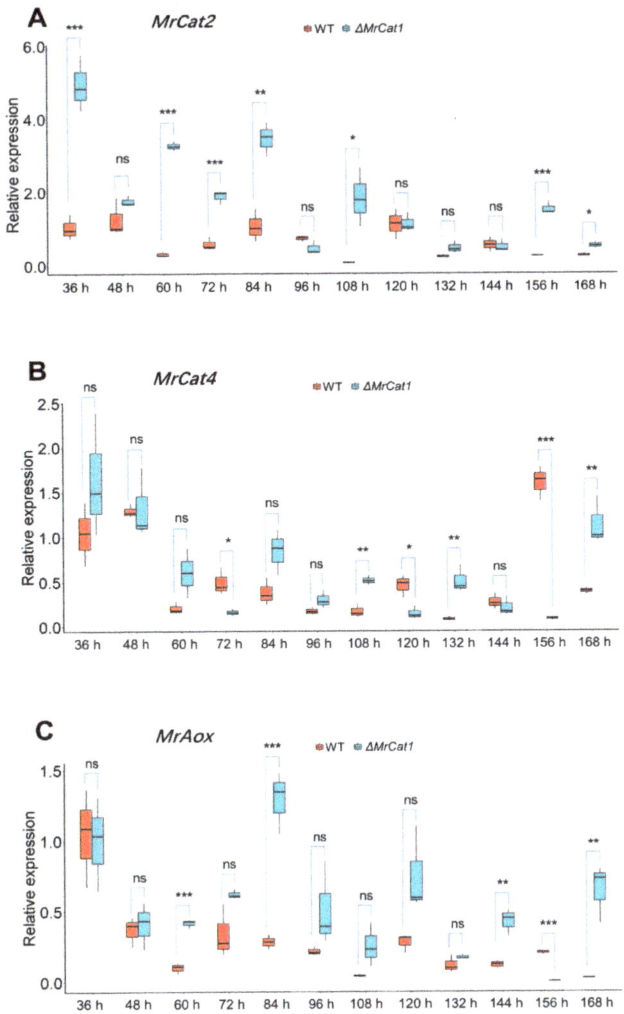

Figure 5. Deletion of *MrCat1* resulted in enhancing the expression of *MrCat2*, *MrCat4*, and *MrAox* during the MS formation of *M. rileyi*. RT-qPCR analysis of *MrCat2* (**A**), *MrCat4* (**B**), and *MrAox* (**C**) expression. Total RNA was isolated during MS development from the WT and Δ*MrCat1* strains. Results are mean relative expression ± SD. * $p < 0.05$, ** $p < 0.01$, *** $p < 0.001$, 'ns' represented no significant difference, compared between WT and Δ*MrCat1* at the same time point.

Similarly to *MrAox*, the transcription levels were higher from the yeast-like cell period to MS maturation (60 h–144 h) and were up-regulated in the later period of secondary mycelial growth (168 h). The most significant compensation occurred at 168 h, where *MrAox* was up-regulated by 22.1-fold. However, transcription levels of the *MrCat4* gene were prominent during the secondary mycelial growth phase and exhibited irregular compensation between the WT and Δ*MrCat1* strains. Taken together, the results indicate that *MrCat2*, *MrCat4*, and *MrAox* genes exhibit significantly higher expression in the Δ*MrCat1* strain compared to the WT strain during the MS initiation period (84 h), which is a key time point for MS initiation (Figure 5A–C). This suggests that in *M. rileyi*, *MrCat1*, *MrCat2*, and *MrAox* may have shared functions during MS formation.

3.6. Deletion of MrCat1 Reduced the Virulence of M. rileyi

To examine the effects of *MrCat1* deletion on the pathogenicity of *M. rileyi*, insect bioassays were conducted to infect *P. litura* larvae. The survival rates and LT_{50} were determined and compared between the WT and Δ*MrCat1* strains. After infection, it was observed that the survival rates of *P. litura* larvae infected by the Δ*MrCat1* strain were significantly higher than those infected by the WT strain at the same concentration (2.5×10^7 sp mL^{-1}, 5.0×10^7 sp mL^{-1}, 1.0×10^8 sp mL^{-1}) (Figure 6A). Moreover, the mortality of the Δ*MrCat1* stain at the concentration of 2.5×10^7 sp mL^{-1} and 5.0×10^7 sp mL^{-1} was less than 50% at the time point of 9 d. However, at the concentration of 1.0×10^8 sp mL^{-1}, the LT_{50} was 6.81 ± 0.52 days. In comparison, the LT_{50} for the WT at the concentrations of 2.5×10^7 sp mL^{-1}, 5.0×10^7 sp mL^{-1}, and 1.0×10^8 sp mL^{-1} were 7.29 ± 0.57 days, 6.56 ± 0.33 days, and 5.57 ± 0.27 days, respectively. The LT_{50} of the Δ*MrCat1* strain was prolonged 1.2 times compared with the WT strain at a concentration of 1.0×10^8 sp mL^{-1}, which exhibited a significant difference ($p < 0.05$; Figure 6B). The LC_{50} of Δ*MrCat1* and WT at the time point of 8 d were 3.077×10^7 sp mL^{-1} and 9.831×10^6 sp mL^{-1}, respectively. However, there was no noticeable difference in the morphology of muscardine cadavers between the WT and the Δ*MrCat1* strains. Furthermore, the insect cadavers were fully covered by fungal spores or mycelia from both the WT and Δ*MrCat1* strains after insect death (Figure 6C). For more detailed information about the bioassay, please refer to Table A2.

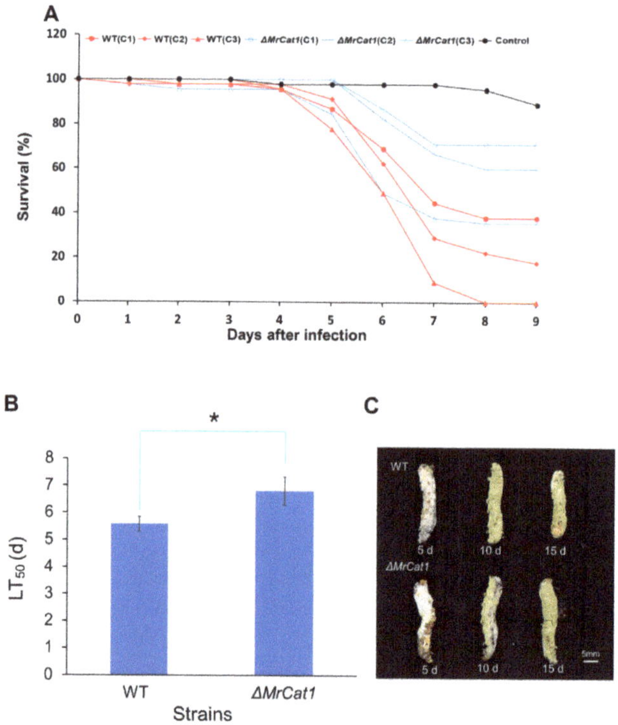

Figure 6. Deletion of *MrCat1* reduced the virulence of *M. rileyi*. (**A**) The survival rates of *P. litura* larvae infected by the WT and Δ*MrCat1* strains measured at different concentrations of spore suspensions (C1: 2.5×10^7 sp mL^{-1}, C2: 5.0×10^7 sp mL^{-1}, C3: 1.0×10^8 sp mL^{-1}). (**B**) LT_{50} of the WT and Δ*MrCat1* strains against *P. litura* larvae under spore concentration of 1×10^8 sp mL^{-1}. * $p < 0.05$, compared between WT and Δ*MrCat1*. (**C**) The morphology of muscardine cadavers infected by WT and Δ*MrCat1*. The *P. litura* larvae were killed by WT and Δ*MrCat1* for 5, 10, and 15 days.

4. Discussion

In this study, a catalase gene (*MrCat1*) was identified and isolated from *M. rileyi*. This gene showed up-regulation during MS formation. Sequence analysis revealed that *MrCat1* shares sequence similarity with monofunctional catalases found in other entomopathogenic fungi such as *M. anisopliae*, *B. bassiana*, and *Magnaporte grisea*. Previous studies have demonstrated the involvement of catalases in virulence and stress response. In these fungi, catalases are also known to play a role in combating oxidative stress, heat stress, hyperosmotic stress, and UV-B radiation in *B. bassiana*, *Cryptococcus neoformans*, and *M. grisea* [22,34,35].

In this study, no significant differences were observed in colony growth or dimorphic switching between yeast and hyphae on solid media when comparing the Δ*MrCat1* mutants and the WT strain. However, under oxidative stress conditions, both colony growth and dimorphic switching were impaired in the Δ*MrCat1* mutants compared to the WT strain. Additionally, the tolerance of conidia to H_2O_2 showed notable variation. These findings align with the documented role of *Cat1* in reactive oxygen species (ROS) metabolism in *B. bassiana* [22], *M. anisopliae* [23], and *M. grisea* [35]. Deletion of the *MrCat1* analog *MgCatB* in *M. grisea* resulted in accelerated hyphal growth but also caused paler pigmentation, reduced biomass, fragile conidia and appressoria, poor sporulation, and impaired melanization [35]. However, in *B. bassiana* and *M. rileyi*, the deletion of *MrCat1* did not lead to these phenotypic changes [22]. The studies have suggested that *MrCat1* in *M. rileyi* is involved in oxidative stress response and plays a role in colony growth, dimorphic switching, and conidial tolerance to H_2O_2. Understanding the specific mechanisms of *MrCat1* and their interactions with microsclerotia (MS) is crucial for future research in this field.

The present study also found that *MrCat1* in *M. rileyi* is significantly up-regulated during the development of mycelial sclerotia (MS) and the early stage of secondary mycelial growth. Deleting *MrCat1* resulted in a 76% reduction in MS production compared to the WT strain, as well as irregular MS morphology. These findings indicate that *Cat1* is involved in both reactive oxygen species (ROS) metabolism and MS formation in *M. rileyi*. Previous studies have established that oxidative stress plays a role in inducing sclerotium differentiation [36–39], with optimal concentrations of H_2O_2 promoting MS formation while high concentrations inhibiting it [13]. In the WT *M. rileyi*, gene expression analysis of *MrCat1*, *MrCat2*, and *MrCat4* during the development of MS revealed that these three catalase genes are up-regulated during sclerotial initiation and MS formation, indicating higher H_2O_2 concentrations during this process. The Δ*MrCat1* mutants displayed deficiencies in both MS proliferation and morphology.

The expression levels of Δ*MrCat1* and *MrCat2* mutants were found to be higher from the spore germinating period to MS initiation (36 to 84 h) compared to the WT strain. This indicates that *MrCat2* is up-regulated as a compensatory mechanism in response to the deletion of *MrCat1* during the early stages of MS formation. This observation is similar to what has been reported in *B. bassiana*, where the disruption of one of the five catalases led to the up-regulation of one or more other catalase genes. This suggests that some of the five catalases were functionally complementary, leading to complicated phenotypic changes [22], which is consistent with the results of the present study. However, the compensatory expression pattern of *MrCat4* in the Δ*MrCat1* mutants is irregular due to the absence of *MrCat1*. In contrast, in *M. grisea*, mutants lacking *MgCatB* do not show overexpression of other catalase or catalase-related antioxidant genes [35]. These findings suggest that the function of *MrCat1* is analogous to that of *BbCatB*, but the compensatory mechanisms may differ between different fungal species.

Interestingly, in the Δ*MrCat1* mutants, it was observed that another gene involved in ROS metabolism, *MrAox*, exhibited over-expression during the yeast-like cell period to MS maturation (60 h–144 h). Previous studies have shown that *Aox* is involved in the metabolism of ROS generated through energy production and metabolism in fungi [40,41]. In *M. rileyi*, defects in *MrAox* have been shown to impair ROS metabolism and MS formation [14]. These findings indicate that reactive oxygen species (ROS) accumulate during microsclerotia (MS) formation. It has been established that catalases are important enzymes for scavenging ROS, as they can convert H_2O_2 to water and molecular oxygen. In *Beauveria bassiana*, the five catalases play distinct roles in regulating tolerance to oxidation. In the current study, it was found that *MrAox* was up-regulated to compensate for the deletion of *MrCat1*. This suggests that *Aox* may function similarly to catalases by regulating ROS metabolism during MS formation. Catalase genes have been identified as essential virulence factors in entomogenous fungi [22,23] and phytopathogenic fungi [35]. In line with previous research, the deletion of *MrCat1* in this study led to a notable decrease in virulence. However, further comprehensive studies are necessary to elucidate the underlying mechanism. Additionally, it is important to conduct further investigations to determine the potential involvement of *MrCat2* or *MrCat4* in virulence and MS formation.

5. Conclusions

The deletion of the *MrCat1* gene had an impact on ROS metabolism and MS development in *M. rileyi*. *MrCat2* and *MrCat4* were up-regulated as compensatory mechanisms in the absence of *MrCat1*. Additionally, the up-regulation of *MrAox* in response to *MrCat1* deletion, which has not been previously reported, suggests its involvement in compensating for the loss of *MrCat1* in *M. rileyi*.

Author Contributions: Conceptualization, Y.S.; methodology, Y.S.; data curation, Y.S.; writing—original draft preparation, Y.S.; validation, X.W., Y.L. and H.J.; visualization, X.W.; formal analysis, Y.L.; investigation, G.T.; writing—review and editing, H.L.; supervision, H.L.; project administration, H.L. All authors have read and agreed to the published version of the manuscript.

Funding: This research received no external funding.

Institutional Review Board Statement: Not applicable.

Informed Consent Statement: Not applicable.

Data Availability Statement: The original contributions presented in the study are included in the article; further inquiries can be directed to the corresponding author.

Conflicts of Interest: The authors declare no conflicts of interest.

Appendix A

Table A1. Primers used for RT-qPCR in this work.

Primers	Sequence 5′-3′
RT-tub-F	GGCAAGGTCGCTATGAAG
RT-tub-R	CTGGATGGAGGTAGAGTTAC
RT-tef-F	GTCATCGTCCTCAACCATC
RT-tef-R	CAGTCTCAACAGCCTTACC
RT-Cat1-F	TTCGTCAACGAAGAGGGTGA
RT-Cat1-R	GTGGAAGTCGGAGTTCTTGC
RT-Cat2-F	ACTGGAAGCTCAACAACCCT
RT-Cat2-R	GCGCATCTGGACATTCTTTA
RT-Cat4-F	GGCGTCCCTGATATTCCTGA
RT-Cat4-R	GGCGATGATATCCGTGTGTG
RT-Aox-F	AATCCCACTACATCCGTGGT
RT-Aox-R	GCTCTCGAGAAAGACGAACC

Table A2. LT_{50}s of the wild-type and Δ*MrCat1* strains against *P. litura* larvae.

Strains	Concentrations of Conidial Suspensions	LT_{50} (d)
WT	2.5×10^7 sp mL^{-1}	7.29 ± 0.57
	5.0×10^7 sp mL^{-1}	6.56 ± 0.33
	1.0×10^8 sp mL^{-1}	5.57 ± 0.27
Δ*MrCat1*	2.5×10^7 sp mL^{-1}	>9 *
	5.0×10^7 sp mL^{-1}	>9 *
	1.0×10^8 sp mL^{-1}	6.81 ± 0.52

* Mortality less than 50% at the time point of 9 d, accurate LT_{50} could not be determined in this case.

References

1. Boucias, D.G.; Tigano, M.S.; Sosa-Gomez, D.R.; Glare, T.R.; Inglis, P.W. Genotypic properties of the entomopathogenic fungus *Nomuraea rileyi*. *Biol. Control* **2000**, *19*, 124–138. [CrossRef]
2. Chen, H.; Yin, Y.P.; Li, Y.; Mahmud, M.S.; Wang, Z.K. Identification and analysis of gene differentially expressed in the *Spodoptera litura* fat body in response to the biocontrol fungus, *Nomuraea rileyi*. *Comp. Biochem. Physiol. Part B Biochem. Mol. Biol.* **2012**, *163*, 203–210. [CrossRef]
3. Devi, P.S.; Prasad, Y.G.; Chowdary, D.A.; Rao, L.M.; Balakrishnan, K. Identification of virulent isolates of the entomopathogenic fungus *Nomuraea rileyi* (F) Samson for the management of *Helicoverpa armigera* and *Spodoptera litura* (identification of virulent isolates of *N. rileyi*). *Mycopathologia* **2003**, *156*, 356–373.
4. Supakdamrongkul, P.; Bhumiratana, A.; Wiwat, C. Characterization of an extracellular lipase from the biocontrol fungus, *Nomuraea rileyi* MJ, and its toxicity toward *Spodoptera litura*. *J. Invertebr. Pathol.* **2010**, *105*, 228–235. [CrossRef]
5. Vega-Aquino, P.; Sanchez-Peña, S.; Blanco, C.A. Activity of oil-formulated conidia of the fungal entomopathogens *Nomuraea rileyi* and *Isaria tenuipes* against lepidopterous larvae. *J. Invertebr. Pathol.* **2010**, *103*, 145–149. [CrossRef] [PubMed]
6. Yin, Y.P.; Shan, H.; Song, Z.Y.; Wang, Z.K. Microsclerotia artificial inductions of *Nomuraea rileyi* CQNr01. *China Agric. Sci.* **2012**, *4523*, 4801–4807. (In Chinese)
7. Jackson, M.A.; Jaronski, S.T. Development of pilot-scale fermentation and stabilisation processes for the production of microsclerotia of the entomopathogenic fungus *Metarhizium brunneum* strain F52. *Biocontrol Sci. Technol.* **2012**, *22*, 915–930. [CrossRef]
8. Jaronski, S.T.; Jackson, M.A. Efficacy of *Metarhizium anisopliae* microsclerotial granules. *Biocontrol Sci. Technol.* **2008**, *18*, 849–863. [CrossRef]
9. Kawasaki, L.; Aguirre, J. Multiple catalase genes are differentially regulated in *Aspergillus nidulans*. *J. Bacteriol.* **2001**, *183*, 1434–1440. [CrossRef]
10. Song, Z.Y.; Yin, Y.P.; Jiang, S.S.; Liu, J.J.; Chen, H.; Wang, Z.K. Comparative transcriptome analysis of microsclerotia development in *Nomuraea rileyi*. *BMC Genom.* **2013**, *141*, 411. [CrossRef]
11. Jiang, S.S.; Yin, Y.P.; Song, Z.Y.; Zhou, G.L.; Wang, Z.K. *Raca* and *cdc42* regulate polarized growth and microsclerotium formation in the dimorphic fungus *Nomuraea rileyi*. *Res. Microbiol.* **2014**, *1653*, 233–242. [CrossRef] [PubMed]
12. Song, Z.Y.; Wei, J.; Yin, Y.P.; Wang, Z.K. Polarity Proteins Mrcdc24 and Mrbem1 Required for Hypha Growth and Microsclerotia Formation in *Metarhizium Rileyi*. *Biocontrol Sci. Techn.* **2016**, *26*, 733–745. [CrossRef]
13. Liu, J.J.; Yin, Y.P.; Song, Z.Y.; Li, Y.; Jiang, S.S.; Shao, C.W.; Wang, Z.K. NADH: Flavin oxidoreductase/NADH oxidase and ROS regulate microsclerotium development in *Nomuraea rileyi*. *World J. Microbiol. Biotechnol.* **2014**, *30*, 1927–1935. [CrossRef] [PubMed]
14. Zhou, G.L.; Song, Z.Y.; Yin, Y.P.; Jiang, W.; Wang, Z.K. Involvement of alternative oxidase in the regulation of hypha growth and microsclerotia formation in *Nomuraea rileyi* CQNr01. *World J. Microbiol. Biotechnol.* **2015**, *31*, 1343–1352. [CrossRef] [PubMed]
15. Fan, L.Q.; Li, B.J.; Wang, J.H.; Li, X.X.; Ma, F.L.; Du, F.; Li, H.L.; Lin, Y.L. Multifunctional regulation of NADPH oxidase in growth, microsclerotia formation and virulence in *Metarhizium rileyi*. *Biotechnol. Lett.* **2023**, *45*, 1441–1455. [CrossRef]
16. Song, Z.Y.; Shen, L.; Yin, Y.P.; Tan, W.Y.; Shao, C.W.; Xu, J.; Wang, Z.K. Role of two *Nomuraea rileyi* transmembrane sensors sho1p and sln1p in adaptation to stress due to changing culture conditions during microsclerotia development. *World J. Microbiol. Biotechnol.* **2015**, *31*, 477–485. [CrossRef] [PubMed]
17. Song, Z.Y.; Zhong, Q.; Yin, Y.P.; Shen, L.; Li, Y.; Wang, Z.K. The high osmotic response and cell wall integrity pathways cooperate to regulate morphology, microsclerotia development, and virulence in *Metarhizium rileyi*. *Sci. Rep.* **2016**, *6*, 38765. [CrossRef] [PubMed]
18. Polidoros, A.N.; Mylona, P.V.; Scandalios, J.G. Transgenic tobacco plants expressing the maize *Cat2* gene have altered catalase levels that affect plant–pathogen interactions and resistance to oxidative stress. *Transgenic Res.* **2001**, *10*, 555–569. [CrossRef]
19. Alscher, R.G.; Erturk, N.; Heath, L.S. Role of superoxide dismutases SODs in controlling oxidative stress in plants. *J. Exp. Bot.* **2002**, *53*, 1331–1341. [CrossRef]
20. Schouten, A.; Tenberge, K.B.; Vermeer, J.; Stewart, J.; Wagemakers, L.; Williamson, B.; van Kan, J.A. Functional analysis of an extracellular catalase of *Botrytis cinerea*. *Mol. Plant Pathol.* **2002**, *3*, 227–238. [CrossRef]

21. Kwok, L.Y.; Schlüter, D.; Clayton, C.; Clayton, C.; Soldati, D. The antioxidant system in *Toxoplasma gondii* and the role of cytosolic catalase in defense against oxidative injury. *Mol. Microbiol.* **2004**, *51*, 47–61. [CrossRef] [PubMed]
22. Wang, Z.L.; Zhang, L.B.; Ying, S.H.; Feng, M.G. Catalases play differentiated roles in the adaptation of a fungal entomopathogen to environmental stresses. *Environ. Microbiol.* **2013**, *152*, 409–418. [CrossRef] [PubMed]
23. Morales Hernandez, C.E.; Padilla Guerrero, I.E.; Gonzalez Hernandez, G.A.; Salazar Solis, E.; Torres Guzman, J.C. Catalase overexpression reduces the germination time and increases the pathogenicity of the fungus *Metarhizium anisopliae*. *Appl. Microbiol. Biotechnol.* **2010**, *87*, 1033–1044. [CrossRef] [PubMed]
24. Wang, N.Y.; Yoshida, Y.; Hasunuma, K. Loss of Catalase-1 (Cat-1) results in decreased conidial viability enhanced by exposure to light in *Neurospora crassa*. *Mol. Genet. Genom.* **2007**, *277*, 13–22. [CrossRef] [PubMed]
25. Michán, S.; Lledías, F.; Hansberg, W. Asexual development is increased in *Neurospora crassa cat-3*-null mutant strains. *Eukaryot. Cell* **2003**, *2*, 798–808. [CrossRef] [PubMed]
26. Hisada, H.; Hata, Y.; Kawato, A.; Abe, Y.; Akita, O. Cloning and expression analysis of two catalase genes from *Aspergillus oryzae*. *J. Biosci. Bioeng.* **2005**, *99*, 562–568. [CrossRef] [PubMed]
27. Garre, V.; Tenberge, K.B.; Eising, R. Secretion of a fungal extracellular catalase by *Claviceps purpurea* during infection of rye: Putative role in pathogenicity and suppression of host defense. *Phytopathology* **1998**, *88*, 744–753. [CrossRef] [PubMed]
28. Pedrini, N.; Juárez, M.P.; Crespo, R.; de Alaniz, M.J. Clues on the role of *Beauveria bassiana* catalases in alkane degradation events. *Mycologia* **2006**, *98*, 528–534. [CrossRef] [PubMed]
29. Su, Y.; Wang, Z.K.; Shao, C.W.; Luo, Y.L.; Wang, L.; Yin, Y.P. An efficient gene disruption method using a positive-negative split-selection marker and Agrobacterium tumefaciens-mediated transformation for *Nomuraea rileyi*. *World J. Microbiol. Biotechnol.* **2018**, *34*, 26. [CrossRef]
30. Wang, Z.; Ye, S.F.; Li, J.J.; Zheng, B.; Bao, M.Z.; Ning, G.G. Fusion primer and nested integrated PCR FPNI-PCR: A new high-efficiency strategy for rapid chromosome walking or flanking sequence cloning. *BMC Biotechnol.* **2011**, *11*, 109. [CrossRef]
31. Kumar, S.; Stecher, G.; Li, M.; Knyaz, C.; Tamura, K. MEGA X: Molecular Evolutionary Genetics Analysis across Computing Platforms. *Mol. Biol. Evol.* **2018**, *35*, 1547–1549. [CrossRef] [PubMed]
32. Gao, Q.; Jin, K.; Ying, S.H.; Zhang, Y.J.; Xiao, G.H.; Shang, Y.F.; Duan, Z.B.; Hu, X.; Xie, X.Q.; Zhou, G.; et al. Genome sequencing and comparative transcriptomics of the model entomopathogenic fungi *Metarhizium anisopliae* and *M. acridum*. *PLoS Genet.* **2011**, *7*, e1001264. [CrossRef] [PubMed]
33. Wang, G.; Liu, Z.G.; Lin, R.M.; Li, E.F.; Mao, Z.C.; Ling, J.; Yang, Y.H.; Yin, W.B.; Xie, B.Y. Biosynthesis of Antibiotic Leucinostatins in Bio-control Fungus *Purpureocillium lilacinum* and Their Inhibition on Phytophthora Revealed by Genome Mining. *PLoS Pathog.* **2016**, *12*, e1005685. [CrossRef] [PubMed]
34. Giles, S.S.; Stajich, J.E.; Nichols, C.; Gerrald, Q.D.; Alspaugh, J.A.; Dietrich, F.; Perfect, J.R. The *Cryptococcus neoformans* catalase gene family and its role in antioxidant defense. *Eukaryot. Cell* **2006**, *5*, 1447–1459. [CrossRef] [PubMed]
35. Skamnioti, P.; Henderson, C.; Zhang, Z.; Robinson, Z.; Gurr, S.J. A novel role for catalase B in the maintenance of fungal cell-wall integrity during host invasion in the rice blast fungus *Magnaporthe grisea*. *Mol. Plant-Microbe Interact.* **2007**, *20*, 568–580. [CrossRef] [PubMed]
36. Georgiou, C.D.; Petropoulou, K.P. The role of ascorbic acid role in the differentiation of sclerotia in sclerotinia minor. *Mycopathologia* **2002**, *154*, 71–77. [CrossRef] [PubMed]
37. Georgiou, C.D.; Zees, A. Lipofuscins and sclerotial differentiation in phytopathogenic fungi. *Mycopathologia* **2002**, *153*, 203–208. [CrossRef] [PubMed]
38. Georgiou, C.D.; Patsoukis, N.; Papapostolou, I.; Zervoudakis, G. Sclerotial metamorphosis in filamentous fungi is induced by oxidative stress. *Integr. Comp. Biol.* **2006**, *46*, 691–712. [CrossRef]
39. Patsoukis, N.; Georgiou, C.D. Thiol redox state and oxidative stress affect sclerotial differentiation of the phytopathogenic fungi sclerotium rolfsii and sclerotinia sclerotiorum. *J. Appl. Microbiol.* **2008**, *104*, 42–50. [CrossRef]
40. Hattori, T.; Honda, Y.; Kino, K.; Kirimura, K. Expression of alternative oxidase gene (*aox1*) at the stage of single-cell conidium in citric acid-producing *Aspergillus niger*. *J. Biosci. Bioeng.* **2008**, *105*, 55–57. [CrossRef]
41. Helmerhorst, E.J.; Stan, M.; Murphy, M.P.; Sherman, F.; Oppenheim, F.G. The concomitant expression and availability of conventional and alternative, cyanide-insensitive, respiratory pathways in *Candida albicans*. *Mitochondrion* **2005**, *5*, 200–211. [CrossRef] [PubMed]

Journal of
Fungi

Article

Oleic Acid and Linoleic Acid Enhances the Biocontrol Potential of *Metarhizium rileyi*

Guang Wang, Xu Zhang, Guangzu Du, Wenqian Wang, Yunhao Yao, Sitong Jin, Haosheng Cai, Yuejin Peng * and Bin Chen *

Yunnan State Key Laboratory of Conservation and Utilization of Biological Resources, College of Plant Protection, Yunnan Agricultural University, Kunming 650201, China; wang_gunag_17@126.com (G.W.); zhxv08@163.com (X.Z.); duguangzu1986@163.com (G.D.); wqwangts@163.com (W.W.); yaoyunh@163.com (Y.Y.); jinsitong1015@163.com (S.J.); cai_hao_sheng@163.com (H.C.)

* Correspondence: 2021053@ynau.edu.cn (Y.P.); chbins@163.com (B.C.)

Abstract: *Metarhizium rileyi* is a wide spread insect fungi with a good biocontrol potentiality to various pests, particularly noctuid insects. However, it is characterized by its slow growth, its sensitivity to abiotic stress, and the slow speed of kill to pests, which hinder its use compared with other entomopathogenic fungi. In this study, the responses of *M. rileyi* to eight types of lipids were observed; among the lipids, oleic acid and linoleic acid significantly promoted the growth and development of *M. rileyi* and enhanced its stress tolerances and virulence. An additional mechanistic study demonstrated that exogenous oleic acid and linoleic acid significantly improved the conidial germination, appressorium formation, cuticle degradation, and cuticle infection, which appear to be largely dependent on the up-regulation of gene expression in growth, development, protective, and cuticle-degrading enzymes. In conclusion, exogenous oleic acid and linoleic acid enhanced the stress tolerances and virulence of *M. rileyi* via protecting conidial germination and promoting cuticle infection. These results provide new insights for the biopesticide development of *M. rileyi*.

Keywords: growth; development; stress tolerances; virulence

1. Introduction

Metarhizium (Nomuraea) *rileyi* (Farlow) Kepler (Hypocreales: Clavicipitaceae) is an entomopathogenic fungus that shows a strong virulence against Nymphalidae, Noctuidae, and other pests [1–5], such as *Spodoptera frugiperda*, *Spodoptera litura*, *Spodoptera cosmioides*, and *Chrysodeixis includens*. *M. rileyi* can control crop pests in different agricultural systems, as shown by greenhouse and field trials [6–8]. However, *M. rileyi* is slow growing, sensitive to abiotic stress, and slow to kill pests, hindering its use compared with other entomopathogenic fungi (*Beauveria bassiana* and *Metarhizium anisopliae*) and decreasing its viability, biological activity, and prevalence under field conditions [8–11]. Therefore, enhancing the resistance of *M. rileyi* to abiotic stresses and improving its speed of killing pests are critical for its industrial-scale use as a bioinsecticide.

Conidia are the main infection and reproductive units of entomopathogenic fungi [12]. Maintaining or improving the viability, biological activity, and abundance of conidia under field conditions is the main strategy for applying entomopathogenic fungi. Many methods of improving the performance of conidia have been reported, such as regulating the culture substrate [13], adding exogenous oil or surfactants [14,15], and adapting to abiotic stress [16]. Oil suspensions are one of the most common types of pesticide formulations. Oil suspension formulations of conidia made from vegetable oil or mineral oil can significantly improve the pathogenicity and abiotic stress tolerance of the conidia [14,15]. Conidial oil suspensions of *M. rileyi* can efficiently control *S. frugiperda* in corn fields [17]. Furthermore, vegetable oil can improve the conidial germination, yield, and thermotolerance of *B. bassiana*, *Simplicillium lanosoniveum*, and *Isaria fumosorosea* [18–20]. Compared to mineral oil, vegetable oil has the function of regulating the growth and development of entomopathogenic fungi [14,15,18–20], suggesting that there is a certain substance in vegetable oil that is beneficial for conidial growth and development.

The main components of vegetable oil are lipids and esters generated from glycerol [21], including palmitic acid (C16:0, 4.6–20.0%), oleic acid (C18:1, 6.2–71.1%), and linoleic acid (C18:2, 1.6–79.0%), which are present in coconut oil, castor oil, olive oil, sunflower oil, and palm oil [21]. In fungal species such as *Cordyceps*, *Metarhizium*, *Beauveria*, and others, lipids can serve as a protective agent for conidia against stresses and signaling molecules for growth and development [10,19,22,23]. Linoleic acid and oleic acid increased the conidial tolerance to heat, drought, and oxidative stresses [10,19,24], which promoted spore germination, the branching of hyphae, pseudohyphal growth, conidial production, and the transcription of growth- and development-related genes [22,23]. Notably, lipids are one of the main components of oxylipins; they change the phospholipid composition, membrane homeostasis, membrane fluidity, and conidial hydrophobicity (an important indicator of the initial cuticle infection) [25–27]. These changes are involved in improving germination, conidiogenesis effects, fungal proliferation, pathogenicity, and stress tolerance [25–28]. Thus, we speculate that lipids may have important functions in the growth, development, virulence, and stress tolerance of entomopathogenic fungi. Despite many previous studies investigating the effect of lipids on the abiotic stress tolerance and conidial yield of *M. rileyi*, the mechanism of lipids in improving conidial quality remains poorly understood.

Therefore, the objectives of this study were as follows: (1) to investigate the effects of eight lipids on the growth and development of *M. rileyi*; (2) to determine the role of oleic acid and linoleic acid in the stress tolerance and pathogenicity of *M. rileyi*; and (3) to evaluate the mechanism through which oleic acid and linoleic acid improve the stress tolerance, growth and development, and pathogenicity of *M. rileyi*.

2. Materials and Methods

2.1. Fungal Strain and Media

M. rileyi strain SZCY201010 was isolated from infected *S. frugiperda* larvae in a corn field in Shizong county, Qujing city, Yunnan province, China, and was cultured on Sabouraud maltose agar plus yeast extract medium (SMAY: 1% peptone, 1% yeast extract, 4% maltose, and 1.5% agarose; Coolaber, Beijing, China), as previously described [29]. The media plates were incubated at 25 °C with a photoperiod consisting of a 12L:12D cycle in a culture chamber. Conidial suspensions were prepared from fungi grown on SMAY plates for 10 days.

2.2. Phenotype Assays

Fungal growth and development: Aliquots of 5 μL of a conidial suspension (1×10^5, 1×10^6, and 1×10^7 conidia/mL) were inoculated onto SMAY with 0.1% lipids, including oleic acid, linoleic acid, stearic acid, eicosane, cetane, palmitic acid, tetracosane, and octacosanol (Aladdin, Shanghai, China; chemical formula provided in Figure S1). The plates were incubated at 25 °C in a 12L:12D cycle within a culture chamber at 25 °C. Daily observations were made of each colony's yeast-like blastospore formation time (YG), hyphal growth time (HG), and sporulation time (ST). The diameters of all the colonies incubated for 5, 7, 9, and 11 days at 25 °C were measured. The experiment was performed in triplicate, with three parallel controls each time. The relative growth rate (GR) was estimated with the formula $(d_{t2} - d_{t1})/(t2 - t1) \times 100$ (dt_1, starting colony diameter; dt_2, end colony diameter).

Fungal stress responses: Aliquots of 5 μL of a conidial suspension (1×10^7 conidia/mL) were inoculated onto SMAY alone (CK) or supplemented with H_2O_2 (3 mM) for oxidative stress; NaCl (0.5 M; Coolaber, Beijing, China), KCl (0.5 M; Coolaber, Beijing, China), glycerol (1 M; Coolaber, Beijing, China), and sorbitol (0.5 M; Coolaber, Beijing, China) for osmotic stress; and calcofluor white (5 μg/mL; Coolaber, Beijing, China) for cell wall stress. The diameters of all the colonies incubated for 10 days under the stresses at 25 °C were measured as described above. The colony response to heat shock was examined by exposing normal 2-day-old SMAY colonies to 42 °C for 1–3 h. The colonies were transferred to 25 °C for 8-day growth recovery, followed by measuring the colony diameters. This experiment was performed in triplicate, with three parallel controls each time. The percentage of relative growth inhibition (RGI) was estimated with the formula $(d_c - d_t)/d_c \times 100$ (d_c, control colony diameter; d_t, stressed colony diameter) and used as an index of hyphal sensitivity to each stress [11].

Conidial germination: Conidial suspensions (1×10^7 conidia/mL, 100 μL) were spread onto SMAY alone (control) or supplemented with 0.190% oleic acid, 0.175% linoleic acid, or different types of stresses. All the treatments were incubated at 25 °C and the germination percentage was determined for 100 conidia. The entire experiment was performed in triplicate.

Yeast-like blastospores: Aliquots of 5 μL of a conidial suspension (1×10^7 conidia/mL) were inoculated onto SMAY alone (control) or supplemented with 0.190% oleic acid or 0.175% linoleic acid. The plates were cultured at 25 °C with a photoperiod of a 12L:12D cycle in a culture chamber. The yeast-like blastospore yield of *M. rileyi* on the SMAY plates was measured at 3, 4, and 5 days. The entire experiment was performed five times.

Conidia production: To evaluate the conidial yield of *M. rileyi*, *M. rileyi* conidia were suspended in a 0.02% Tween 80 solution (1×10^6 and 1×10^7 conidia/mL). Aliquots of a 100 μL conidial suspension were spread onto SMAY plates with varying concentrations (0.01, 0.05, 0.1, 0.2, and 0.4%) of oleic acid and linoleic acid. The plates were cultured at 25 °C with a photoperiod of a 12L:12D cycle in a culture chamber. When the color of the colony changed from pale or olive green to malachite green, the conidial yield of *M. rileyi* on the SMAY plates was evaluated at 10 d. Three replicates were used for each treatment.

2.3. Fungal Virulence-Related Phenotypes

Simulation of host cuticle conidial germination: To view conidial germination on water agar plates (1.5% agarose), conidial suspensions (1×10^7 conidia/mL, 100 μL) were spread onto the plates. Disinfected and sterilized *Locusta migratoria manilensis* hindwings and water agar plates prepared in high-temperature and -pressure facilities were used to mimic the host cuticle and oligotrophic conditions, respectively. A conidial suspension (1×10^7 conidia/mL) was sprayed onto the locust hindwings. All the treatments were incubated at 25 °C and the germination percentage was determined for 100 conidia. Three replicates were used for each treatment.

Appressorium induction: Appressorium formation assays were carried out according to previous reports [30]. Conidia were harvested in 0.01% (vol/vol) Triton X-100 from 10 d old fungal cultures on SMAY. A 20 μL conidial suspension (5×10^8 conidia/mL) was added to 4 mL of MM-Gly (6 g/L $NaNO_3$, 0.52 g/L KCl, 0.52 g/L $MgSO_4 \cdot 7H_2O$, 0.25 g/L KH_2PO_4, and 10 mL glycerol; Coolaber, Beijing, China) in plastic plates (diameter = 6 cm) and incubated at 27 °C for 48 or 72 h for appressorium induction. We selected 100 conidia to calculate the appressorium formation rate and performed three biological replicates.

Cuticle-degrading enzyme: Extracellular enzyme activity assays were performed according to the method described in previous reports by Gebremariam et al. [31], and appropriate modifications were made. The conidia were made into a 1×10^7 conidia/mL suspension, alone or supplemented with 0.190% (*v*/*v*) oleic acid or 0.175% (*v*/*v*) linoleic acid. Aliquots of 5 μL of the conidial suspension were inoculated onto 1/8 SMAY supplemented with 1.2% powdered skim milk (*v*/*v*; Yili, Huhehaote, China), 0.2% olive oil (*v*/*v*; Sangon Biotech, Shanghai, China), and 0.5% colloidal chitin (*w*/*w*; coolaber, Beijing, China) to measure the protease, chitinase, or lipase activity, respectively. The plates were incubated at 25 °C for 14 days. The determination of enzyme activity was based on the transparent hydrolysis zone around the colony. The diameter of the colonies and hydrolysis zones was measured using the cross method. The enzymatic index (EI) was calculated using the following formula: EI = (hydrolysis zone diameter)/(colony growth diameter). The entire experiment was performed five times.

2.4. Bioassays for Fungal Virulence

The *S. frugiperda* larvae that were used for virulence determination were 3rd, 4th, and 5th instars. A Tween 80 solution (0.02%) was used alone as the control (CK) or supplemented with 0.190% (*v*/*v*) oleic acid (OA) or 0.175% (*v*/*v*) linoleic acid (LA). The conidia were made into a 1×10^8 conidia/mL suspension comprising *M. rileyi* conidia alone (Mr) or supplemented with 0.190% (*v*/*v*) oleic acid (Mr+OA) or 0.175% (*v*/*v*) linoleic acid (Mr+LA). Healthy 3rd-, 4th-, and 5th-instar larvae of the same size were selected, put into the prepared conidial suspension for 15 s, and then removed; the excess water on the surface of the insect body was removed with aseptic filter paper. Subsequently, the larvae were placed individually into a disinfected and sterilized 12-well cell culture plate with 75% alcohol. A fresh artificial diet (125 g of soybean flour, 225 g of corn flour, 225 g of corn leaf flour, 40 g of yeast extract, 20 g of casein, 0.6 g of cholesterol, 3 g of choline chloride, 6 g of sorbic acid, 6 g of methyl p-hydroxybenzoate, 36 g of agarose, 0.1 g of inositol, 7 g of vitamin C, and 1300 mL of water) was added to the holes where the larvae were placed. The above experiment was performed three times, with three treatments for each experiment and 30–36 larvae for each treatment. After the treatment, the larvae were fed the artificial diet and maintained in an artificial climate chamber at 25 °C with a relative humidity of 75% and a photocycle of 16L:8D. The number of dead insects was recorded each day, and the dead insects were kept in a Petri dish to determine whether they were infected with the test strain.

The blastospore growth was observed to examine the time it took to penetrate the cuticle. The hemolymph of five 4th-instar larvae in the Mr, Mr+OA, and Mr+LA infection groups was collected and diluted 1:1 in a sterile physiological solution (0.14 M NaCl, 0.1 M glucose, 25 mM sodium citrate, and 30 mM citric acid). The hyphal bodies were photographed and observed under a microscope every 12 h from 48 to 132 h post-inoculation of the conidia. The average of five fields was taken as one biological replicate. Five replicates were used for each treatment.

2.5. CAT, POD, and SOD Activity Assays

M. rileyi was cultured separately in the SMY culture media (1% peptone, 1% yeast extract, and 4% maltose), alone or supplemented with 0.190% (*v*/*v*) oleic acid or 0.175% (*v*/*v*) linoleic acid. These liquid culture media were inoculated with a final concentration of 10^7 conidia/mL and cultured for 6 days at 25 °C on a rotary with a speed of 200 rpm. The mycelium was filtered and washed four times with PBS. A BCA protein content assay kit, catalase (CAT) activity assay kit, peroxidase (POD) activity assay kit, and superoxide dismutase (SOD) activity assay kit (Boxbio Science & Technology Co., Ltd., Beijing, China) were used to determine the protein concentration and total enzymatic activity (U/mg protein). Three replicates were used for each treatment.

2.6. Quantitative Real-Time Reverse-Transcription Polymerase Chain Reaction (RT-qPCR)

Aliquots of 100 μL of the conidial suspension (1×10^7 conidia/mL) were smeared onto SMAY alone (CK) or supplemented with 0.190% oleic acid or 0.175% linoleic acid and incubated at 25 °C for 5 days for the determination of gene expression. The total RNA was isolated using the TRIzol reagent (BBI Life Sciences, Shanghai, China) by following the manufacturer's instructions. The first-strand cDNA was synthesized from the total RNA using a HiScript III All-in-One RT SuperMix Perfect for qPCR (Vazyme, Nanjing, China). The RT-qPCR analysis was carried out in 96-well 0.2 mL block plates using a QuantStudio™ 7 Flex system (Thermo Scientific, Wilmington, DE, USA). Each reaction contained 1.0 μL of the template (1 ng/μL), 10.0 μL of 2 × Taq Pro Universal SYBR qPCR Master Mix (Vazyme, Nanjing, China), and 0.4 μL of each primer (10 μM), and 8.2 μL of EDPC-ddH_2O, with a final volume of 20 μL. The RT-qPCR condition was pre-denaturation at 95 °C for 30 s, and then 40 cycles at 95 °C for 5 s and at 60 °C for 20 s. After each reaction, a melting curve analysis (denatured at 95 °C for 15 s, annealed at 60 °C for 60 s, and denatured at 95 °C for 15 s) was conducted to ensure the consistency and specificity of the amplified product. Three biological replicates were used for each treatment in the RT-qPCR analyses. The β-tubulin gene (TUB) and transcription elongation factor (TEF) genes were employed as internal references [32,33]. The $2^{-\Delta\Delta CT}$ technique was utilized to determine the relative expression of every gene [34]. Table S1 contains a list of primers utilized in the RT-qPCR tests.

2.7. Statistical Analysis

All the data are presented as the mean ± standard deviation. The data were analyzed for the normality of their distribution using the Shapiro–Wilk test and for homoscedasticity using Levene's test. When these assumptions were not fulfilled, the data were subjected to a log or square root transformation. If the data met the assumptions of normality and homogeneity, two groups were analyzed using an unpaired Student's *t*-test and data among the various groups were analyzed using a one-way analysis of variance (ANOVA) followed by Tukey's post hoc test. If the data were not normal and/or not homogenous, two groups were analyzed using a Wilcox test and the data among the various groups were analyzed using the non-parametric Kruskal–Wallis test, followed by the Games–Howell test. The survival data were subjected to a Kaplan–Meier survival log-rank analysis. The median lethal time (LT_{50}) for certain treatments was calculated using linear regression in Microsoft Excel, version 2019. $p < 0.05$ was considered to indicate statistical significance.

3. Results

3.1. Effects of Lipids Addition on M. rileyi Growth and Development

The growth and development rhythm of *M. rileyi*, including conidial germination, yeast-like blastospore production, mycelial growth, and sporulation, was determined (Figure S1). We used a heat map clustering analysis based on the yeast-like blastospore formation time, hyphal growth time, sporulation time, and relative growth rate data to compare the effects of different lipids. As shown in Figure 1A, oleic acid and linoleic acid clustered into one branch, while other lipids (SA, eicosane, cetane, PA, tetracosane, and octacosanol) clustered into another. Exogenous oleic acid and linoleic acid significantly increased the colony diameter in the conidial suspensions with concentrations of 10^5, 10^6, and 10^7 conidia/mL (Figure 1B; Table S2). Next, we measured the conidial production of *M. rileyi* at different concentrations of oleic acid and linoleic acid and used curve fitting ($y = c + bx + ax^2$) to determine the optimal concentration, where y is the conidial production and x is the concentration. The optimal concentration of oleic acid was 0.199% and 0.190% for concentrations of 10^6 and 10^7 conidia/mL, respectively (Figure 1C; Table S3). The optimal concentration of linoleic acid was 0.224% and 0.175% for concentrations of 10^6 and 10^7 conidia/mL, respectively (Figure 1D; Table S3). Thus, the concentrations of 0.190% oleic acid and 0.175% linoleic acid were selected for further study.

3.2. Exogenous Oleic Acid and Linoleic Acid Enhanced the Growth and Development of M. rileyi

The growth and the development of *M. rileyi* on SMAY alone or SMAY with 0.190% oleic acid or 0.175% linoleic acid were observed. Exogenous oleic acid and linoleic acid significantly enhanced the conidial germination compared with the CK group (Figure 2A). The GT_{50} of the CK group was 54.2 ± 3.3 h, which was significantly higher than that of the exogenous oleic acid group (40.5 ± 1.6 h; Student's *t*-test, $p = 0.0030$) and linoleic acid group (34.0 ± 0.8 h; Student's *t*-test, $p = 0.0005$; Figure 2B). In addition, these exogenous fatty acids (OA and LA) significantly reduced the yeast-like blastospore formation time (Figure 2C; Wilcox test, $p = 0.0254$), hyphal growth time (Figure 2D; Wilcox test, $p = 0.0254$), and sporulation time (Figure 2E; Student's *t*-test, $p = 0.0245$). The number of yeast-like blastospores in all the groups increased over time, and it was significantly increased in the exogenous oleic acid group after 4 (Wilcox test, $p = 0.0088$), 5 (Wilcox test, $p = 0.0088$), and 6 (Wilcox test, $p = 0.0090$) days of incubation (Figure 2F). Exogenous linoleic acid also significantly increased the number of yeast-like blastospores at 4 (Student's *t*-test, $p < 0.0000$), 5 (Wilcox test, $p = 0.0090$), and 6 (Student's *t*-test, $p < 0.0000$) days of incubation (Figure 2F). After 10 days of cultivation, exogenous oleic acid (Student's *t*-test, $p = 0.0074$) and linoleic acid (Student's t-test, $p = 0.0130$) notably raised the conidia production compared with the CK group (Figure 2G). Furthermore, we determined the expression of growth- and development-related genes (*MrPbs2*, *MrMsn2*, *MrSwi6*, *MrNsdD*, and *MrSte12*) [23]. Exogenous oleic acid significantly up-regulated the expression of *MrPbs2* (Student's t-test, $p = 0.0055$), *MrMsn2* (Student's t-test, $p = 0.0024$), *MrSwi6* (Student's t-test, $p = 0.0041$), *MrNsdD* (Student's t-test, $p < 0.0000$), and *MrSte12* (Student's t-test, $p = 0.0005$; Figure 2H). Similarly, the expression of *MrPbs2* (Student's t-test, $p = 0.0004$), *MrMsn2* (Student's t-test, $p = 0.0006$), *MrSwi6* (Student's t-test, $p = 0.0047$), *MrNsdD* (Student's t-test, $p < 0.0000$), and *MrSte12* (Student's t-test, $p = 0.0028$) was significantly increased in the linoleic acid supplementation group (Figure 2H).

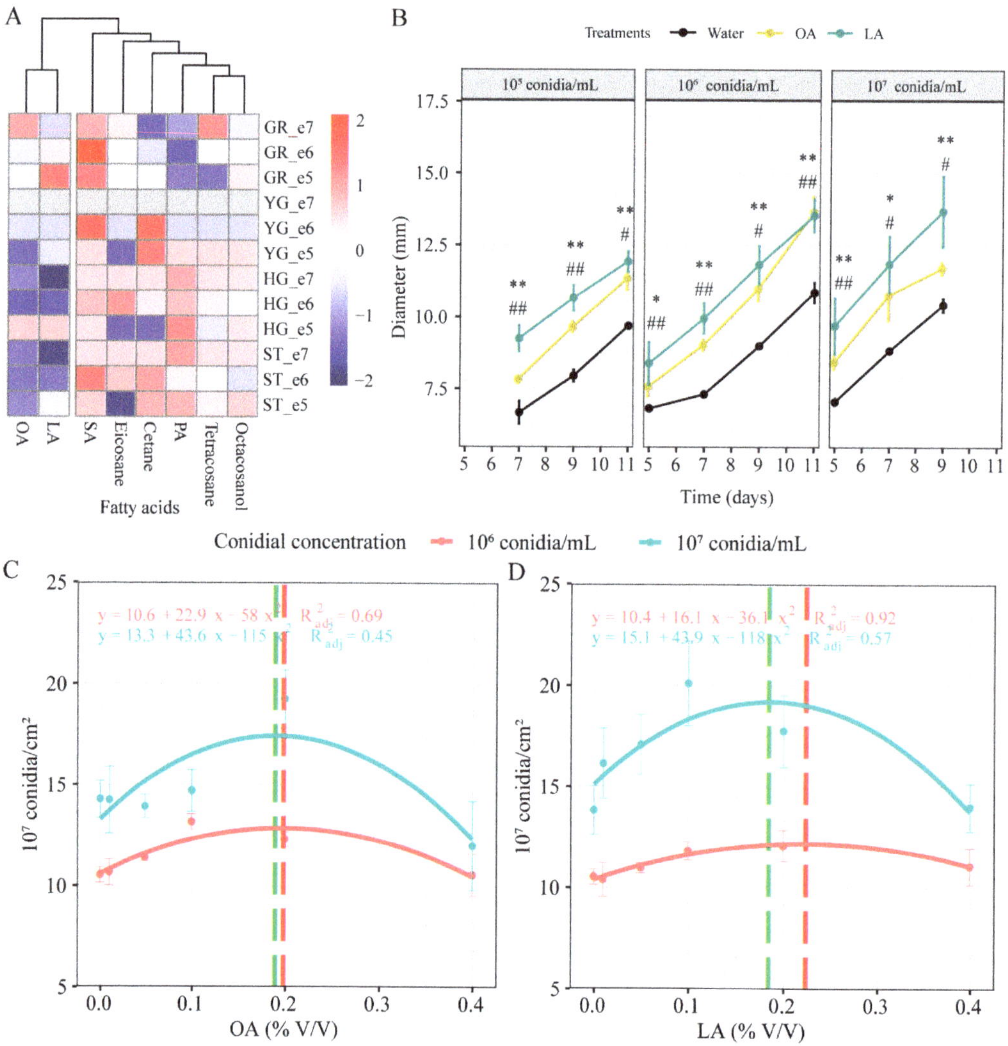

Figure 1. Effects of lipids addition on *M. rileyi* growth and development. (**A**) Cluster heat map and clustering through the level of growth and development. Red and blue boxes indicate increased and decreased levels, respectively. LA, linoleic acid; OA, oleic acid; SA, stearic acid; PA, palmitic

acid; GR, relative growth rate; YG, yeast cell formation time; HG, hyphal growth time; ST, sporulation time; e5, 10^5 conidia/mL; e6, 10^6 conidia/mL; e7, 10^7 conidia/mL. (**B**) Colony diameter. The conidial production of *M. rileyi* at different concentrations of oleic acid (**C**) and linoleic acid (**D**) was determined and curve fitting ($y = c + bx + ax^2$) was used to determine the optimal concentration. The conidial yield for the 10^6 conidia/mL suspension is shown by the red line and the conidial yield for the 10^7 conidia/mL suspension is shown by the green line. GT_{50}, conidial median germination time. The data are presented as the mean ± standard deviation. SDs of the means are from three independent replicates. * $p < 0.05$ and ** $p < 0.01$ when the OA group is compared to the water group. # $p < 0.05$ and ## $p < 0.01$ when the LA group is compared to the water group.

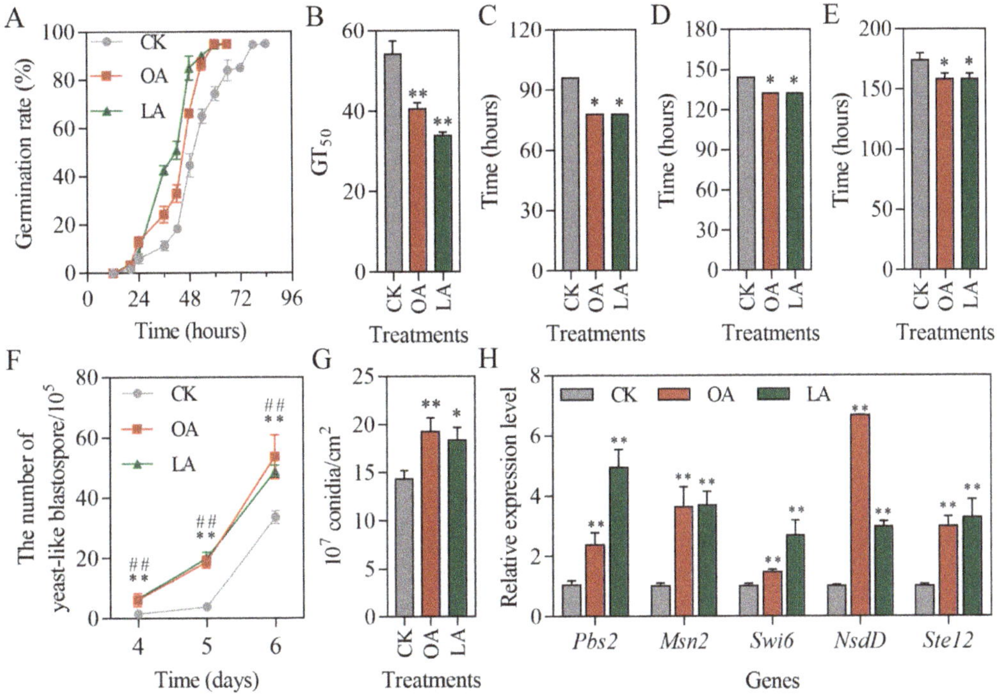

Figure 2. Exogenous oleic acid and linoleic acid enhanced the growth and development of *M. rileyi*. (**A**) Conidial germination curve. (**B**) Conidial median germination time. (**C**) Yeast-like blastospore formation time. (**D**) Hyphal growth time. (**E**) Sporulation time. (**F**) The number of yeast-like blastospores after 4, 5, and 6 days. (**G**) The conidial yield at 10 days. (**H**) The expression of growth- and development-related genes. LA, linoleic acid; OA, oleic acid; Pbs2, mitogen-activated protein kinase kinase Pbs2; Msn2, zinc finger DNA-binding protein Msn2; Swi6, cell cycle box-binding transcription factor MrSwi6; NsdD, GATA-type transcription factor NsdD; Ste12, transcription factor Ste12; TEF, transcription elongation factor; TUB, β-tubulin. The data are presented as the mean ± standard deviation. A-E, G, and H: SDs of the means from three independent replicates. F: SDs of the means from five independent replicates. * $p < 0.05$ and ** $p < 0.01$ when the OA group is compared to the control (CK) group; ## $p < 0.01$ when the LA group is compared to the control (CK).

3.3. Exogenous Oleic Acid and Linoleic Acid Enhanced the Multiple Stress Tolerances of M. rileyi

Exogenous oleic acid and linoleic acid resulted in increased tolerances of *M. rileyi* to various stress cues during 10-day colony growth on SMAY plates exposed to chemical stressors and a 42 °C heat shock for 1–3 h during the period of incubation (Figure 3). During the 10 days of incubation (Figure 3A), the *M. rileyi* colonies in all the treatment groups produced olive green conidia on SMAY with 1 h and 2 h of heat shock, H_2O_2,

KCl, NaCl, and CFW, as well as pale yellow conidia and a white mycelium on SBT and GLY, respectively. After 3 h of heat shock, the *M. rileyi* in the exogenous oleic acid and linoleic acid groups grew a white mycelium, while that in the CK group only grew a white mycelium in the center of the colony. We also evaluated the RGI on the 10th day of incubation (Figure 3B); compared to the CK group after 1 h of HS (8.7 ± 1.2%), 3 h of HS (26.4 ± 1.4%), GLY (27.9 ± 0.4%), and CFW (3.5 ± 1.6%), exogenous oleic acid significantly reduced the RGI in the 1 h (0.1 ± 1.1%; Student's *t*-test, $p = 0.0018$) and 3 h (21.1 ± 1.1%; Student's *t*-test, $p = 0.0125$) heat shock, GLY (21.0 ± 0.5%; Student's *t*-test, $p = 0.0001$), and CFW (0.2 ± 0.3%; Student's *t*-test, $p = 0.0474$) groups. Compared to the CK group after 1 h of HS (8.7 ± 1.2%), H_2O_2 (21.3 ± 0.6%), NaCl (31.9 ± 1.3%), GLY (27.9 ± 0.4%), and CFW (3.5 ± 1.6%), exogenous linoleic acid significantly reduced the RGI in the 1 h HS (3.5 ± 0.8%; Student's *t*-test, $p = 0.0071$), H_2O_2 (10.0 ± 1.6%; Student's *t*-test, $p = 0.0007$), NaCl (28.0 ± 1.5%; Student's *t*-test, $p = 0.0487$), GLY (22.4 ± 0.20%; Student's *t*-test, $p < 0.0000$), and CFW (0.0 ± 0.0%; Student's *t*-test, $p = 0.0386$) groups.

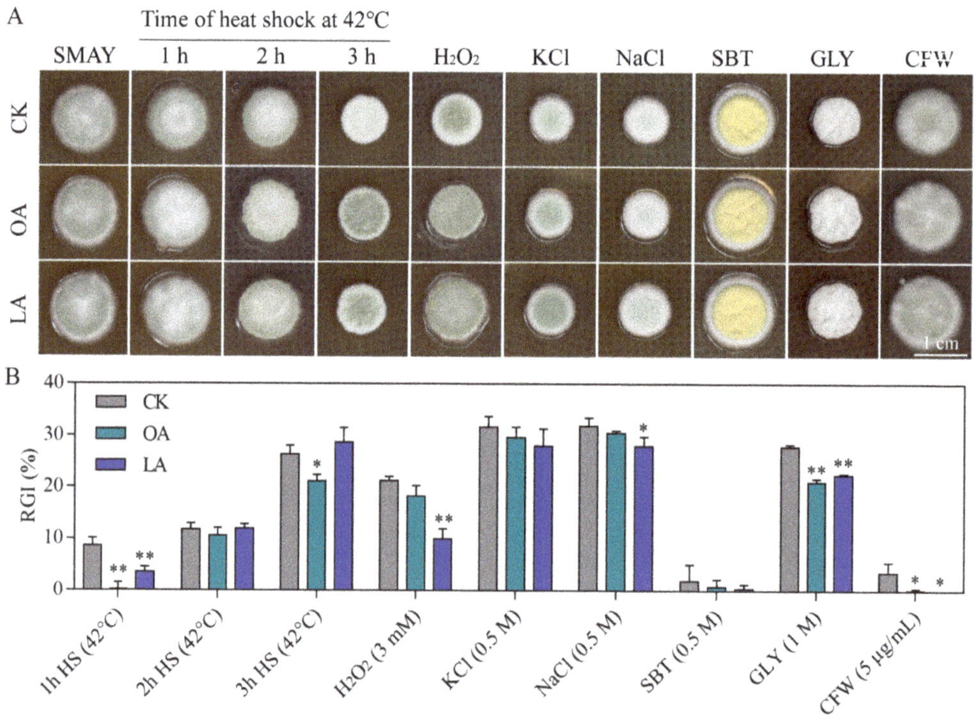

Figure 3. Exogenous oleic acid and linoleic acid enhanced multiple stress tolerances of *M. rileyi*. Images (**A**) and relative growth inhibition percentages (**B**) of fungal colonies incubated at 25 °C for 10 days on SMAY plates or supplemented with the indicated concentrations of H_2O_2, KCl, NaCl, sorbitol (SBT), glycerol (GLY), and calcofluor white (CFW) and of SMAY colonies incubated at 25 °C for 8 days of growth recovery after 2-day-old colonies were exposed to a 42 °C heat shock (HS) for 1, 2, or 3 h. Scale bars = 1.0 cm; LA, linoleic acid; OA, oleic acid; RGI, relative growth inhibition; HS, heat shock. The data are presented as the mean ± standard deviation. SDs of the means are from three independent replicates. * $p < 0.05$ and ** $p < 0.01$ compared to the control (CK) response.

3.4. Exogenous Oleic Acid and Linoleic Acid Protected Conidial Germination of M. rileyi and Improved Stress Tolerance

The germination dynamics of *M. rileyi* conidia under H_2O_2, glycerol, and NaCl stresses are shown in Figure 4A, 4B, and 4C, respectively. Exogenous oleic acid and linoleic acid protected the conidial germination of *M. rileyi* under H_2O_2, glycerol, and NaCl stresses. The conidial median germination time (GT_{50}) under H_2O_2 (Student's *t*-test, $p = 0.0005$), glycerol (Student's *t*-test, $p = 0.0037$), and NaCl (Student's *t*-test, $p < 0.0000$) stresses was significantly reduced by exogenous oleic acid compared to that of the CK group (Figure 4D). The linoleic acid groups also significantly reduced the GT_{50} under H_2O_2 (Student's *t*-test, $p = 0.0002$), glycerol (Student's *t*-test, $p = 0.0002$), and NaCl (Student's *t*-test, $p < 0.0000$) stresses (Figure 4D). To further explore the potential functions underlying oleic acid and linoleic acid against stresses, we investigated the activity of protective enzymes (CAT, POD, and SOD) in *M. rileyi* induced by exogenous oleic acid and linoleic acid. The activity of CAT (Figure 4E; Student's *t*-test, $p = 0.0007$) and POD (Figure 4F; Student's *t*-test, $p = 0.0265$) in *M. rileyi* was increased in the oleic acid group compared to that in the CK group. The activity of CAT (Student's *t*-test, $p = 0.0158$), POD (Student's *t*-test, $p = 0.0182$), and SOD (Figure 4G; Student's *t*-test, $p = 0.0215$) in *M. rileyi* was increased in the linoleic acid group compared to that in the CK group. The expression of *MrCat1* (Wilcox test, $p = 0.0495$), *MrCat2* (Student's *t*-test, $p = 0.0006$), *MrCat3* (Wilcox test, $p = 0.0495$), *MrSod1* (Student's *t*-test, $p = 0.0015$), *MrSod2* (Wilcox test, $p = 0.0495$), *MrSod3* (Wilcox test, $p = 0.0495$), and *MrSod4* (Wilcox test, $p = 0.0495$) was significantly increased in the oleic acid group compared to that in the CK group (Figure 4H,I). Similarly to oleic acid group, the expression of *MrCat1* (Wilcox test, $p = 0.0495$), *MrCat2* (Student's *t*-test, $p < 0.0000$), *MrCat3* (Student's *t*-test, $p = 0.0007$), *MrSod1* (Student's *t*-test, $p = 0.0006$), *MrSod2* (Wilcox test, $p = 0.0495$), *MrSod3* (Wilcox test, $p = 0.0495$), and *MrSod4* (Wilcox test, $p = 0.0495$) was significantly increased in the linoleic acid group compared to that in the CK group (Figure 4H,I). Moreover, we measured the expression of the chitin synthase family genes, as they are strongly associated with the stress tolerances of entomopathogenic fungi [35,36]. As shown in Figure 4J, the expression of *MrChs1* (Student's *t*-test, $p = 0.0003$, $p = 0.0001$), *MrChs4* (Student's *t*-test, $p = 0.0013$), *MrChs5* (Student's *t*-test, $p = 0.0004$), *MrChs6* (Student's *t*-test, $p = 0.0403$), and *MrChs7* (Student's *t*-test, $p = 0.0098$) was significantly up-regulated in the oleic acid group compared to that in the CK group (Figure 4J). The expression of *MrChs1* (Student's *t*-test, $p = 0.0001$), *MrChs3* (Student's *t*-test, $p = 0.0010$), *MrChs4* (Student's *t*-test, $p = 0.0002$), *MrChs5* (Student's *t*-test, $p = 0.0002$), *MrChs6* (Student's *t*-test, $p = 0.0002$), and *MrChs7* (Student's *t*-test, $p = 0.0066$) was also significantly up-regulated in the linoleic acid group compared to that in the CK group (Figure 4J). The expression of *MrChs2* (Student's *t*-test, $p = 0.0361$) was down-regulated with linoleic acid supplementation (Figure 4J).

3.5. Exogenous Oleic Acid and Linoleic Acid Improved the Virulence of M. rileyi

The laboratory bioassays revealed a high virulence for the 0.190% oleic acid and 0.175% linoleic acid treatments against the 3rd instars of *S. frugiperda*, with death beginning on the first day and survival reaching $34.3 \pm 1.3\%$ and $35.2 \pm 1.3\%$ after 6 days (Figure 5A,B; Table S4), respectively. The virulence of *M. rileyi* against the 3rd, 4th, and 5th instars of *S. frugiperda* was determined, with survival reaching $13.6 \pm 3.7\%$, $60.0 \pm 1.8\%$, and $68.7 \pm 2.9\%$ (Figure 5; Table S4), respectively, 10 days after the treatment using suspensions of 1×10^8 conidia/mL. The LT_{50} of the 3rd instars treated with *M. rileyi* was 5.7 ± 0.1 d. However, the death rate of the 4th and 5th instars due to *M. rileyi* did not reach 50%, so the LT_{50} could not be calculated. The combined treatment of *M. rileyi* and oleic acid against the 3rd, 4th, and 5th instars of *S. frugiperda* resulted in survival rates of $0.9 \pm 1.3\%$, $45.6 \pm 2.3\%$, and $51.5 \pm 4.3\%$, which were significantly lower than those resulting from the 0.190% oleic acid treatment alone or the *M. rileyi* suspensions alone (Figure 5A,C,E; Table S4). Similarly, the combined treatment of *M. rileyi* and linoleic acid against the 3rd, 4th, and 5th instars of *S. frugiperda* resulted in survival rates of $0.9 \pm 1.3\%$, $47.8 \pm 3.4\%$, and $51.5 \pm 4.3\%$, which were significantly lower than those resulting from the 0.175%

linoleic acid treatment alone or the *M. rileyi* suspensions alone (Figure 5B,D,F; Table S4). The LT_{50} for 4th-instar *S. frugiperda* with *M. rileyi*+OA and *M. rileyi*+LA was 7.6 ± 0.2 d and 7.6 ± 0.3 d, respectively.

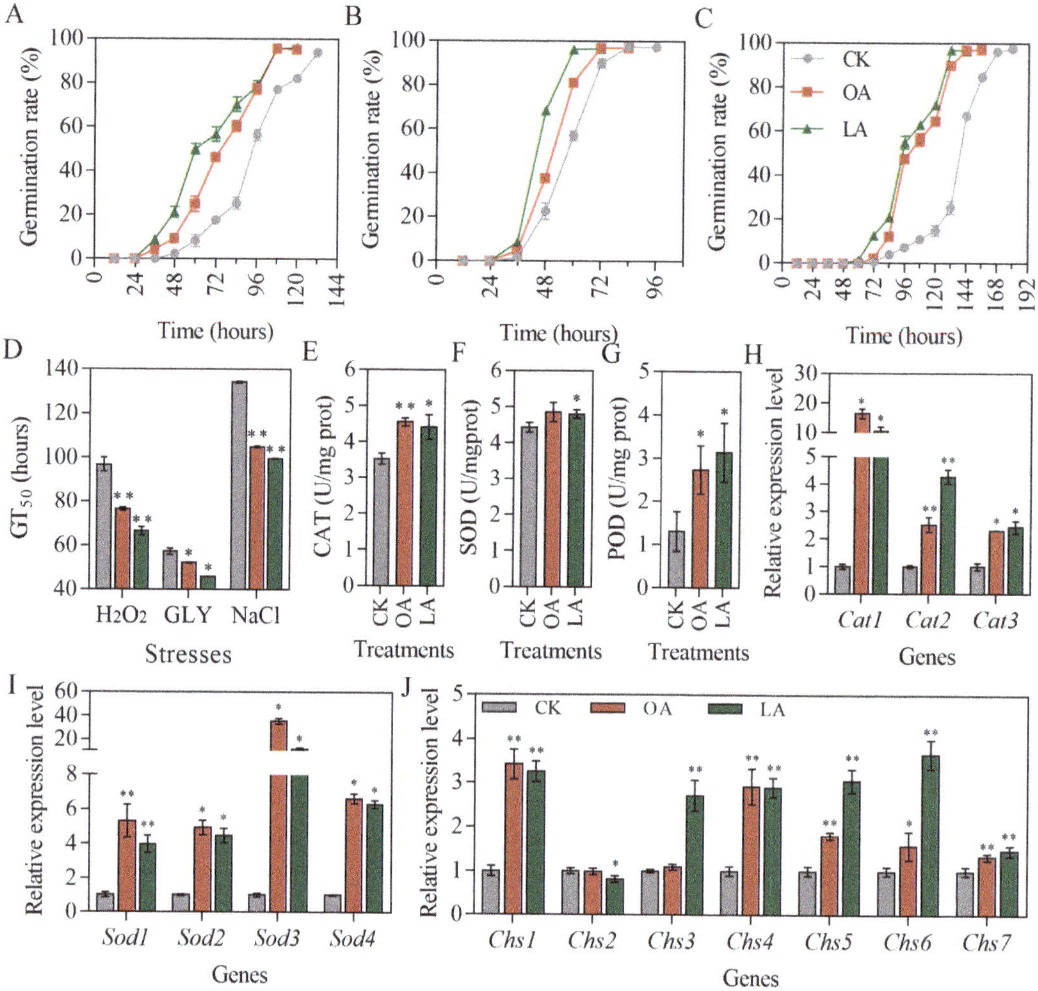

Figure 4. Exogenous oleic acid and linoleic acid protected conidial germination of *M. rileyi* and improved potential for stress tolerance. Conidial germination curve with H_2O_2 (**A**), GLY (**B**), and NaCl (**C**) stresses. (**D**) Conidial median germination time. The activity of CAT (**E**), SOD (**F**), and POD (**G**). The expression of genes encoding the CAT (**H**) and SOD (**I**) isoenzymes. (**J**) The expression of genes encoding the chitin synthase isoenzymes. LA, linoleic acid; OA, oleic acid; GT_{50}, conidial median germination time; CAT, catalase; SOD, superoxide dismutase; POD, peroxidase; Chs, chitin synthase. The data are presented as the mean ± standard deviation. SDs of the means are from three independent replicates. * $p < 0.05$ and ** $p < 0.01$ when compared to the control (CK).

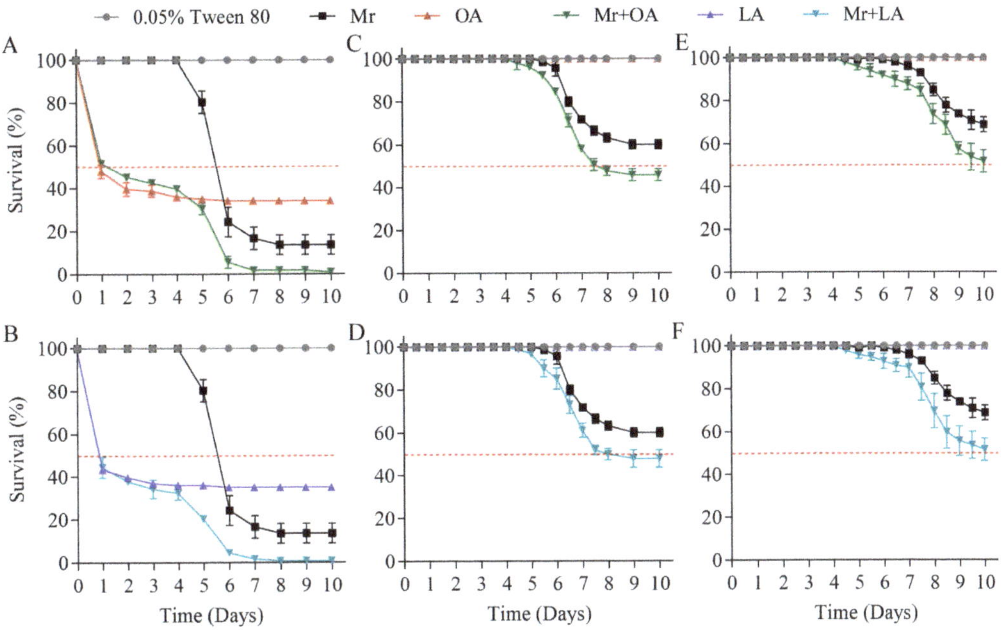

Figure 5. Exogenous oleic acid and linoleic acid improved the virulence of *M. rileyi*. (**A**,**B**) The survival of 3rd instars of *S. frugiperda* exposed to *M. rileyi*. (**C**,**D**) The survival of 4th instars of *S. frugiperda* exposed to *M. rileyi*. (**E**,**F**) The survival of 5th instars of *S. frugiperda* exposed to *M. rileyi*. LA, linoleic acid with a 0.175% concentration; The dashed line represents a survival rate of 50%; OA, oleic acid with a 0.190% concentration; Mr, *M. rileyi* conidial suspension with 10^8 conidia/mL; Mr+OA, combined treatment of *M. rileyi* and oleic acid; Mr+LA, combined treatment of *M. rileyi* and linoleic acid. The data are presented as the mean ± standard deviation. SDs of the means are from three independent replicates.

3.6. Exogenous Oleic Acid and Linoleic Acid Increased Fungal Virulence-Related Phenotypes

The conidial germination on water agar plates was significantly increased in the oleic acid (Student's t-test, $p = 0.0032$) and linoleic acid (Student's t-test, $p = 0.0007$) groups compared with the CK group (Figure 6A). The exogenous oleic acid and linoleic acid groups showed conidial germination on locust hindwings, but the CK group did not show conidial germination (Figure 6B). Exogenous oleic acid significantly improved the appressorium formation rates at 48 (Wilcox test, $p = 0.0369$) and 72 h (Student's t-test, $p < 0.0000$) and accelerated appressorium formation (Figure 6C,D). Linoleic acid also significantly improved the appressorium formation rates at 48 (Wilcox test, $p = 0.0369$) and 72 h (Student's t-test, $p = 0.0009$) and accelerated appressorium formation (Figure 6C,D). Moreover, we detected cuticle-degradation-related extracellular enzyme activity and gene expression. As shown in Figure 6E,F, the exogenous oleic acid (Student's t-test, $p = 0.0178$) and linoleic acid (Student's t-test, $p < 0.0000$) groups significantly increased the extracellular protease activity. However, no differences were identified for extracellular lipase or chitinase. The expression of *MrEL* (extracellular lipase gene; Wilcox test, $p = 0.0495$) and *MrSL* (secreted lipase gene; Wilcox test, $p = 0.0495$) was significantly up-regulated in the oleic acid and linoleic acid groups compared to that in the CK group (Figure 6G). The expression of protease-related genes *MrPr1* (Student's t-test, $p = 0.0079$) and *MrPr2* (Student's t-test, $p = 0.0081$) was significantly increased after linoleic acid supplementation (Figure 6H). Exogenous oleic acid only significantly up-regulated *MrPr1* expression (Student's t-test, $p = 0.0090$). The expression of *MrChit1* (Student's t-test, $p = 0.0030$), *MrChit2* (Wilcox test, $p = 0.0495$), *MrChit3* (Student's t-test, $p < 0.0000$), and *MrChit5* (Wilcox test, $p = 0.0495$) was significantly up-regulated in the oleic acid group compared to that in the CK group, and the expression of *MrChit4* was significantly decreased (Student's *t*-test, $p = 0.0308$; Figure 6I). Similarly, linoleic acid also significantly up-regulated the expression of *MrChit1* (Student's *t*-test, $p = 0.0022$), *MrChit2* (Wilcox test, $p = 0.0495$), *MrChit3* (Student's *t*-test, $p = 0.0020$), *MrChit4* (Student's *t*-test, $p = 0.0043$), and *MrChit5* (Wilcox test, $p = 0.0495$) compared to the CK group (Figure 6I).

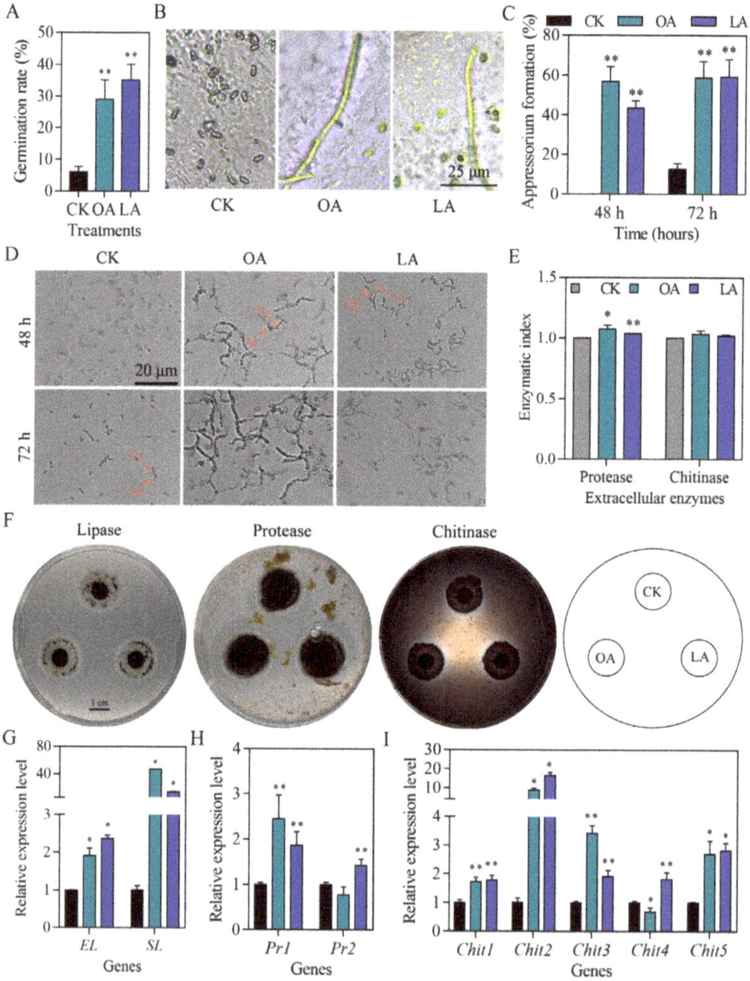

Figure 6. Exogenous oleic acid and linoleic acid increased fungal virulence-related phenotypes. Conidial germination rate on water agar plates (**A**) and locust hindwings (**B**). Appressorium formation rates (**C**) and images (**D**) at 48 and 72 h. The red arrow represents the appressorium. Ap, appressorium; Co, conidia. Extracellular enzymatic index (**E**) and images (**F**) at 14 days. (**G**) The expression of extracellular lipase gene and secreted lipase gene. (**H**) The expression of genes encoding the protease isoenzymes. (**I**) The expression of genes encoding the chitinase isoenzymes. LA, linoleic acid; OA, oleic acid; Chit, chitinase; Pr, protease; EL, extracellular lipase; SL, secreted lipase. The data are presented as the mean ± standard deviation. A, C, and G-I: SDs of the means are from three independent replicates. E: SDs of the means are from five independent replicates. * $p < 0.05$ and ** $p < 0.01$ compared to the control (CK).

3.7. Exogenous Oleic Acid and Linoleic Acid Enhanced Cuticle Infection of *M. rileyi*

Exogenous oleic acid and linoleic acid improved the virulence-related phenotypes and increased the cuticle-degradation-related enzyme activity and gene expression in *M. rileyi*, suggesting that exogenous oleic acid and linoleic acid enhance host cuticle infection by *M. rileyi*. After fungal invasion, the host hemocytes flocked together and formed melanic dots. Then, the time for host melanic dots to be created and blastospores to be generated during *M. rileyi* infection was observed. The times of melanic dot formation were 96, 72, and 72 h in the CK, oleic acid, and linoleic acid groups, respectively. The CK, oleic acid, and linoleic acid treatments started to generate blastospores from the host hemocoel at 108, 84, and 96 h, respectively (Figure 7A). The number of blastospores gradually increased from 72 to 132 h, and significantly increased in the oleic acid and linoleic acid treatment groups at multiple time points compared to the CK group (Figure 7B; Table S5). Finally, we observed 30 larvae in each treatment group at 132 h. The infection rates in the CK, oleic acid, and linoleic acid groups were 47, 67, and 63%, respectively (Figure 7C).

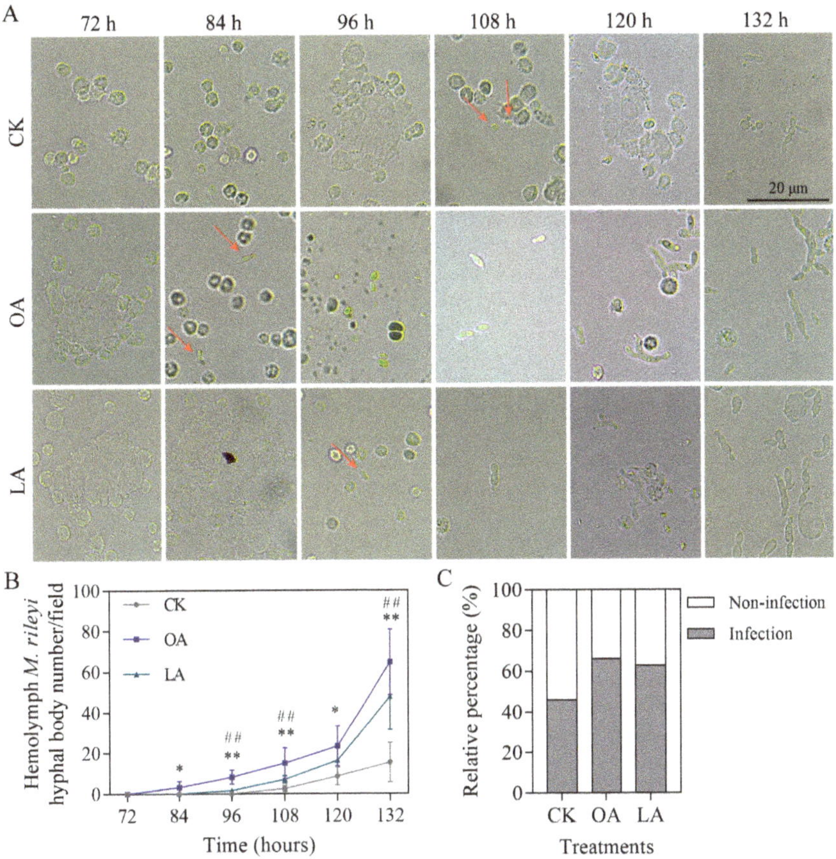

Figure 7. Exogenous oleic acid and linoleic acid enhanced cuticle infection of *M. rileyi*. (**A**) Hemocyte encapsulation was examined under a microscope at an interval of 12 h. Scale bars = 20 μm. The red arrow represents the fungal blastospore in the hemolymph. (**B**) The number of *M. rileyi* hyphal bodies in the hemolymph from 72 to 132 h. (**C**) The relative percentage of infection at 132 h. LA, linoleic acid; OA, oleic acid. The data are presented as the mean ± standard deviation. SDs of the means are from five independent replicates. * $p < 0.05$ and ** $p < 0.01$ when OA is compared to the control (CK); ## $p < 0.01$ when LA is compared to the control (CK).

4. Discussion

Fatty acids have previously been identified as protectors of *M. rileyi* conidia against abiotic stresses, such as dehydration and thermotolerance [10,23,37]. In this study, exogenous oleic acid and linoleic acid protected the *M. rileyi* conidial germination and improved various abiotic stress tolerances, including the tolerance to oxidative, osmotic, cell wall, and heat shock stresses. However, fatty acids can also inhibit the spore germination, hyphal growth, and pathogenicity of entomopathogenic fungi (e.g., *B. bassiana* and *Conidiobolus coronatus*) [38–40]. The mechanism for this involves a lipid-droplet protein (*BbLar1*) that maintains the intracellular homeostasis of fatty acids and is crucial to the fungal tolerance to linoleic acid stress, which significantly contributes to fungal compatibility with fatty acids [41]. We found that exogenous oleic acid and linoleic acid significantly increased the activity of protective enzymes and the expression of genes encoding the CAT and SOD isoenzymes. The cumulative evidence proves that the protective enzymes of fungi are crucial to resisting biological and abiotic stress processes [36,42,43], implying that exogenous oleic acid and linoleic acid protect the conidial germination of *M. rileyi* and improve its stress tolerance by increasing the expression of protective enzymes. The cell wall integrity pathway senses hypotonic conditions or multi-stresses, including heat shock, oxidative, osmotic, and cell wall stresses, by controlling cell wall biosynthetic enzymes and the expression of cell-wall-related genes [36,44]. In this study, the transcription level of chitin synthesis genes was significantly up-regulated by the oleic acid and linoleic acid treatments, suggesting that oleic acid and linoleic acid may be signal molecules of the cell wall integrity pathway. These results indicate that fatty acids may function as protector agents of biological and abiotic stresses for *M. rileyi*.

Oil suspension formulations of conidia significantly improve the pathogenicity of entomopathogenic fungi [14,15], among which fatty acids may play an important role [26]. In this study, exogenous oleic acid and linoleic acid significantly improved the virulence of *M. rileyi* against the 4th and 5th instars of *S. frugiperda*. Exogenous oleic acid and linoleic acid significantly elevated the conidial germination on water agar plates and locust hindwings and the appressorium formation rates. After adhering to the cuticle of insects, a fungal conidium germinates and differentiates into a specialized infection structure called an appressorium [12,30,45]. Appressorium-mediated cuticle penetration is accompanied by the combined action of cuticle-degrading enzymes (protease, lipase, and chitinase) [31,46]. Our results also showed that the protease activity was significantly elevated after the oleic acid and linoleic acid treatments. Similarly, protease genes were significantly up-regulated in the oleic acid and linoleic acid groups. No differences were found for extracellular lipase or chitinase. However, the expression of the extracellular lipase gene, secreted lipase gene, and chitinase genes was significantly up-regulated in the oleic acid and linoleic acid groups. This may have been due to the insufficient sensitivity of the plate enzyme activity detection method. Furthermore, vegetable oil has the function of regulating the growth and development of entomopathogenic fungi [14,15,18–20]. Triacylglycerides is one of the main components of oil suspension, and is hydrolyzed into fatty acids and glycerol by lipases [47]. In addition, exogenous fatty acids also can enter cells through fatty acid-binding protein. Degradation of fatty acids by the β-oxidation pathway results in the formation of acetyl-CoA, which enters the tricarboxylic acid (TCA) cycle for the production of ATP. TCA cycle intermediates may be used as key sources of carbon molecules for biosynthesis of nucleotides, amino acids, and lipids [48]. Notably, the oleic acid and linoleic acid treatments improved the conidial germination rate on water agar plates and locust hindwings and increased the growth and development of *M. rileyi* [23], suggesting that exogenous oleic acid and linoleic acid provide energy substance for *M. rileyi*'s growth and development [23,30]. Combined with the results of this study, we speculate that oleic acid and linoleic acid may improve the ability of *M. rileyi* to infect *S. frugiperda* via enhancing the process of conidial germination, appressorium formation, and cuticle penetration.

When the conidia attach to the host cuticle under the appropriate environmental conditions and signals, they germinate and invade the host [49]. Lipids are one of the main components of the body wall, and they may be signaling molecules for conidial germination [22]. Oleic acid and linoleic acid have a generic structure of lipo-chitooligosaccharides (LOCs), and LOCs are perceived at the plasma membrane by receptor-like kinases with extracellular LysM domains [50,51]. For example, LOCs act as regulatory signals of fungal growth and development, and an LOC treatment increased the differentially expressed genes of *A. fumigatus* from 91 to 152 at 30 min and 130 min, respectively [22]. Fungi possess a significant number of LysM-containing proteins [22], suggesting that oleic acid and linoleic acid can be recognized by LysM-containing proteins. Furthermore, fatty acids can be oxidized to produce oxylipins through enzymatic and non-enzymatic reactions [52]. Oxylipin may act as ligands through specific G-protein-coupled receptors (GPCRs), and GPCRs are abundant in filamentous fungi [53,54]. For example, oleic acid promotes GPCR membrane docking and activity as well as related signaling molecules [55]; oleic acid increases the expression of growth- and development-related genes (*Mrap1*, *MrNsdD*, *MrPbs2*, *MrSwi6*, *MrSte12*, and *MrMsn2*) in *M. rileyi* [23]; and oxylipin increases the level of chitin in filamentous fungi [54]. In this study, exogenous oleic acid and linoleic acid up-regulated the expression of growth- and development-related genes, CAT genes, SOD genes, chitin synthase family genes, and cuticle-degrading enzyme-related genes, suggesting that oleic acid and linoleic acid function as signaling molecules for *M. rileyi* conidia. Notably, fatty acid is converted to polyunsaturated fatty acid [56]. Polyunsaturated fatty acids can be metabolized to multiple oxylipins by the cyclooxygenase, lipoxygenase, and/or cytochrome P450 pathways [57]. Oxylipins induce fungal cellular differentiation, including lateral branching, appressorium formation, the yeast-to-hyphal transition, sexual development, and spore release [52,54], suggesting that oleic acid- and linoleic acid-derived oxylipins play a critical role in the regulation of growth and development in *M. rileyi*.

Our and others' previous work has demonstrated that oleic acid and linoleic acid increase the conidial tolerance to heat, drought, oxidative, osmotic, and cell wall stresses [10,19,24]. These abiotic stresses strongly limit the application of entomopathogenic fungi as biological insecticides [14,58]. Conidial oil suspensions of *M. rileyi*, *M. anisopliae*, and *B. bassiana* can efficiently prevent and control pests in fields [17,59,60]. These results indicate that oleic acid and linoleic acid can be added to oil suspensions of conidia.

5. Conclusions

Exogenous oleic acid and linoleic acid improve the stress tolerances and virulence of *M. rileyi*. Their favorable biological characteristics appear to be largely mediated through protecting conidial germination and promoting cuticle infection. These results provide new insights for the development of biopesticides based on *M. rileyi*.

Supplementary Materials: The following supporting information can be downloaded at: https://www.mdpi.com/article/10.3390/jof10080521/s1, Figure S1: Chemical structures for oleic acid, linoleic acid, stearic acid, eicosane, cetane, palmitic acid, tetracosane, and octacosanol. Figure S2: Growth and development of *M. rileyi*, including conidial germination at 48 h, yeast-like blastospore production at 4 days, mycelial growth at 5 days, and sporulation at 7 days; Table S1: qRT-PCR primers used in this study; Table S2: Colony diameter for all the conidial suspension concentrations; Table S3: Conidial production of *M. rileyi* at different concentrations of oleic acid and linoleic acid; Table S4: The mortality rate of *M. rileyi* against the 3rd, 4th, and 5th instars of *S. frugiperda*; Table S5: The number of *M. rileyi* hyphal bodies in the hemolymph from 72 to 132 h.

Author Contributions: Conceptualization, G.W. and Y.P.; methodology, B.C.; software, X.Z.; validation, G.W., X.Z. and Y.P.; formal analysis, G.D. and W.W.; investigation, H.C.; data curation, Y.Y.; writing—original draft preparation, G.W.; writing—review and editing, Y.P. and B.C.; visualization, G.W.; supervision, S.J. and B.C.; project administration, G.D.; funding acquisition, B.C. All authors have read and agreed to the published version of the manuscript.

Funding: This research was supported by the Yunnan Fundamental Research Projects (grant number 202401AS070089), the National Natural Science Foundation of China (grant number 32060616), and the Yunnan Provincial Education Department Scientific and Technological Innovation Team (grant number 2022[69]).

Institutional Review Board Statement: Not applicable.

Informed Consent Statement: Not applicable.

Data Availability Statement: The original contributions presented in this study are included in the article; further inquiries can be directed to the corresponding authors.

Conflicts of Interest: The authors declare no conflicts of interest.

References

1. Wang, L.; Wang, J.; Zhang, X.; Yin, Y.; Li, R.; Lin, Y.; Deng, C.; Yang, K.; Liu, X.; Wang, Z. Pathogenicity of *Metarhizium rileyi* against *Spodoptera litura* larvae: Appressorium differentiation, proliferation in hemolymph, immune interaction, and reemergence of mycelium. *Fungal Genet. Biol.* **2021**, *150*, 103508. [CrossRef] [PubMed]
2. Loureiro, E.D.S.; Lima, A.R.D.; Pessoa, L.G.A.; Dias, P.M.; Assis, L.F. Virulência de *Metarhizium rileyi* (Ascomycota:Clavicipitaceae) a *Spodoptera cosmioides* (Lepidoptera: Noctuidae). *Res. Soc. Dev.* **2020**, *9*, 186973962. [CrossRef]
3. Barros, S.; Pitta, R.; Lopes, R.; Almeida, E.; Ferreira, F. Susceptibility of *Spodoptera frugiperda* and *Chrysodeixis includens* (Lepidoptera: Noctuidae) to infections caused by *Metarhizium rileyi*. *Pesqui. Agropecu. Trop.* **2020**, *50*, e61713. [CrossRef]
4. Devi, P.S.V.; Prasad, Y.G. Nomuraea Rileyi—A potential mycoinsecticide. In *Biocontrol Potential and Its Exploitation in Sustainable Agriculture: Volume 2: Insect Pests*; Upadhyay, R.K., Mukerji, K.G., Chamola, B.P., Eds.; Springer: Boston, MA, USA, 2001; pp. 23–38.
5. Zhang, Z.; Sui, L.; Tian, Y.; Lu, Y.; Xia, X.; Liu, W.; Cheng, K.; Li, Q.; Shi, W. *Metarhizium rileyi* with broad-spectrum insecticidal ability confers resistance against phytopathogens and insect pests as a phytoendophyte. *Pest Manag. Sci.* **2024**, *80*, 3246–3257. [CrossRef]
6. Faria, M.; Souza, D.A.; Sanches, M.M.; Schmidt, F.G.V.; Oliveira, C.M.; Benito, N.P.; Lopes, R.B. Evaluation of key parameters for developing a *Metarhizium rileyi*-based biopesticide against *Spodoptera frugiperda* (Lepidoptera: Noctuidae) in maize: Laboratory, greenhouse, and field trials. *Pest Manag. Sci.* **2022**, *78*, 1146–1154. [CrossRef]
7. Barros, S.K.A.; de Almeida, E.G.; Ferreira, F.T.R.; Barreto, M.R.; Lopes, R.B.; Pitta, R.M. Field efficacy of *Metarhizium rileyi* applications against *Spodoptera frugiperda* (Lepidoptera: Noctuidae) in maize. *Neotrop. Entomol.* **2021**, *50*, 976–988. [CrossRef] [PubMed]
8. Grijalba, E.P.; Espinel, C.; Cuartas, P.E.; Chaparro, M.L.; Villamizar, L.F. *Metarhizium rileyi* biopesticide to control *Spodoptera frugiperda*: Stability and insecticidal activity under glasshouse conditions. *Fungal Biol.* **2018**, *122*, 1069–1076. [CrossRef]
9. Edelstein, J.; Lecuona, R.; Trumper, E. Selection of culture media and in vitro assessment of temperature-dependent development of *Nomuraea rileyi*. *Neotrop. Entomol.* **2004**, *33*, 737–742. [CrossRef]
10. Sánchez-Rey, L.E.; Moreno-Sarmiento, N.; Grijalba-Bernal, E.P.; Quiroga-Cubides, G. Physiological response of *Metarhizium rileyi* with linoleic acid supplementation. *Fungal Biol.* **2024**, *128*, 1827–1835. [CrossRef] [PubMed]
11. Wang, G.; Xu, S.; Chen, L.; Zhan, T.; Zhang, X.; Liang, H.; Chen, B.; Peng, Y. Gut microbial diversity reveals differences in pathogenicity between *Metarhizium rileyi* and *Beauveria bassiana* during the early stage of infection in *Spodoptera litura* larvae. *Microorganisms* **2024**, *12*, 1129. [CrossRef]
12. Shin, T.Y.; Lee, M.R.; Park, S.E.; Lee, S.J.; Kim, W.J.; Kim, J.S. Pathogenesis-related genes of entomopathogenic fungi. *Arch. Insect Biochem.* **2020**, *105*, e21747. [CrossRef] [PubMed]
13. Kim, J.J.; Xie, L.; Han, J.H.; Lee, S.Y. Influence of additives on the yield and pathogenicity of conidia produced by solid state cultivation of an *Isaria javanica* isolate. *Mycobiology* **2014**, *42*, 346–352. [CrossRef] [PubMed]
14. Paixão, F.R.S.; Muniz, E.R.; Barreto, L.P.; Bernardo, C.C.; Mascarin, G.M.; Luz, C.; Fernandes, É.K.K. Increased heat tolerance afforded by oil-based conidial formulations of *Metarhizium anisopliae* and *Metarhizium robertsii*. *Biocontrol Sci. Technol.* **2017**, *27*, 324–337. [CrossRef]
15. Saeed, N.; Wakil, W.; Farooq, M.; Shakeel, M.; Arain, M.S.; Shakeel, Q. Evaluating the combination of *Metarhizium anisopliae* and an enhanced form of diatomaceous earth (Grain-Guard) for the environmentally friendly control of stored grain pests. *Environ. Monit. Assess.* **2020**, *192*, 210. [CrossRef] [PubMed]
16. Wang, F.; Sethiya, P.; Hu, X.; Guo, S.; Chen, Y.; Li, A.; Tan, K.; Wong, K.H. Transcription in fungal conidia before dormancy produces phenotypically variable conidia that maximize survival in different environments. *Nat. Microbiol.* **2021**, *6*, 1066–1081. [CrossRef] [PubMed]
17. Gu, M.; Tian, J.; Lou, Y.; Ran, J.; Mohamed, A.; Keyhani, N.O.; Jaronski, S.; Wang, G.; Chen, X.; Zang, L.-S.; et al. Efficacy of *Metarhizium rileyi* granules for the control of *Spodoptera frugiperda* and its synergistic effects with chemical pesticide, sex pheromone and parasitoid. *Entomol. Gen.* **2023**, *6*, 1211–1219. [CrossRef]
18. Rangel, D.E.N.; Acheampong, M.A.; Bignayan, H.G.; Golez, H.G.; Roberts, D.W. Conidial mass production of entomopathogenic fungi and tolerance of their mass-produced conidia to UV-B radiation and heat. *Fungal Biol.* **2023**, *127*, 1524–1533. [CrossRef]

19. Kim, J.S.; Je, Y.H.; Roh, J.Y. Production of thermotolerant entomopathogenic *Isaria fumosorosea* SFP-198 conidia in corn-corn oil mixture. *J. Ind. Microbiol. Biot.* **2010**, *37*, 419–423. [CrossRef]
20. Mishra, S.; Kumar, P.; Malik, A. Preparation, characterization, and insecticidal activity evaluation of three different formulations of *Beauveria bassiana* against *Musca domestica*. *Parasitol. Res.* **2013**, *112*, 3485–3495. [CrossRef] [PubMed]
21. Nitbani, F.O.; Tjitda, P.J.P.; Nurohmah, B.A.; Wogo, H.E. Preparation of fatty acid and monoglyceride from vegetable oil. *J. Oleo Sci.* **2020**, *69*, 277–295. [CrossRef]
22. Rush, T.A.; Puech-Pagès, V.; Bascaules, A.; Jargeat, P.; Maillet, F.; Haouy, A.; Maës, A.Q.; Carriel, C.C.; Khokhani, D.; Keller-Pearson, M.; et al. Lipo-chitooligosaccharides as regulatory signals of fungal growth and development. *Nat. Commun.* **2020**, *11*, 3897. [CrossRef]
23. Xue, R.; Du, G.; Chen, C.; Chen, B.; Peng, Y. Oleic acid improves the conidial production and quality of *Metarhizium rileyi* as a biocontrol agent. *Biocontrol Sci. Technol.* **2023**, *33*, 758–771. [CrossRef]
24. Jasińska, A.; Różalska, S.; Rusetskaya, V.; Słaba, M.; Bernat, P. Microplastic-induced oxidative stress in metolachlor-degrading filamentous fungus *Trichoderma harzianum*. *Int. J. Mol. Sci.* **2022**, *23*, 12978. [CrossRef]
25. Peng, Y.; Chen, B. Role of cell membrane homeostasis in the pathogenicity of pathogenic filamentous fungi. *Virulence* **2024**, *15*, 2299183. [CrossRef] [PubMed]
26. Peng, Y.J.; Wang, J.J.; Lin, H.Y.; Ding, J.L.; Feng, M.G.; Ying, S.H. HapX, an indispensable bzip transcription factor for iron acquisition, regulates infection initiation by orchestrating conidial oleic acid homeostasis and cytomembrane functionality in mycopathogen *Beauveria bassiana*. *mSystems* **2020**, *5*, e00695-20. [CrossRef]
27. Peng, Y.J.; Zhang, H.; Feng, M.G.; Ying, S.H. SterylAcetyl hydrolase 1 (BbSay1) links lipid homeostasis to conidiogenesis and virulence in the entomopathogenic fungus *Beauveria bassiana*. *J. Fungi* **2022**, *8*, 292. [CrossRef]
28. Gessler, N.N.; Filippovich, S.Y.; Bachurina, G.P.; Kharchenko, E.A.; Groza, N.V.; Belozerskaya, T.A. Oxylipins and oxylipin synthesis pathways in fungi. *Appl. Biochem. Microbiol.* **2017**, *53*, 628–639. [CrossRef]
29. Peng, Y.; Yao, Y.; Pang, J.; Di, T.; Du, G.; Chen, B. Genetic diversity and virulence variation of *Metarhizium rileyi* from infected *Spodoptera frugiperda* in corn fields. *Microorganisms* **2024**, *12*, 264. [CrossRef]
30. Wang, L.; Lai, Y.; Chen, J.; Cao, X.; Zheng, W.; Dong, L.; Zheng, Y.; Li, F.; Wei, G.; Wang, S. The ASH1-PEX16 regulatory pathway controls peroxisome biogenesis for appressorium-mediated insect infection by a fungal pathogen. *Proc. Natl. Acad. Sci. USA* **2023**, *120*, e2217145120. [CrossRef] [PubMed]
31. Gebremariam, A.; Chekol, Y.; Assefa, F. Extracellular enzyme activity of entomopathogenic fungi, *Beauveria bassiana* and *Metarhizium anisopliae* and their pathogenicity potential as a bio-control agent against whitefly pests, *Bemisia tabaci* and *Trialeurodes vaporariorum* (Hemiptera: Aleyrodidae). *BMC Res. Notes* **2022**, *15*, 117. [CrossRef]
32. Fan, L.; Li, B.; Wang, J.; Li, X.; Ma, F.; Du, F.; Li, H.; Lin, Y. Multifunctional regulation of NADPH oxidase in growth, microsclerotia formation and virulence in *Metarhizium rileyi*. *Biotechnol. Lett.* **2023**, *45*, 1441–1455. [CrossRef] [PubMed]
33. Li, Y.; Wang, Z.; Liu, X.; Song, Z.; Li, R.; Shao, C.; Yin, Y. Siderophore biosynthesis but not reductive iron assimilation is essential for the dimorphic fungus *Nomuraea rileyi* conidiation, dimorphism transition, resistance to oxidative stress, pigmented microsclerotium formation, and virulence. *Front. Microbiol.* **2016**, *7*, 931. [CrossRef] [PubMed]
34. Livak, K.J.; Schmittgen, T.D. Analysis of relative gene expression data using real-time quantitative PCR and the 2(-Delta Delta C(T)) Method. *Methods* **2001**, *25*, 402–408. [CrossRef] [PubMed]
35. Zhang, J.; Jiang, H.; Du, Y.; Keyhani, N.O.; Xia, Y.; Jin, K. Members of chitin synthase family in *Metarhizium acridum* differentially affect fungal growth, stress tolerances, cell wall integrity and virulence. *PLoS Pathog.* **2019**, *15*, e1007964. [CrossRef]
36. Wang, D.; Lv, C.; Guan, Y.; Ni, X.; Wu, F. Dsk2 involves in conidiation, multi-stress tolerance and thermal adaptation in *Beauveria bassiana*. *Microbiol. Rep.* **2021**, *13*, 384–393. [CrossRef] [PubMed]
37. Sánchez, L.; León, D.; Grijalba, E.; Moreno, N.; Cuartas-Otálora, P.; Quiroga-Cubides, G. Selection of inductors for thermotolerant conidia production of the entomopathogenic fungus *Metarhizium rileyi* Nm017. *Biocontrol Sci. Technol.* **2022**, *32*, 891–898. [CrossRef]
38. Keyhani, N.O. Lipid biology in fungal stress and virulence: Entomopathogenic fungi. *Fungal Biol.* **2018**, *122*, 420–429. [CrossRef]
39. Hou, J.; Ding, J.L.; Peng, Y.J.; Feng, M.G.; Ying, S.H. Genome-wide identification of BCS1 domain-containing proteins reveals the mitochondrial bcs1 essential for growth, stress response, and virulence of the filamentous entomopathogenic fungus *Beauveria bassiana*. *Microbiol. Res.* **2023**, *267*, 127262. [CrossRef] [PubMed]
40. Boguś, M.I.; Czygier, M.; Gołebiowski, M.; Kedra, E.; Kucińska, J.; Mazgajska, J.; Samborski, J.; Wieloch, W.; Włóka, E. Effects of insect cuticular fatty acids on in vitro growth and pathogenicity of the entomopathogenic fungus *Conidiobolus coronatus*. *Exp. Parasitol.* **2010**, *125*, 400–408. [CrossRef]
41. Hou, J.; Zhang, H.; Ding, J.L.; Feng, M.G.; Ying, S.H. Transcriptomic investigation reveals a physiological mechanism for *Beauveria bassiana* to survive under linoleic acid stress. *iScience* **2023**, *26*, 106551. [CrossRef]
42. Pang, J.; Peng, Y.; Di, T.; Du, G.; Chen, B. Virulence of *Metarhizium rileyi* is determined by its growth and antioxidant stress and the protective and detoxifying enzymes of *Spodoptera frugiperda*. *Insects* **2023**, *14*, 260. [CrossRef] [PubMed]
43. Paixão, F.R.S.; Huarte-Bonnet, C.; Ribeiro-Silva, C.S.; Mascarin, G.M.; Fernandes, É.K.K.; Pedrini, N. Tolerance to abiotic factors of microsclerotia and *Mycelial Pellets* from *Metarhizium robertsii*, and molecular and ultrastructural changes during microsclerotial differentiation. *Front. Fungal Biol.* **2021**, *2*, 654737. [CrossRef] [PubMed]

44. Jiang, H.; Chi, Z.; Liu, G.L.; Hu, Z.; Zhao, S.Z.; Chi, Z.M. Melanin biosynthesis in the desert-derived *Aureobasidium melanogenum* XJ5-1 is controlled mainly by the CWI signal pathway via a transcriptional activator Cmr1. *Curr. Genet.* **2020**, *66*, 173–185. [CrossRef] [PubMed]
45. Ortiz-Urquiza, A.; Keyhani, N.O. Action on the surface: Entomopathogenic fungi versus the insect cuticle. *Insects* **2013**, *4*, 357–374. [CrossRef] [PubMed]
46. Wang, C.; Wang, S. Insect pathogenic fungi: Genomics, molecular interactions, and genetic improvements. *Annu. Rev. Entomol.* **2017**, *62*, 73–90. [CrossRef] [PubMed]
47. Brzozowski, A.M.; Derewenda, U.; Derewenda, Z.S.; Dodson, G.G.; Lawson, D.M.; Turkenburg, J.P.; Bjorkling, F.; Huge-Jensen, B.; Patkar, S.A.; Thim, L. A model for interfacial activation in lipases from the structure of a fungal lipase-inhibitor complex. *Nature* **1991**, *351*, 491–494. [CrossRef] [PubMed]
48. Donnelly, R.P.; Finlay, D.K. Glucose, glycolysis and lymphocyte responses. *Mol. Immunol.* **2015**, *68*, 513–519. [CrossRef] [PubMed]
49. Fan, Y.; Ortiz-Urquiza, A.; Garrett, T.; Pei, Y.; Keyhani, N.O. Involvement of a caleosin in lipid storage, spore dispersal, and virulence in the entomopathogenic filamentous fungus, *Beauveria bassiana*. *Environ. Microbiol.* **2015**, *17*, 4600–4614. [CrossRef] [PubMed]
50. Feng, F.; Sun, J.; Radhakrishnan, G.V.; Lee, T.; Bozsóki, Z.; Fort, S.; Gavrin, A.; Gysel, K.; Thygesen, M.B.; Andersen, K.R.; et al. A combination of chitooligosaccharide and lipochitooligosaccharide recognition promotes arbuscular mycorrhizal associations in *Medicago truncatula*. *Nat. Commun.* **2019**, *10*, 5047. [CrossRef]
51. Kelly, S.; Radutoiu, S.; Stougaard, J. Legume LysM receptors mediate symbiotic and pathogenic signalling. *Curr. Opin. Plant Biol.* **2017**, *39*, 152–158. [CrossRef]
52. Tsitsigiannis, D.I.; Keller, N.P. Oxylipins as developmental and host-fungal communication signals. *Trends Microbiol.* **2007**, *15*, 109–118. [CrossRef] [PubMed]
53. Xu, J.R.; Peng, Y.L.; Dickman, M.B.; Sharon, A. The dawn of fungal pathogen genomics. *Annu. Rev. Phytopathol.* **2006**, *44*, 337–366. [CrossRef] [PubMed]
54. Niu, M.; Steffan, B.N.; Fischer, G.J.; Venkatesh, N.; Raffa, N.L.; Wettstein, M.A.; Bok, J.W.; Greco, C.; Zhao, C.; Berthier, E.; et al. Fungal oxylipins direct programmed developmental switches in filamentous fungi. *Nat. Commun.* **2020**, *11*, 5158. [CrossRef] [PubMed]
55. Lopez, S.; Bermudez, B.; Montserrat-de la Paz, S.; Jaramillo, S.; Varela, L.M.; Ortega-Gomez, A.; Abia, R.; Muriana, F.J. Membrane composition and dynamics: A target of bioactive virgin olive oil constituents. *Biochim. Biophys. Acta* **2014**, *1838*, 1638–1656. [CrossRef] [PubMed]
56. Du, A.; Wang, Z.; Huang, T.; Xue, S.; Jiang, C.; Qiu, G.; Yuan, K. Fatty acids in cancer: Metabolic functions and potential treatment. *MedComm-Oncology* **2023**, *2*, e25. [CrossRef]
57. Eccles, J.A.; Baldwin, W.S. Detoxification Cytochrome P450s (CYPs) in families 1-3 produce functional oxylipins from polyunsaturated fatty acids. *Cells* **2022**, *12*, 82. [CrossRef] [PubMed]
58. Rangel, D.E.; Braga, G.U.; Fernandes, É.K.K.; Keyser, C.A.; Hallsworth, J.E.; Roberts, D.W. Stress tolerance and virulence of insect-pathogenic fungi are determined by environmental conditions during conidial formation. *Curr. Genet.* **2015**, *61*, 383–404. [CrossRef]
59. Barbieri, A.; Rico, I.B.; Silveira, C.; Feltrin, C.; Dall Agnol, B.; Schrank, A.; Lozina, L.; Klafke, G.M.; Reck, J. Field efficacy of *Metarhizium anisopliae* oil formulations against *Rhipicephalus microplus* ticks using a cattle spray race. *Ticks Tick Borne Dis.* **2023**, *14*, 102147. [CrossRef]
60. Brito, E.S.; de Paula, A.R.; Vieira, L.P.; Dolinski, C.; Samuels, R.I. Combining vegetable oil and sub-lethal concentrations of Imidacloprid with *Beauveria bassiana* and *Metarhizium anisopliae* against adult guava weevil *Conotrachelus psidii* (Coleoptera: Curculionidae). *Biocontrol Sci. Technol.* **2008**, *18*, 665–673. [CrossRef]

Journal of *Fungi*

MDPI

Article

A Countermeasure Strategy against Peramine Developed by *Chilesia rudis* in the Endophyte–Ryegrass–Herbivore Model

Manuel Chacón-Fuentes [1,*], Daniel Martínez-Cisterna [2,3], Marcelo Lizama [4,5], Valeria Asencio-Cancino [6], Ignacio Matamala [4,7] and Leonardo Bardehle [2,7,*]

1 Agriaquaculture Nutritional Genomic Center, CGNA, Las Heras 350, Temuco 4780000, Chile
2 Centro de Investigación Biotecnológica Aplicada al Medio Ambiente (CIBAMA), Universidad de La Frontera, Av. Francisco Salazar 01145, Casilla 54-D, Temuco 4811230, Chile
3 Programa de Doctorado en Ciencias de Recursos Naturales, Universidad de La Frontera, s/n, P.O. Box 58-D, Temuco 4780000, Chile
4 Programa de Doctorado en Ciencias Agroalimentarias y Medioambiente, Facultad de Ciencias Agropecuarias y Medioambiente, Universidad de La Frontera, Av. Francisco Salazar 01145, Casilla 54-D, Temuco 4811230, Chile
5 Escuela de Universidad de Valladolid Doctorado (ESDUVa), Departamento Producción Vegetal y Recursos Forestales, Escuela Técnica Superior de Ingenierías Agrarias, Campus "La Yutera", Avda. de Madrid, 50, 34004 Palencia, Spain
6 Carrera de Ingenieria en Recursos Naturales, Facultad de Ciencias Agropecuarias y Medioambiente, Universidad de La Frontera, Temuco 4811230, Chile
7 Departamento de Producción Agropecuaria, Facultad de Ciencias Agropecuarias y Medioambiente, Universidad de La Frontera, Temuco 4811230, Chile
* Correspondence: manuel.chacon@cgna.cl (M.C.-F.); leonardo.bardehle@ufrontera.cl (L.B.)

Abstract: Exploitation of the symbiotic relationship between endophytic fungi and ryegrass is a crucial technique for reducing the incidence of insect pests. This is primarily due to the production of alkaloids, such as peramine, by the fungi. This alkaloid has been reported as both a deterrent and toxic to a variety of insects. However, insects have developed various strategies to counteract plant defenses. One of the most studied methods is their ability to sequester toxic compounds from plants. In this study, we examined the feeding preferences and adaptation to peramine in *Chilesia rudis*, a native Chilean larva. Using a no-choice assay, we assessed larval feeding preferences and mass gain on seven experimental lines and two commercial cultivars of endophyte-infected and non-infected ryegrass. Pupal development time and adult performance were evaluated post-assay. Additionally, we measured peramine content in larval carcasses, feces, and ryegrass leaves. Jumbo was the most preferred cultivar with 32 mm^2 of leaf tissues consumed. The longest pupal development time was observed in L161 and ALTO AR1, both at 28 days. Wing length in adults was greatest in the Jumbo and L163 cultivars, measuring 1.25 cm and 1.32 cm, respectively. Peramine concentrations were detected in the bodies of *C. rudis*. In conclusion, this larva can adapt to endophyte-infected ryegrass and develop counter-adaptation mechanisms to mitigate the effects of peramine.

Keywords: native herbivore; detoxification; alkaloids; fungus; Erebidae; Epichloë

1. Introduction

Due to their sessile nature, plants cannot flee from stress and hence have developed sophisticated strategies to survive both biotic and abiotic stresses [1]. Plants have developed various defense mechanisms, such as producing toxic secondary metabolites, to defend themselves against herbivores and pathogens, minimizing damage and preserving their overall fitness. Oftentimes, secondary metabolites from a single class of compounds, e.g., alkaloids, occur in complex mixtures within plant tissues [2,3]. Alkaloids are defensive substances synthesized by plants to cope with attacks from organisms, animals, or insects. Alkaloids have been shown to influence the feeding behavior of insect herbivores in a variety of ways, acting as attractants, phagostimulants, deterrents, or repellents [4,5]. However, chemical defense strategies are never 100% effective, as insects have developed several ways to cope with the presence of secondary metabolites produced by their host plants. One such strategy is sequestration, which involves the selective uptake, transport, modification, storage, and deployment of plant secondary metabolites for the insect's own defense. These strategies result from a counter-adaptation process generated by the coevolution between plants and their herbivores [6].

Fungal organisms known as endophytes can inhabit their host plants without causing damage. Endophytic fungi have been reported to increase yield production, improve nutrient uptake from the soil, and enhance the ability to cope with insect pests in *Lolium perenne* [7]. Four families of natural alkaloids are produced as a result of this symbiosis: ergot alkaloids, pyrrolizidines (lolines), indole-diterpenes (lolitrems), and pyrrolopyrazine (peramine). Individual alkaloids within these classes often have specific toxic effects depending on the type of herbivore (i.e., vertebrate vs. invertebrate). Lolines and peramine are well known for their insecticidal or insect-deterring effects [8,9]. Peramine functions exclusively as an anti-invertebrate defensive metabolite; it is not acutely toxic or a contact poison but deters feeding in several insect species [10], for example, the Argentine stem weevil (*Listronotus bonariensis*, Coleoptera: Curculionidae) [11]. Rowan [12] showed that peramine at concentrations from 0.1 to 10 μg/g deters both larvae and adults of the Argentine stem weevil, *L. bonariensis*. This symbiotic interaction between ryegrass and endophytes has also been associated with the control of the fall armyworm (*Spodoptera frugiperda*, Lepidoptera: Noctuidae); black cutworm (*Agrotis ipsilon*, Lepidoptera: Noctuidae); green bug (*Chinavia hilaris*, Hemiptera: Pentatomidae) and bluegrass billbug (*Stenophorus parvulus*, Coleoptera: Curculionidae), these being among the most relevant [13–15]. Additionally, peramine is distributed throughout all plant fluids and is water-soluble, offering protection to the entire plant, not just the areas colonized by the endophyte. Recent findings show that this alkaloid can accumulate in higher trophic levels, such as in predaceous ladybird beetles (*Adalia bipunctata* or *Harmonia axyridis*, Coleoptera: Coccinellidae) that feed on aphids reared on endophyte-infected grass [16]. Furthermore, Parra et al. [17] reported that extracts of feces containing peramine from animals fed on pastures infected with endophyte fungus increased the larval mortality (90%) of horn flies (*Haematobia irritans*, Diptera: Muscidae), reducing their survival compared to a control group fed on endophyte-free pastures.

The use of alkaloids to combat insects can trigger an arms race between plants and their herbivores. Plants evolve defense mechanisms, and insects develop countermeasures such as detoxification, sequestration mechanisms and passive accumulation [18,19]. This adaptation and co-evolutionary process suggests that insects can detoxify toxic metabolites, allowing them to cope with the defensive systems of their host plants. According to Nishida [20] and Opitz and Müller [21], these chemical compounds are diverse and are used by various lepidopterans for processes such as host selection, defense, and even mating. For example, species in the Papilionidae family have been associated with plants containing aristolochic acids, which are sequestered by larvae and transferred to eggs by adults, providing defense against predators [22]. Additionally, the genus *Battus* (Lepidoptera: Papilionidae) has been studied for its ability to use these compounds for host selection and defense [23,24]. Particularly, the Erebidae family has been studied in regard to some sequestration mechanisms of plant chemical defense. Nevertheless, there is no report about the storage, sequestration or detoxification mechanism of peramine by lepidopterans Erebidae.

In this context, *Chilesia rudis* (Lepidoptera: Erebidae) is a native, polyphagous, and univoltine larva from Chile, known as one of the most serious pests of grasslands due to its severe defoliating behavior [25–27]. In its larval stage, this insect is a severe and abundant defoliator of various plants, including introduced pastures grasses, ornamental plants, and native species such as *Ugni molinae* [28]. Previous studies by Nickisch-Rosenegk and Wink [29] have shown that pyrrolizidine alkaloids are sequestered by the larvae and adults of six European erebid moth species (*Spilosoma lubricipeda, Arctia caja, Phragmatobia fuliginosa, Callimorpha dominula, Diacrisia sannio,* and *Tyria jacobaeae*). Therefore, the main objective of this research was to evaluate the production of peramine due to endophyte fungus infection in seven experimental lines and two cultivars of ryegrass and to analyze the feeding preference and countermeasures to peramine of the native herbivorous *C. rudis*.

2. Materials and Methods

2.1. Plant Material

Ryegrass samples, including seven experimental lines (LE161, LE162, LE163, LE164, LE165, LE166, and LE167), as well as the commercial cultivars ALTO AR1 (endophyte-infected, denoted as E+) and JUMBO (endophyte-free, denoted as E−), were sourced from the plots of *L. perenne* at the Maquehue Experimental Station (University of La Frontera), located in Temuco, Chile [7]. During early summer (December 2023), tillers were manually harvested and carefully placed into paper bags for transportation to the Chemical Ecology Laboratory at the University of La Frontera. Upon arrival, leaves designated for bioassays were promptly utilized, while those earmarked for chemical analyses were stored in bags at −20 °C until subjected to analysis using high-performance liquid chromatography (HPLC).

2.2. Larval Rearing

The original population of *Chilesia rudis* larvae (30–40 mm) was collected from natural grassland in Temuco in 2020 [28]. Once the parental population was obtained, it was allowed to develop in acrylic boxes (30 × 30 × 30 cm) with cloth on each side to allow ventilation. The larvae were raised on the general-purpose lepidopteran diet (soy-wheat germ diet with vitamin and agar) (Frontier, agricultural Sciences, Newark, DE, USA) in 30 mL plastic cups. The cups were kept with a photoperiod of 16:8 (L:D), 60% R.H. and a temperature of 25 °C [8,30]. Briefly, 19 g of Agar was added to 875 mL of cold water. The mixture was brought to a full boil for 1 min with occasional stirring. Subsequently, the agar solution was transferred to the blender container, and the Dry Mix containing 144 g of ingredients including sucrose, soy flour (50%), stabilized wheat germ, salt mix, USDA vitamin premix, fiber, sorbic acid, methyl paraben, and ascorbic acid, was added. The mixture was blended for approximately 30 s or until thoroughly mixed. Finally, the prepared diet was dispensed immediately, and any unused diet was stored in the refrigerator for later use. The first generation of *C. rudis* (2021) was kept in rearing throughout its generation, which gave way to the offspring or F2 generation (2022), which were again raised in the same way as the parental population. This F2 population gave rise to the F3 population (2023), with which the experiments were conducted. For all experiments, larvae in their final instar were used.

2.3. Evaluation, Isolation and Characterization of Endophytic Fungus in *Lolium perenne*

The endophyte infection levels of each experimental line and cultivar were assessed using the method described by Dombrowski et al. [31] and Chacón-Fuentes et al. [32] prior to commencing the experiments. In summary, 120 tillers were harvested per cultivar or experimental line and the inner epidermis of a leaf sheath was carefully peeled off and mounted on a glass slide. Two drops of rose Bengal stain (Merck KGaA, Darmstadt, Germany) were applied to each sample and covered with a cover slip after 1–2 min. Subsequently, microscopic examination (40×) was conducted to determine the presence or absence of typical fungal mycelium, from which the infection percentage was calculated. Positive endophyte infection was determined by the presence of fungal hyphae in the assessed tillers. For endophyte isolation, *L. perenne* tillers underwent disinfection using 3% sodium hypochlorite, followed by rinsing with sterile distilled water. The tillers were then cultured on potato dextrose agar (PDA) supplemented with penicillin and streptomycin (100 $\mu g\ g^{-1}$) for 3 weeks at 23 °C. *Epichloë* sp. presence in *L. perenne* was confirmed through molecular techniques following the methodology proposed by Chacón-Fuentes et al. [7]. Fungal tissue from petri dishes containing strains cultured from *L. perenne* was macerated under sterile conditions with liquid nitrogen. The obtained macerate was processed using the NucleoSpin Soil DNA isolation Kit (Macherey-Nagel) to extract non-denatured DNA, with purity index values analyzed using Gen5. Subsequently, a PCR reaction was conducted to associate the obtained DNA with the *Epichloë* sp. fungus identity.

2.4. No-Choice Assay

This experiment evaluated the consumption of *C. rudis* on seven experimental lines and two cultivars of ryegrass. A single *C. rudis* larva was placed in a petri dish (94 × 16 mm) with a hole in the top covered with mesh for air interchange. Dishes contained one fresh fully expanded leaf (10 cm wide × 1 cm long) per treatment (each treatment corresponded to each single experimental line or cultivar). The consumption was evaluated by scanning the leaves and calculating the foliar area removed by the larva with the software ImageJ (version 1.42), the ryegrass leaves were replaced every 12 h [28]. Furthermore, weight gain of the larvae was measured at the end of the experiment and calculated according to the following formula: Wf − Wi, where Wf is final weight and Wi corresponds to the initial weight. The bioassay was evaluated every 12 h over 7 days. At 7 days, the experiment was concluded. All larvae remained in the final instar without showing any evidence of molting or entering the pupal stage. Twenty replicates were carried out for each of the nine treatments.

2.5. Pupal Development Time and Adult Performance Evaluation

To assess the impact of peramine on the performance of *C. rudis,* the pupal development time was evaluated, and the same individuals from the no-choice assays were utilized. Briefly, pupal development time was determined by calculating the number of days from larvae to pupae (from original Petri dish). Furthermore, upon emergence, adult *C. rudis* were evaluated for various parameters: (1) total length of adults (cm), (2) male wing length (cm), and female wing length of adults, and (3) total weight of the last larval instar (g). Measurements were taken using a digital caliper and an analytical balance following the methodology outlined by Wang et al. [33].

2.6. Carcasses, Feces and Guts Peramine Extraction

In parallel, a new group of *C. rudis* larvae (20 larvae per experimental line or cultivar) were fed with leaves from the two commercial cultivars and the seven experimental lines separately. These larvae were allowed to feed, and upon reaching their final larval stage, they were sacrificed. Larvae were deprived of food for 3 days and dissected. The guts of the larvae were removed carefully using a scalpel (N° 11) under an optical glass to exclude any influence of remaining plant material. The carcasses of the larvae were stored in a paper bag and placed in an oven at 40 °C over three days or until it reached a constant weight. The insect material was ground to obtain a dry powder. A total of 25 mg of this material was placed in an amber Eppendorf tube (2 mL) and 1 mL of a mixture of methanol/water/formic acid (70:29.5:0.5 *v*/*v*) (chromatographic grade, Sigma-Aldrich, St. Louis, MO, USA) was added. Then, tubes were centrifuged at 6000 rpm for 3 min. The supernatant was collected in an amber chromatographic vial (1.7 mL) and analyzed by HPLC. Feces of the insects were collected each day and stored in the dark at −20 °C until chemical analyses by HPLC. These samples were freeze-dried for 24 h and then ground following the process explained above for larval carcasses. The guts obtained from the larval carcass protocol were also subjected to the same larval tissue extraction to determine if there was any remaining peramine inside them. Finally, the extracts of carcasses, feces and guts were filtered through 0.22 μm membranes before their injection into the HPLC [34].

2.7. Plant Material Extraction

Leaves of the seven experimental lines and the two commercial cultivars of *L. perenne* were dried in an oven for 3 days at 60 °C. Then, the samples were milled in a grinder until they reached a particle size of 2 mm and placed in a polypropylene tube in darkness.

Next, 0.5 g of milled material was first extracted with 3 mL of methanol and then with 3 mL of chloroform (HPLC grade, Sigma-Aldrich, St. Louis, MO, USA) for 30 min at room temperature. Subsequently, 3 mL of hexane and 3 mL of water (HPLC grade, Sigma-Aldrich, St. Louis, MO, USA) were added, and the mixture was centrifuged at 6000× *g* for 5 min. Afterward, 1 mL of the aqueous phase was passed through a CCX column (UCT 2731 Bartram Road, Bristol, PA, USA). In the same column, 1 mL of HPLC water and finally 0.5 mL of 0.5% formic acid were used to elute the peramine from the column. Samples were collected in amber vials, filtered through a 0.22 μm membrane, and analyzed by HPLC [28].

2.8. HPLC-DAD Analysis

Twenty microliters of samples were injected into a Shimadzu HPLC system (LC-20A Prominence, Kyoto, Japan) equipped with a C-18 column (250 × 4.6 mm I.D.; particle size 5 μm) maintained at 25 °C. A peramine standard (purchased from AgResearch, Hamilton, New Zealand) was monitored at 280 nm. The analysis was conducted using an isocratic mobile phase composed of 80% guanidinium buffer (A) and 20% acetonitrile (B) at a flow rate of 1 mL/min, with each HPLC analysis completed in approximately 10 min. The mobile phase was prepared by dissolving 1.44 g of guanidinium carbonate in 800 mL of HPLC water, adding formic acid (1.3 mL/L), and 200 mL of HPLC-grade acetonitrile (Merck KGaA, Darmstadt, Germany). Samples were prepared in triplicate, and the amount of peramine was determined from a calibration curve generated by injecting commercial standards in a serial dilution from 100 ppm to 0.01 ppm [35].

2.9. Survival Rate Test

A new group of 20 larvae per experimental line or commercial cultivar was used to evaluate larval survival. Each larva was placed individually in a separate Petri dish containing a whole leaf from the endophyte-free cultivar Jumbo (10 × 1 cm). Subsequently, 100 μL of peramine at concentrations of 1, 5, 10, and 100 mg/L were added to each leaf. This method was replicated twenty times for each of the seven experimental lines and two commercial cultivars. Then, the larvae were allowed to feed for two days, with the leaves being replaced every 12 h. After the two-day period, mortality was assessed by counting the deceased individuals, and a percentage was calculated based on the total number of surviving individuals versus those that died [36].

2.10. Statistical Analysis

The statistical software Statistix 10 (Tallahassee, Florida) was utilized to analyze the presence of endophyte in the seven experimental lines and two commercial cultivars, as well as the effect of peramine on no-choice assays (leaf consumption), pupal development time, adult performance, survival rate tests, and the presence of peramine in leaves, carcasses, feces, and guts of *C. rudis*. Briefly, to analyze the presence of endophytic fungi, a chi-square test was conducted for each experimental line or commercial cultivar. In addition, an ANOVA followed by a Bonferroni test was conducted to analyze differences in the presence of endophyte among the different experimental lines and cultivars. To analyze the no-choice assays, survival rate range, and evaluation of morphological characteristics, as well as peramine content, an analysis of variance (ANOVA) followed by Tukey's multiple comparison test was conducted. Additionally, bivariate Pearson correlations were calculated to estimate specific relationships among variables. Values of $p \leq 0.05$ were considered significant. Results are presented as means with their corresponding standard errors.

3. Results

3.1. Evaluation, Isolation and Characterization of Endophytic Fungus in Lolium perenne

The experimental results presented in Table 1 depict the endophyte infection rates across various line experimental cultivars along with JUMBO (E−) and ALTO AR1 (E+). Among the experimental lines, L161 and ALTO AR1 showcased complete endophyte colonization, with 100% infection rates, whereas L162, L164, L166, and L165 exhibited significant endophyte presence, ranging from 74% to 91.6%. Conversely, L163 and L167 exhibited lower endophyte colonization rates, with values of 10% and 11%, respectively. As expected, JUMBO (E−) showed no endophyte colonization, while ALTO AR1 (E+) showed complete colonization, mirroring the findings observed in L161 ($F_{8,104}$ = 2.35, $p < 0.001$).

Table 1. Percentage of endophyte fungal infection in tillers of the several experimental lines and commercial cultivars of ryegrass. Different letters indicate significant differences based on ANOVA followed by Bonferroni test. * mean significant differences into every experimental line or cultivars based on Chi-square test ($p \leq 0.05$).

Line Experimental	N° of Plants Tested	(E+)	(E−)	Endophyte (%)
L161	120	120	0	100 *a
L162	120	89	31	74 *d
L163	120	12	108	10 *e
L164	120	110	10	91.6 *b
L165	120	100	20	83.3 *c
L166	120	110	10	91.6 *b
L167	120	13	107	11 *e
JUMBO (E−)	120	0	120	0 *f
ALTO AR1 (E+)	120	120	0	100 *a

Table 2 provides the results of a PCR analysis specifically targeting the Tub2 intron to detect the presence of the endophytic fungus *Epichloë* sp. The results indicate the confirmed presence or absence of the endophytic fungus in the analyzed samples. The columns "IS-RS-5′; IS-NS3" and "IS-tub2w-5′; IS-tub2w-3" represent the testing conditions used. In the column corresponding to "IS-RS-5′ and IS-NS3" as well as in "IS-tub2w-5′ and IS-tub2w-3", all samples, including experimental lines L161 to L167, as well as the commercial cultivar ALTO AR1 (E+), show the confirmed presence of the endophytic fungus, indicated by the symbol "+". Meanwhile, the commercial cultivar JUMBO (E−) exhibits the absence of the endophytic fungus.

Table 2. PCR analysis of Tub2 Intron for presence of *Epichloë* sp. endophyte. "+" means the presence of endophyte, while "−" means the absence of endophyte.

Experimental Lines or Commercial Cultivars	PCR (Tub2 Intron)	
	IS-RS-5′; IS-NS3′	IS-tub2w-5′; IS-tub2w-3′
L161	+	+
L162	+	+
L163	+	+
L164	+	+
L165	+	+
L166	+	+
L167	+	+
JUMBO (E−)	−	−
ALTO AR1 (E+)	+	+

3.2. No-Choice Assay

Leaf area consumed from the seven experimental lines by *C. rudis* larvae on grasses, measured in cm^2 in a no-choice assay, reveals significant variations among different experimental lines (Figure 1). Overall, the commercial cultivar JUMBO (E−) exhibits the highest level of consumption, averaging 34 ± 2.0 mm^2 (leaf-removing area), closely followed by the experimental line L163 with 30 ± 3.0 mm^2, while the commercial cultivar ALTO AR1 (E+) and the experimental line L164 record the lowest consumption levels, averaging 4 ± 0.4 mm^2 each ($F_{8,1011}$ = 17.25, $p < 0.001$). These discrepancies underscore the importance of different experimental lines in grass susceptibility to larval performance and suggest the need for further investigation to understand the contributing factors to these variations.

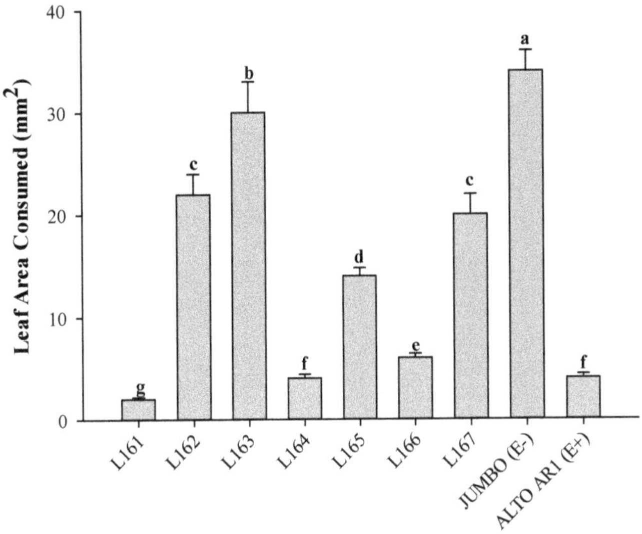

Figure 1. Leaf area consumed from the seven experimental lines and two cultivars of ryegrass by *C. rudis* in a no-choice test. Different letters indicate significant differences according to Tukey's test ($p \leq 0.05$). The data correspond to the mean with its corresponding standard error.

3.3. Pupal Development Time and Adult Performance Evaluation and HPLC-DAD Analysis

Table 3 shows the relationship between peramine in leaves and several analyses of larvae performance and content of peramine in tissues of larvae. For example, when analyzing the pupal development time ($F_{8,171} = 19.91$, $p < 0.001$) and peramine concentration in leaves ($F_{8,18} = 1377$, $p < 0.001$), it is observed that larvae feeding from leaves from experimental lines and the commercial cultivar with lower peramine concentration in the leaves, such as JUMBO (E−) (0.0 ± 0.0 μg/g), have a shorter pupal development time (23 ± 1.4 days), while larvae feeding from leaves from experimental lines or the commercial cultivar with a higher concentration, such as ALTO AR1 (E+) (184.1 ± 29.8 μg/g), have a longer pupal development time (28 ± 1.2 days). This relationship suggests that the concentration of peramine might be related to the pupal development time, resulting in an increase of up to 5 days in the larva-to-pupa cycle for this species when it is subjected to a diet with the presence of peramine (ALTO AR1 (E+)). In regard to the amount of peramine in leaves related to the length of *C. rudis* larvae, it is observed that experimental lines and commercial cultivar with higher concentrations of peramine in the leaves, such as L161 (179.3 ± 24.5 μg/g) and ALTO AR1 (E+) (184.1 ± 29.8 μg/g), show significantly smaller larvae length than other experimental lines such as L163 and L167 as well as the commercial cultivar JUMBO (E−) ($F_{8,171} = 142.8$, $p < 0.001$). For example, L163 has a larval length of 2.9 ± 0.2 cm, L167 has a length of 3.1 ± 0.1 cm, and JUMBO (E−) has a length of 3.2 ± 0.3 cm, while ALTO AR1 (E+) has a length of 1.4 ± 0.1 cm. This observation suggests that the concentration of peramine in the leaves has a direct relationship with the decrease in larval length. Furthermore, the weight of the larva decreases as the amount of peramine increases, with the lowest amount of peramine (JUMBO (E−)) corresponding to a higher larval weight (0.7 ± 0.1 g) and the highest amount of peramine.

Table 3. Peramine concentration in leaves, carcasses, feces and guts, pupal development time and performance of the adult of *C. rudis* after no-choice test for each experimental line and commercial cultivar of ryegrass. Different letters per column indicate significant differences ($p \leq 0.05$) based on ANOVA test followed by Tukey's test.

Experimental lines and Cultivars	Peramine in Leaves (μg/g)	Pupal Devel-opment Time (days)	Length (cm)	Weight (g)	Wing Length ♀ (cm)	Wing Length ♂ (cm)	Peramine in Carcasses (μg/g)	Peramine in Feces (μg/g)	Peramine in Guts (μg/g)
L161	179.3 ± 24.5^{a}	28 ± 1.1^{a}	1.7 ± 0.1^{e}	0.5 ± 0.0^{b}	0.17 ± 0.0^{b}	0.63 ± 0.1^{e}	12.1 ± 1.0^{a}	31.7 ± 4.2^{a}	0.0 ± 0.0^{e}
L162	99.8 ± 17.3^{d}	25 ± 1.2^{b}	1.8 ± 0.1^{d}	0.6 ± 0.1^{b}	0.14 ± 0.0^{d}	0.99 ± 0.1^{c}	8.2 ± 0.9^{c}	13.7 ± 2.7^{d}	0.8 ± 0.1^{c}
L163	15.2 ± 4.9^{e}	24 ± 1.0^{c}	2.9 ± 0.2^{b}	0.7 ± 0.0^{a}	0.13 ± 0.0^{d}	1.32 ± 0.2^{a}	1.6 ± 0.8^{d}	1.1 ± 0.3^{e}	0.0 ± 0.0^{e}
L164	160.8 ± 39.9^{b}	26 ± 2.1^{b}	1.9 ± 0.2^{d}	0.5 ± 0.0^{b}	0.15 ± 0.0^{c}	0.70 ± 0.1^{d}	11.3 ± 2.1^{a}	27.9 ± 4.2^{b}	0.2 ± 0.0^{d}
L165	127.0 ± 34.2^{c}	25 ± 2.0^{b}	2.4 ± 0.1^{c}	0.6 ± 0.0^{b}	0.16 ± 0.0^{c}	1.19 ± 0.2^{b}	9.5 ± 0.9^{b}	19.7 ± 2.3^{c}	1.5 ± 0.1^{b}
L166	146.4 ± 21.2^{c}	26 ± 1.5^{b}	2.0 ± 0.1^{d}	0.5 ± 0.1^{b}	0.12 ± 0.0^{e}	0.80 ± 0.1^{d}	10.1 ± 0.1^{b}	24.0 ± 2.1^{b}	1.7 ± 0.1^{a}
L167	17.1 ± 6.1^{e}	24 ± 1.1^{c}	3.1 ± 0.1^{a}	0.8 ± 0.1^{a}	0.15 ± 0.0^{c}	1.20 ± 0.2^{b}	2.2 ± 0.2^{d}	1.1 ± 0.4^{e}	0.0 ± 0.0^{e}
JUMBO (E−)	0.0 ± 0.0^{f}	23 ± 1.4^{c}	3.2 ± 0.3^{a}	0.7 ± 0.1^{a}	0.20 ± 0.0^{a}	1.25 ± 0.2^{b}	0.0 ± 0.0^{e}	0.0 ± 0.0^{f}	0.0 ± 0.0^{e}
ALTO AR1 (E+)	184.1 ± 29.8^{a}	28 ± 1.2^{a}	1.4 ± 0.1^{f}	0.3 ± 0.0^{c}	0.17 ± 0.0^{b}	0.43 ± 0.1^{f}	10.5 ± 1.1^{a}	32.8 ± 4.4^{a}	1.3 ± 0.1^{b}

ALTO AR1 (E+) corresponded to a lower larval weight (0.3 ± 0.0 g). On the other hand, for both male ($F_{8,135} = 63.1$, $p < 0.001$) and female ($F_{8,27} = 16.55$, $p < 0.001$) adults, there was a trend of decreasing wing length with an increase in the amount of peramine, with the lowest amount of peramine (JUMBO (E−)) corresponding to longer wing lengths and the highest amount of peramine (ALTO AR1 (E+)) corresponding to shorter wing lengths. In addition, when comparing the concentration of peramine in leaves with that found in the carcasses, feces, and guts of the larvae, a general trend was observed where experimental lines and commercial cultivars with higher concentrations of peramine in the leaves also tended to have higher concentrations in carcasses, feces, and guts ($F_{8,171} = 981.1$, $p < 0.001$; $F_{8,171} = 20{,}062$, $p < 0.001$; $F_{8,171} = 2484$ $p < 0.001$, respectively). For example, the experimental line L161, with a high concentration of peramine in the leaves (179.3 ± 24.5 μg/g), also shows a high concentration in the carcass (12.1 ± 1.0 μg/g) and feces (31.7 ± 4.2 μg/g), contrary to what happens in other experimental lines such as L167, which presents 17.1 ± 6.1 μg/g of peramine in leaves and therefore the concentration of this alkaloid in the carcass and feces is 2.2 ± 0.2 μg/g and 1.1 ± 0.4 μg/g, respectively. Complementing this, the Pearson correlation revealed various correlations among the studied variables. For instance, a strong positive correlation was observed between the concentration of peramine in the leaves and its presence in the carcass and feces of the larvae, with correlation coefficients of 0.976 and 0.9937, respectively. Additionally, significant correlations were found between the concentration of peramine in the carcass and feces (0.9498), as well as between the concentration of peramine in feces and intestines (0.5086). On the other hand, notable negative correlations were found between total length and the concentration of peramine in leaves, carcasses, and feces, with correlation coefficients of −0.9128, −0.8348, and −0.8537, respectively. These correlations suggested both positive and negative relationships among the studied variables, indicating potential complex interactions in the biology and development of *Chilesia rudis* larvae. Finally, these observations suggest possible complex interactions between the concentration of peramine in the leaves and various aspects of the development and physiology of *Chilesia rudis*.

3.4. Survival Rate Test

The results of the mortality assay on *C. rudis* larvae using different concentrations of peramine reveal that as the concentration of peramine increases, the percentage of larval survival decreases (Figure 2). The lowest concentration of 1 mg/L shows high survival rates of 100%, while the highest concentration of 100 mg/L results in the lowest survival rate, recording 45%. This pattern suggests a significant dose–response relationship, where higher concentrations of peramine correlate with greater larval mortality ($F_{3,76} = 9.53$, $p < 0.001$).

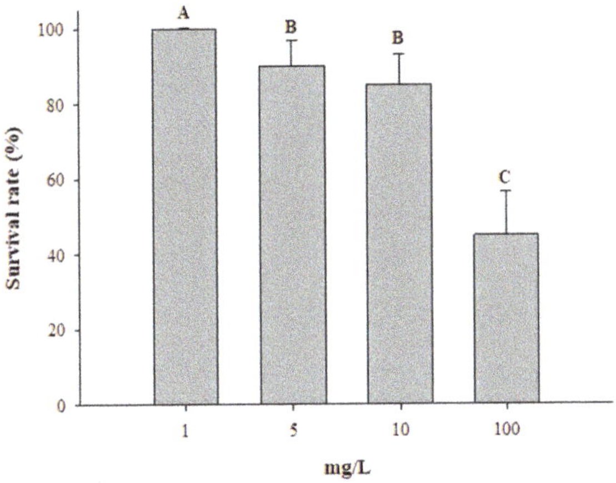

Figure 2. Survival rate for C. rudis larvae feeding on leaves of JUMBO (E−) subjected to several doses of peramine. Different letters indicate significant differences ($p \leq 0.05$) based on ANOVA test following by Tukey's test.

4. Discussion

The experimental findings outlined in the study provide a comprehensive analysis of the presence and infection rates of *Epichloë* sp. in *L. perenne* across various experimental lines and commercial cultivars. This incorporation of endophytic fungal infection in ryegrass has facilitated the control of several pests, such as *L. bonariensis*, due to specific strains that produce peramine. However, our results indicate that the concentration of peramine in ryegrass is not directly correlated with the percentage of endophyte infection. This discrepancy arises because various endophyte strains independently produce different groups of alkaloids. For example, Guerre [37] identified *Epichloë festucae* var. *lolii* as one of the endophytic fungi responsible for producing lolitrem B in ryegrass, while studies by Schardl et al. [38] reported various fungi capable of producing peramine, such as *N. gansuense* and *E. typhina*. The effects of these alkaloids generally relate to their toxicity, deterrence, or mortality on insects.

As plant defense mechanisms evolve, they also prompt insects to develop countermeasures [39]. Within this framework, our no-choice experiment results demonstrated that the cultivar JUMBO was more preferred by the herbivore *C. rudis* compared to other cultivars and experimental lines. Interestingly, despite this preference, larvae showed similar weight gain across all treatments. This suggests that while peramine may act as a deterrent, *C. rudis* might be either storing, sequestering, or passively accumulating this alkaloid to mitigate its toxicity. Chacón-Fuentes et al. [28] reported an example of potential detoxification, sequestration, or passive accumulation in *C. rudis*. They found that when larvae consumed artificial diets containing phenolic compounds like kaempferol, rutin, myricetin, and gallic acid at concentrations of 10 or 100 mg/g, it led to reduced consumption percentages and increased larval mass without affecting survival. This study underscores the role of sequestration and passive accumulation as significant defense strategies in herbivores, particularly within the Lepidoptera order [29,40–42].

Research has highlighted a link between the sequestration of specific compound classes and the level of dietary specialization observed in insects. For instance, our survival assays underscore the potential role of peramine as a potent deterrent against herbivory in ryegrass, highlighting its ecological significance in plant–insect interactions and suggesting its potential application in pest management strategies. However, this effect is observed at high concentrations. Interestingly, at concentrations ranging from 1 to 10 mg/L, mortality was not affected, consistent with findings reported by Chacón-Fuentes et al. [28], suggesting a possible passive accumulation of compounds by the larvae as an adaptation method to the alkaloids present in these grasses. In complement to our results, Lampert and Bowers [43] and Lampert et al. [44] compared the sequestration of iridoid glycosides between the specialist larva *Junonia coenia* and other herbivores (*Anartria jatrophae* and *Vanessa cardui*); the specialist sequestered nearly 12% more iridoid glycosides than other species. Therefore, this phenomenon has been identified as a result of evolutionary relationships between insects and phytochemicals [45].

In summary, our results indicate that the concentration of peramine in various experimental lines and cultivars of ryegrass ranged from 15.2 μg/g to 184.1 μg/g. These values are higher than concentrations reported by Spiering et al. [46], who found values ranging from 14 to 34 μg/g in various parts of ryegrass. Studies by Fuchs et al. [16] showed that ryegrass infected with endophytes produced concentrations of peramine around 0.06 μg/g, while their herbivore, *Rhapalosiphum padi*, was able to accumulate 0.19 μg/g in their tissues. Generally, the peramine content in ryegrass is influenced by endophyte strain, environmental factors, and host age [47]. The presence of fungus did not significantly affect the total length of insects, with Jumbo (E−) showing lengths of 3.2 ± 0.3 cm and Alto AR1 (80% endophyte) showing lengths of 1.40 ± 0.1 cm. Furthermore, insect mass was also not significantly affected when comparing cultivars JUMBO and Alto AR1. Our results suggest that the pyrrolopyrazine alkaloid peramine acted as a deterrent for *C. rudis* but was not toxic and did not significantly affect larval weight due to the ability of larvae to store peramine within their tissues and eliminate a percentage of the alkaloid through feces. Future research will focus on determining the ecological role of sequestered compounds or passive accumulation by *C. rudis* and evaluating other alkaloids that may be sequestered from ryegrass, such as PAs. Finally, these results offer insights into the variability of endophyte colonization among different experimental lines, indicating potential candidates for further investigation and breeding programs aimed at enhancing endophyte resistance in crops.

Author Contributions: Conceptualization, M.C.-F. and L.B.; methodology, M.C.-F., D.M.-C., V.A.-C., I.M., M.L. and L.B.; software, M.C.-F.; validation, L.B.; formal analysis, M.C.-F.; investigation, L.B.; resources, L.B.; data curation, M.C.-F.; writing—original draft preparation, M.C.-F. and L.B.; writing—review and editing, L.B.; visualization, M.C.-F. and L.B.; supervision, L.B. All authors have read and agreed to the published version of the manuscript.

Funding: The APC was funded by Universidad de La Frontera.

Institutional Review Board Statement: Not applicable.

Informed Consent Statement: Not applicable.

Data Availability Statement: Data are contained within the article.

Acknowledgments: Financial support for this research was supplied by FONDECYT (Project 11130715). Additional support (partial) was provided by Dirección de Investigación, Universidad de La Frontera and FONDECYT INICIACIÓN 11240860. Additional support (partial) was provided by Dirección de Investigación, Universidad de La Frontera.

Conflicts of Interest: The authors declare no conflicts of interest. The funders had no role in the design of the study; in the collection, analyses, or interpretation of data; in the writing of the manuscript; or in the decision to publish the results.

References

1. Al-Khayri, J.M.; Rashmi, R.; Toppo, V.; Chole, P.B.; Banadka, A.; Sudheer, W.N.; Nagella, P.; Shehata, W.F.; Al-Mssallem, M.Q.; Alessa, F.M.; et al. Plant Secondary Metabolites: The Weapons for Biotic Stress Management. *Metabolites* **2023**, *13*, 716. [CrossRef] [PubMed]
2. Wang, X.; Yan, W.; Liu, W.; Chen, H.; Yuan, Q.; Wang, Z.; Liu, H. Endophytic *Beauveria bassiana* of Tomato Resisted the Damage from Whitefly *Bemisia tabaci* by Mediating the Accumulation of Plant-Specialized Metabolites. *J. Agric. Food Chem.* **2023**, *71*, 13244–13254. [CrossRef] [PubMed]
3. Hoogshagen, M.; Hastings, A.P.; Chavez, J.; Duckett, M.; Pettit, R.; Pahnke, A.A.; Agrawal, A.A.; de Roode, J.C. Mixtures of Milkweed Cardenolides Protect Monarch Butterflies against Parasites. *J. Chem. Ecol.* **2023**, *50*, 52–62. [CrossRef] [PubMed]

4. Shan, Z.; Zhou, S.; Shah, A.; Arafat, Y.; Arif Hussain Rizvi, S.; Shao, H. Plant Allelopathy in Response to Biotic and Abiotic Factors. *Agronomy* **2023**, *13*, 2358. [CrossRef]
5. Erb, M.; Robert, C.A.M. Sequestration of plant secondary metabolites by insect herbivores: Molecular mechanisms and ecological consequences. *Curr. Opin. Insect Sci.* **2016**, *14*, 8–11. [CrossRef]
6. Petschenka, G.; Agrawal, A. How herbivores coopt plant defenses: Natural selection, specialization, and sequestration. *Curr. Opin. Insect Sci.* **2016**, *14*, 17–24. [CrossRef] [PubMed]
7. Chacón-Fuentes, M.; Martínez-Cisterna, D.; Vera, W.; Ortega-Klose, F.; Reyes, C.; Matamala, I.; Bardehle, L. Feeding Performance of Argentine Stem Weevil Is Reduced by Peramine from Perennial Ryegrass Infected with Endophyte Fungus. *Insects* **2024**, *15*, 410. [CrossRef] [PubMed]
8. Lin, W.; Gao, C.; Wang, J.; Xu, W.; Wang, M.; Li, M.; Ma, B.; Tian, P. Effects of Drought Stress on Peramine and Lolitrem B in Epichloë-Endophyte-Infected Perennial Ryegrass. *Life* **2022**, *12*, 1207. [CrossRef] [PubMed]
9. Shymanovich, T.; Musso, A.M.; Cech, N.B.; Faeth, S.H. *Epichloë* endophytes of *Poa alsodes* employ alternative mechanisms for host defense: Insecticidal versus deterrence. *Arthropod-Plant Interact.* **2019**, *13*, 79–90. [CrossRef]
10. Nelli, M.R. Total Synthesis of Peramine, a Defensive Alkaloid Produced by Endophytic Fungi of Cool Season Grasses Possessing Anti-Insect Properties. *J. Nat. Prod.* **2016**, *71*, 1370–1377.
11. Nelli, M.R.; Scheerer, J.R. Synthesis of peramine, an anti-insect defensive alkaloid produced by endophytic fungi of cool season grasses. *J. Nat. Prod.* **2016**, *79*, 1189–1192. [CrossRef] [PubMed]
12. Rowan, D.D.; Dymock, J.J.; Brimble, M.A. Effect of fungal metabolite peramine and analogs on feeding development of Argentine stem weevil (*Listronotus bonariensis*). *J. Chem. Ecol.* **1990**, *16*, 1683–1695. [CrossRef] [PubMed]
13. Young, C.; Hume, D.; McCulley, R. Fungal endophytes of tall fescue and perennial ryegrass: Pasture friend or foe? *J. Anim. Science.* **2013**, *91*, 2379–2394. [CrossRef] [PubMed]
14. Keogh, R.G.; Tapper, B.A.; Fletcher, R.H. Distributions of the fungal endophyte *Acremonium lolii*, and of the alkaloids Lolitrem B and Peramine, within perennial ryegrass. *N. Z. J. Agricutural Res.* **1996**, *39*, 121–127. [CrossRef]
15. Siegel, M.R.; Bush, L.P. Defensive chemicals in grass-fungal endophyte associations. *Recent Adv. Phytochem.* **1996**, *30*, 81–118.
16. Fuchs, B.; Krischke, M.; Mueller, M.J.; Krauss, J.J. Peramine and Lolitrem B from Endophyte-Grass Associations Cascade Up the Food Chain. *Chem. Ecol.* **2013**, *39*, 1385–1389. [CrossRef] [PubMed]
17. Parra, L.; Mutis, A.; Chacón, M.; Lizama, M.; Rojas, C.; Catrileo, A.; Rubilar, O.; Tortella, G.; Birkett, M.; Quiroz, A. Horn fly larval survival in cattle dung is reduced by endophyte infection of tall fescue pasture. *Pest Manag. Sci.* **2016**, *72*, 1328–1334. [CrossRef] [PubMed]
18. Faeth, S.H.; Saari, S. Fungal grass endophytes and arthropod communities: Lessons from plant defence theory and multitrophic interactions. *Fungal Ecol.* **2012**, *5*, 364–371. [CrossRef]
19. Zust, T.; Agrawal, A.A. Population growth and sequestration of plant toxins along a gradient of specialization in four aphid species on the common milkweed, *Asclepias syriaca. Funct. Ecol.* **2016**, *30*, 547–556. [CrossRef]
20. Nishida, R. Sequestration of defensive substances from plants by lepidoptera. *Annu. Rev. Entomol.* **2002**, *47*, 57–92. [CrossRef]
21. Opitz, S.E.W.; Müller, C. Plant chemistry and insect sequestration. *Chemoecology* **2009**, *1*, 71. [CrossRef]
22. Weintraub, J.D. Host plant association patterns and phylogeny in the tribe Troidini (Lepidoptera: Papilionidae). In *Swallowtail Butterflies: Their Ecology and Evolutionary Biology*; Scriber, J.M., Tsubaki, Y., Lederhouse, R.C., Eds.; Scientific Publishers: Gainesville, FL, USA, 1995; pp. 307–316.
23. Pinto, C.F.; Troncoso, A.J.; Urzua, A.; Niemeyer, H.M. Use of volatiles of Aristolochia chilensis (aristolochiaceae) in host searching by fourth-instar larvae and adults of battus Polydamas archidamas (Lepidoptera: Papilionidae: Troidini). *Eur. J. Entomol.* **2009**, *106*, 63–68. [CrossRef]
24. Pinto, C.F.; Urzúa, A.; Niemeyer, H.M. Sequestration of aristolochic acids from meridic diets by larvae of *Battus Polydamas archidamas* (Papilionidae: Troidini). *Eur. J. Entomol.* **2011**, *108*, 41–45. [CrossRef]
25. Ángulo, A.; Ruiz, V. *Maenas rudis* (Butler): Cuncuna colorada de prados y jardines; biología y estados inmaduros (Lepidoptera: Arctiidae). *Boletín Soc. Biológica Concepción* **1975**, 139–147.
26. Chacón-Fuentes, M.; Parra, L.; Rodriguez-Saona, C.; Seguel, I.; Ceballos, R.; Quiroz, A. Domestication in murtilla (*Ugni molinae*) reduced defensive flavonol levels but increased resistance against a native herbivorous insect. *Environ. Entomol.* **2015**, *44*, 627–637. [CrossRef] [PubMed]
27. Chacón-Fuentes, M.; Parra, L.; Lizama, M.; Seguel, I.; Urzúa, A.; Quiroz, A. Plant flavonoid content modified by domestication. *Environ. Entomol.* **2017**, *46*, 1080–1089. [CrossRef]
28. Chacon-Fuentes, M.; Bardehle, L.; Seguel, I.; Rubilar, F.; Martínez-Cisterna, D.; Quiróz, A. Domestication of Plants of *Ugni molinae* Turcz (Myrtaceae) Interferes in the Biology of *Chilesia rudis* (Lepidoptera: Erebidae) Larvae. *Molecules* **2021**, *26*, 2063. [CrossRef] [PubMed]
29. Nickisch-Rosenegk, E.; Wink, M. Sequestration of pyrrolizidine alkaloids in several arctiid moths (Lepidoptera: Arctiidae). *J. Chem. Ecol.* **1993**, *19*, 1189–1903. [CrossRef] [PubMed]
30. Dickel, F.; Freitak, D.; Mappes, J. Long-term prophylactic antibiotic treatment: Effects on survival, immunocompetence and reproduction success of *Parasemia plantaginis* (Lepidoptera: Erebidae). *J. Insect Sci.* **2016**, *16*, 46. [CrossRef]
31. Dombrowski, J.; Baldwin, J.; Azevedo, M.; Banowetz, G. A sensitive PCR-based assay to detect *Neotyphodium* fungi in seed and plant tissue of tall, fescue and ryegrass species. *Crop Sci.* **2006**, *46*, 1064–1070. [CrossRef]

32. Chacon-Fuentes, M.; Martínez-Cisterna, D.; Reyes, C.; Vera, W.; Fincheira, P.; Lizama, M.; Quiroz, A.; Bardehle, L. Infection of perennial ryegrass (*Lolium perenne*) by an endophyte fungus (*Neotyphodium lolii*) decreases the abundance and diversity of predators and parasitoids. *Rev. Bras. De Entomol.* **2023**, *67*, e20230012. [CrossRef]
33. War, A.R.; Paulraj, M.G.; Ahmad, T.; Buhroo, A.A.; Hussain, B.; Ignacimuthu, S. Mechanisms of plant defense against insect herbivores. *Plant Signal. Behav.* **2012**, *7*, 1306–1320. [CrossRef] [PubMed]
34. Kholghahmadi, M.; Karimi-Malati, A.; Jalali Sendi, J. Ecophysiological responses of individually and group reared *Cydalima perspectalis* (Lepidoptera: Crambidae) to alkaloid-containing host plants. *Environ. Entomol.* **2023**, *52*, 426–435. [CrossRef] [PubMed]
35. Ait Lhaj, R.; Saffaj, T.; Belmir, H.; Ihssane, B. The uncertainty profile used for full validation of the HPLC Method to determine 22 azo amines in fabrics. *J. AOAC Int.* **2023**, *106*, 813–824. [CrossRef] [PubMed]
36. Chacón-Fuentes, M.; Bardehle, L.; Lizama, M.; Seguel, I.; Quiroz, A. Restoration of flavonols and isoflavonoids in Ugni molinae subjected to a reciprocal transplant experiment in a domestication framework. *Chem. Ecol.* **2019**, *35*, 115–127. [CrossRef]
37. Guerre, P. Ergot alkaloid produced by endophytic fungi of the genus Epichloe. *Toxins* **2015**, *7*, 773–790. [CrossRef] [PubMed]
38. Schardl, C.L.; Young, C.A.; Hesse, U.; Amyotte, S.G.; Andreeva, K.; Calie, P.J.; Fleetwood, D.J.; Haws, D.C.; Moore, N.; Oeser, B.; et al. Plant-symbiotic fungi as chemical engineers: Multi-genome analysis of the clavicipitaceae reveals dynamics of alkaloid loci. *PLoS Genet* **2013**, *9*, e1003323. [CrossRef] [PubMed]
39. Shymanovich, T.; Faeth, S.H. Environmental factors affect the distribution of two Epichloë fungal endophyte species inhabiting a common host grove bluegrass (*Poa alsodes*). *Ecol. Evol.* **2019**, *9*, 6624–6642. [CrossRef]
40. Newton, E.; Bullock, J.M.; Hodgson, D. Bottom-up effects of glucosinolate variation on aphid colony dynamics in wild cabbage populations. *Ecol. Entomol.* **2009**, *34*, 614–623. [CrossRef]
41. Kos, M.; Broekgaarden, C.; Kabouw, P.; Lenferink, K.O.; Poelman, E.H.; Vet, L.E.M.; Dicke, M.; Van Loon, J.J.A. Relative importance of plant mediated bottom-up and top-down forces on herbivore abundance on Brassica oleracea. *Funct. Ecol.* **2011**, *25*, 113–1124. [CrossRef]
42. Cibils-Stewart, X.; Putra, R.; Islam, T.; Fanna, D.J.; Wuhrer, R.; Mace, W.J.; Johnson, S.N. Silicon and Epichloë-endophyte defences in a model temperate grass diminish feeding efficiency and immunity of an insect folivore. *Funct. Ecol.* **2023**, *37*, 3177–3192. [CrossRef]
43. Lampert, E.C.; Bowers, M.D. Host plant influences on iridoid glycoside sequestration of generalist and specialist caterpillars. *J. Chem. Ecol.* **2010**, *36*, 1101–1104. [CrossRef]
44. Lampert, E.C.; Dyer, L.A.; Bowers, M.D. Dietary specialization and the effects of plant species on potential multitrophic interactions of three species of nymphaline caterpillars. *Entomol. Exp. Appl.* **2014**, *153*, 207–216. [CrossRef]
45. Abe, F.; Yamauchi, T.; Honda, K.; Omura, H.; Hayashi, N. Sequestration of phenanthroindolizidine alkaloids by an Asclepiadaceae-feeding danaid butterfly, *Ideopsis similis*. *Phytochemistry* **2001**, *56*, 697–701. [CrossRef] [PubMed]
46. Spiering, M.J.; Wilkinson, H.H.; Blankenship, J.D.; Schardl, C.L. Expressed sequence tags and genes associated with loline alkaloid expression by the fungal endophyte *Neotyphodium uncinatum*. *Fungal Genet. Biol.* **2002**, *36*, 242–254. [CrossRef]
47. Leuchtmann, A.; Schmidt, D.; Bush, L.P. Different levels of protective alkaloids in grasses with stroma-forming and seed-transmitted *Epichloë/Neotyphodium* endophytes. *J. Chem. Ecol.* **2000**, *26*, 1025–1036. [CrossRef]

Journal of *Fungi*

MDPI

Correction

Correction: Preisegger et al. Visualizing Oral Infection Dynamics of *Beauveria bassiana* in the Gut of *Tribolium castaneum*. *J. Fungi* 2025, *11*, 101

Lautaro Preisegger [1], Juan Cruz Flecha [1], Fiorella Ghilini [2], Daysi Espin-Sánchez [1], Eduardo Prieto [2,3], Héctor Oberti [4,5], Eduardo Abreo [4], Carla Huarte-Bonnet [1], Nicolás Pedrini [1,*] and Maria Constanza Mannino [1,*]

1 Instituto de Investigaciones Bioquímicas de La Plata (INIBIOLP), CCT La Plata Consejo Nacional de Investigaciones Científicas y Técnicas (CONICET), Universidad Nacional de La Plata (UNLP), Calles 60 y 120, La Plata 1900, Argentina; lpreisegger@med.unlp.edu.ar (L.P.); juancruzflecha16@gmail.com (J.C.F.); despinsanchez@med.unlp.edu.ar (D.E.-S.); carlahb@biol.unlp.edu.ar (C.H.-B.)
2 Instituto de Investigaciones Fisicoquìmicas Teóricas y Aplicadas (INIFTA), CCT La Plata Consejo Nacional de Investigaciones Científicas y Técnicas (CONICET), Universidad Nacional de La Plata (UNLP), Diagonal 113 y 64 S/N, La Plata 1900, Argentina; fiorelag@inifta.unlp.edu.ar (F.G.); edprieto@quimica.unlp.edu.ar (E.P.)
3 Instituto de Ciencias de la Salud, Universidad Nacional Arturo Jauretche (ICS-UNAJ), Avenida Calchaqui, Florencio Varela 6200, Argentina
4 Laboratorio de Bioproducción, Plataforma de Bioinsumos, Instituto Nacional de Investigación Agropecuaria, Estación Experimental Wilson Ferreira Aldunate, Ruta 48 Km 10, Canelones 90000, Uruguay; hoberti@inia.org.uy (H.O.); eabreo@inia.org.uy (E.A.)
5 Departamento de Biotecnología, Universidad ORT Uruguay, Montevideo 10129, Uruguay
* Correspondence: npedrini@med.unlp.edu.ar (N.P.); mmannino@med.unlp.edu.ar (M.C.M.)

In the original article [1], the omission of two critical references led to a shift in the citation numbering throughout the discussion section, which significantly impacts the interpretation of the text. The specific changes and their implications are detailed below:

Adding References 27 and 28: These references were omitted in the editing process, they were in the original manuscript and when it was sent to review if it was correct, they were missing from the list, hence the reference numbering was wrong. Now, they are all included and the numbering should be 27. Kilaru; 28. Bernasconi and 29. Oberti, since that is the correct referencing according to what is written in the main text in the manuscript.

Reference 27: Kilaru, S.; Schuster, M.; Studholme, D.; Soanes, D.; Lin, C.; Talbot, N.J.; Steinberg, G. A codon-optimized green fluorescent protein for live cell imaging in Zymoseptoria tritici. *Fungal Genet Biol.* **2015**, *79*, 125–131. https://doi.org/10.1016/j.fgb.2015.03.022.

Reference 28: Bernasconi, A.; Alassimone, J.; McDonald, B.A.; Sánchez-Vallet, A. Asexual reproductive potential trumps virulence as a predictor of competitive ability in mixed infections. *Environ. Microbiol.* **2022**, *24*, 4369–4381. https://doi.org/10.1111/1462-2920.16018.

Volume and page was added in reference 29 as one of the sources was a preprint and has since been published.

Reference 29: Oberti, H.; Sessa, L.; Oliveira-Rizzo, C.; Di Paolo, A.; Sánchez-Vallet, A.; Seidl, M.F.; Abreo, E. Novel genomic features in entomopathogenic fungus *Beauveria bassiana ILB308*: Accessory genomic regions and putative virulence genes involved in the infection process of soybean pest Piezodorus guildinii. *Pest. Manag. Sci.* **2025**, *81*, 2323–2336. https://doi.org/10.1002/ps.8631.

Reference 27 should be Kilaru as is shown in the updated version. Oberti should be reference 29, since the techniques cited were published in Kilaru (Ref. 27) and Bernasconi (Ref. 28) and in Oberti et al., 2025 (Ref. 29).

These references provide essential background and methodology details that are foundational to the discussion of the results. Their omission not only diminishes the scientific rigor of the article but also disrupts the logical flow of the argument presented.

The omission has caused all subsequent citations to be misnumbered, starting from Section 2.1.1. This misnumbering altered the meaning of several key points in the discussion, as the numerical citations pointed to incorrect sources, potentially misleading readers.

The authors state that the scientific conclusions are unaffected. This correction was approved by the Academic Editor. The original publication has also been updated.

Reference

1. Preisegger, L.; Flecha, J.C.; Ghilini, F.; Espin-Sánchez, D.; Prieto, E.; Oberti, H.; Abreo, E.; Huarte-Bonnet, C.; Pedrini, N.; Mannino, M.C. Visualizing Oral Infection Dynamics of *Beauveria bassiana* in the Gut of *Tribolium castaneum*. *J. Fungi* **2025**, *11*, 101. [CrossRef] [PubMed]

MDPI AG
Grosspeteranlage 5
4052 Basel
Switzerland
Tel.: +41 61 683 77 34

Journal of Fungi Editorial Office
E-mail: jof@mdpi.com
www.mdpi.com/journal/jof

www.ingramcontent.com/pod-product-compliance
Lightning Source LLC
Chambersburg PA
CBHW061451210926
51850CB00008B/1677
9783725867424